Qualit
Assuran

Applying Methodologies for
Launching New Products, Servic
and Customer Satisfaction

Quality Assurance

Applying Methodologies for Launching New Products, Services, and Customer Satisfaction

D. H. Stamatis

CRC Press
Taylor & Francis Group
Boca Raton London New York

CRC Press is an imprint of the
Taylor & Francis Group, an **informa** business

CRC Press
Taylor & Francis Group
6000 Broken Sound Parkway NW, Suite 300
Boca Raton, FL 33487-2742

Printed on acid-free paper
Version Date: 20150305

International Standard Book Number-13: 978-1-4987-2868-3 (Hardback)

Library of Congress Cataloging-in-Publication Data

Stamatis, D. H., 1947-
Quality assurance : applying methodologies for launching new products, services, and customer satisfaction / author, D. H. Stamatis.
pages cm
Includes bibliographical references and index.
ISBN 978-1-4987-2868-3 (alk. paper)
1. Quality assurance. I. Title.

TS156.6.S718 2015
658.5'62--dc23 2015004539

Visit the Taylor & Francis Web site at
http://www.taylorandfrancis.com

and the CRC Press Web site at
http://www.crcpress.com

To my friend

Αθανασιου Καλοζουμη

(Thanasi)

Contents

List of Figures

List of Tables

Preface

Since the early 1980s, everyone has talked about continual improvement (CI). However, to talk about it is not enough. It must be engrained in an organization with a three-prong approach. That approach is based on the following: (1) a consistent philosophy of improvement by management, (2) a receptive organizational culture, and (3) the entire culture of the organization must be willing to make decisions based on measurement and data.

These three basic approaches are necessary for the organization to develop an attitude for doing *things right the first time*, whether it will be for a product or a service. This attitude must be incorporated into the processes of any organization and create products or services for the marketplace that will *delight* the customers rather than just satisfy them. It is interesting to note here that even though we all know these fundamental principles, products and services fail or at least miss their launching deadlines, and for some reason even though we claimed that we were short on time, now after the failure we find the *time to do it again*. The question is why? We think that the answer is one or more of the following five categories. They are

1. Design and development. We rush to bring products and services into market before they are ready for mass production and/or marketing.
2. Process design. We design products and services with specific requirements and not robustness. We design and we expect the customer will tell us how to improve the products and services once they hit the market.
3. Problem solving. We focus on fixing the problems after the fact and even then we fail to use a systematic (standardized) and consistent approach for the organization.
4. Share knowledge systematically. We preach corporate knowledge but we do not systematically use it for the benefit of the organization. Quite often, once the launch is complete, we do not even hold debriefing meetings to evaluate Things Gone Wrong (TGW) or Things Gone Right (TGR). In fact, in most cases, the launch team is replaced by a new one and we start the process all over again.
5. Leaders are not involved with the development of others. We practice and encourage silo development within our organizations. Leaders focus on financial metrics and they seem to spend very little time on mentoring upcoming talent. We forget that one of the most important tasks of leadership is to build and train new leaders for the future.

To circumvent these deficiencies, the concept of CI must be an active philosophy in the management of any organization. In fact, that philosophy must be interpreted through a strategy and implemented in such a way that *superb results* will be accomplished. If the strategy is great, everything else will work to positive results. Unfortunately, most organizations circumvent a good strategy and depend on managing broken processes to have good results with additional procedures and/or requirements.

We are not suggesting that there is a specific policy within organizations that encourages cheating in any way, shape, or form. Absolutely not! However, we do suspect that many organizations, due to time and cost constraints, *sometimes* look the other way. After all, they assume acts that have welcomed benefits are good and the sooner they reach the market the better. We suggest that this kind of thought is 100% wrong. When this happens, organizations must look at the bigger picture internally. It is a system problem! The problem is not that individuals cut corners, or cheat, or coerce someone to do something unethical. The problem is that the idea of *doing it right the first time* is only a slogan and many organizations have developed the ridiculous notion that you can hide or reinvent reality by creating new requirements and policies and trying to enforce them. This approach has not worked and it will not work.

To be sure, one of the risks we take in a free society is to allow ourselves and others to make mistakes. Why? Because no one is perfect! One can never be certain when something is a mistake, oversight or not. So what do we do? Technology is emerging and rapidly growing in all industries and service organizations. The opportunity exists for all organizations to grow as well. However, for that growing to take place, the same organizations must recognize the consumer's needs and wants and align them to their priorities. One good way to do that is to follow the seven principles:

1. Research/plan: Research for innovations that will delight the consumer. Always scan/survey the consumers to find their needs and wants. Remember that current wants become needs in the future (Kano model).
2. Assure: The consumer has to be assured that the product you are selling is reliable (engineer's language) and durable (customer's language) because we want products that last.
3. Explain: Why a particular feature and innovation are important and how they will facilitate the customer's safety and/or comfort.
4. Prioritize: Product performance; it should work better if it is really good.
5. Demonstrate: The product is environmentally safe and friendly and its performance is better than a competitor's product or service.

6. Confirm: The material used in the product is the best, so the quality metric is what we want and expect.
7. Show: How you value and deliver product safety.

Based on 35 years of experience dealing with OEMs and suppliers this book covers the essential items for doing *the right thing the first time* especially when launching a good product and/or service to the customer. The requirements have been split into two sections. Section I covers the original equipment manufacturer (OEM)/customer requirements and Section II covers supplier requirements. In both sections, we do not take a *deep dive* in the content because each chosen topic requires its own book. We have chosen, however, to identify key indicators and methodologies that will help attain excellent performance, delivery, and cost for both the customer and the supplier. In other words, if you follow these methodologies and indicators, the *job will get done right the first time*. Specifically, each chapter covers the following:

Chapter 1: Timing (Product Development). The significance of timing is discussed. Emphasis is placed on setting the milestones of deliverables.

Chapter 2: Basic OEM Quality System. The significance of a customer quality system is explained and the need for such a system to exist before a customer makes demands on a supplier for quality products/services.

Chapter 3: Manufacturing (Facility) Site Assessment (MSA). The significance of a customer knowing what to expect from a supplier and how it should validate its requirements in the supplier's facility.

Chapter 4: Sourcing. The significance of having the right supplier with which to do business and a discussion of how to go about selecting the right supplier.

Chapter 5: Segmentation. The significance of categorizing suppliers with which to do business depending on their own capability, capacity, and quality requirements.

Chapter 6: Supplier Development. The significance of supplier development by the customer, so that the supplier can meet the customer's requirements.

Chapter 7: Role of Supplier Technical Engineer. The significance of the customer designating a specific individual to make sure that the supplier is meeting all the requirements of the customer. It defines several roles and responsibilities that the supplier technical engineer must do for success.

Chapter 8: Commitment to Continual Improvement. This chapter discusses the significance of the continual improvement philosophy and how it affects the supplier and customer relationship.

Chapter 9: Lean Manufacturing. This chapter introduces the philosophy of Lean and several methods of reducing waste.

Chapter 10: Quality Operating System (QOS). This chapter discusses one of the key systems in pursuing quality. It focuses on key processes and key indicators of these processes.

Chapter 11: Certification to the International Standard and Customer-Specific Requirements. This chapter gives a cursory review of the fundamental principles of the ISO standard and how the supplier can use it to improve quality.

Chapter 12: Statistical Process Control (SPC). This chapter gives a cursory review of the SPC fundamentals and explains the need for the supplier to have a system in place to evaluate its processes. Without SPC, *the voice of the process* will never be known.

Chapter 13: Problem-Solving Methodology. This chapter discusses the role of problem-solving methodology in any organization. It also addresses several key points about definition of the problem and the characteristics of problem solving and the process improvement cycle.

Chapter 14: Internal Audits. This chapter explains the validation process of fulfilling the requirements of both standards and customer requirements by explaining the concept of an audit and specifically elaborates on several types that may be used by either the customer or the supplier.

Chapter 15: Poka-Yoke. This chapter extends the discussion of Chapter 13 by providing a specific approach to problem solving—the mistake-proofing approach. This discussion covers the rationale, method, and several approaches to mistake proofing.

Chapter 16: Measurement System Analysis (MSA). This chapter discusses the rationale for measurement, approaches to measurement by differentiating attribute and measurable data, and provides a mini-tutorial on gauge R&R via Minitab®.

Chapter 17: Supplier's Perspective of APQP. This chapter discusses the fundamental reasons for the existence of APQP and emphasizes the significance of the milestones. It further discusses the measuring effectiveness of each milestone.

Chapter 18: Supplier's Perspective on the PPAP. This chapter discusses the elements of the PPAP process, the rationale for them, and the PSW significance. It also delineates the levels of the PPAP as well as the significance of capacity as part of the PPAP process for some customers.

Chapter 19: Capacity Analysis. This chapter discusses the need for a capacity study and explains the components of it as well as giving a rationale for incorporating overal equipment effectiveness (OEE) as part of the capacity evaluation. The chapter also discusses the required OEE, demonstrated OEE, shared loading analysis, and total allocation.

Chapter 20: Geometric Dimensioning and Tolerancing (GD&T). This chapter gives a cursory overview of GD&T and focuses primarily on the most common characteristics: form, profile, orientation, location, and runout. It also discusses the concepts of tolerances, dimensions, datums, maximum material condition (MMC), least material condition (LMC), and feature of size (FoS).

Chapter 21: 8D. This chapter is also an extension of Chapter 13. It discusses a very specific approach to problem solving, which is very popular in many industries. Specifically, it discusses the individual steps of the method but it also covers an evaluation process for effectiveness at the completion of all the steps by giving the reader a series of questions to make sure that the key areas of the methodology have been covered.

Chapter 22: Miscellaneous. This chapter covers some unique topics that most organizations need to practice. They are very important and many customers demand that their supplier practice them. Specifically, the topics are production system, supplier request engineering approval (SREA), failure mode and effect analysis (FMEA), warranty, Weibull distribution, and Six Sigma.

Epilogue: Here we close the book with a discussion of some essential items that both customers and suppliers must keep in mind in order to continue the journey to flawless launches.

Appendix I: A typical detailed *expectation list* of a site assessment by the customer to the supplier's facility is given.

Appendix II: A typical PPAP checklist is given to make sure that all elements are covered appropriately.

Appendix III: Mistake proofing.

Appendix IV: Auditing requirements, criteria, and evaluation.

Appendix V: Many Six Sigma and Lean explanations of specific concepts and typical forms/tools are given to help suppliers with their data or selection methodologies to pursue continual improvement.

The book concludes with a glossary and selected bibliography.

Acknowledgments

No book is the result of one person. Many individuals both directly and indirectly contribute in subtle ways that make the formal book worth printing. This book is the result of many years contemplating what is necessary to have a launch that is as flawless as possible. Recognizing that nothing is perfect and that the unexpected always finds its way into any given situation, I decided that I would try to identify the key indicators to guide an organization through the cultural change that is necessary to pursue a way of life that expects *doing it right the first time.*

I concluded that this cultural change takes two parties. The first one is the customer, and the second one is the supplier. The more I looked at these two, the more I realized I was not too far from the target of improvement. My experience as an academician, trainer, and consultant over the last 35 years guided me in the direction to pursue this endeavor. In the process, I met many individuals I cannot possibly give credit to but for whom I am grateful for their suggestions and discussions about the key indicators that make an organization perform at a level that is indeed exceptional.

In addition to these silent names, there are also individuals who encouraged me to pursue the project, reviewed rough drafts, gave me advice, helped me focus on the points of interest and content in individual chapters, and helped with typing, printing, drawing graphs, software, and so on. High on the list are the following:

ASME for giving me permission to reprint some of the symbols and definitions from the ASME Y14.5 2009 standard. Specifically, figures C.5, C.6, and C.7. Special thanks also go to Ivette Estevez, CAP administrator, ASME, for helping in the process of granting the permission.

G. Kontos, PhD; R. Munro, PhD; and J.J. Yu, PhD; for their relentless encouragement and support. They were always there to bounce thoughts, ideas, and content. I appreciate their effort.

C. Stamatis and S. Stamatis, PhD, my sons, for helping with software and computer work. I do not think that this book would have been completed without their constant help—especially when problems arose due to power failures.

K. Kukor (Freudenberg-NOK), E. Voskuhl (DANA), R. Horner (DANA), R. Landreth (Stant), C. Sanchez (Federal Mogul), A. Barnett (Quantiles), J. Methot (Methot Associates), S. Mitchell (SJM Associates), J. Lapolla (General Physics), C. Cunitz (MPC), and K. Little (Bosch) for their advice on supplier issues/perspectives as well as ideas on specific content that makes suppliers more efficient and effective in their roles to satisfy the customer.

E. Liasi, PhD (Ford), R. Jiang (Ford, China), J. Tassis (Ford), Y. Yin (Ford), A. Katakis (GM), B. Mooridian (GM), J. Kapur, PhD (Six Sigma Solutions),

G. Polychronis (Chrysler), K. Shankar (Parker), M. Lutz (Kostal), A. Allen (Weirton), M. Mihas (Chrysler), S. Gikas (Chrysler), N. Papazoglou (Chrysler), and B. Anagnostou (GM) for their detailed suggestions and advice to make the requirements of the original equipment manufacturers (customers) more concise and easily understood.

G. Kissonergis (Process Product Consulting), N. Simons (Performance Innovation), V. Simmers (Quality Performance), J. Worley (Worley Associates), and A. Woodley (Woodley Consulting) for reviewing and recommending suggestions to improve flow and content.

Whereas all their suggestions and advice were considered, the final product is the author's responsibility. I have incorporated many of their ideas and suggestions throughout the chapters.

I also thank the editors of the book for designing it in such a way that made the final product look professional and easier to read.

Furthermore, I thank all my clients (both original equipment manufacturers and suppliers) that have given me the opportunity to learn from their processes and transfer that learning into the ideas of this book.

My list would not be complete without recognizing the most important supporter, editor, motivator, and life partner, my wife, Carla. As always, she has been there to motivate me and encourage me to go on even at the lowest points of my writing. Thank you so much, Carla Jeanne. Without you, this project would not have been completed.

Author

D. H. Stamatis is the president of Contemporary Consultants Co., in Southgate, Michigan. He is a specialist in management consulting, organizational development, and quality science. He has taught project management, operations management, logistics, mathematical modeling, economics, management, and statistics at both graduate and undergraduate levels at Central Michigan University, the University of Michigan, Anhui University (Bengbu, China), the University of Phoenix, and the Florida Institute of Technology.

With over 30 years of experience in management, quality training, and consulting, Dr. Stamatis has serviced numerous private sector industries including but not limited to steel, automotive, general manufacturing, tooling, electronics, plastics, food, U.S. Navy, U.S. Department of Defense, pharmaceutical, chemical, printing, healthcare, and medical device.

He is a certified quality engineer through the American Society of Quality Control, certified manufacturing engineer through the Society of Manufacturing Engineers, certified Master Black Belt through IABLS, Inc., and a graduate of BSI's ISO 900 Lead Assessor training program.

Dr. Stamatis has written over 70 articles, presented many speeches, and participated in both national and international conferences on quality. He is a contributing author on several books and the sole author of 52 books.

Introduction

The world of quality has been involved with *quality* ideas since prehistoric times, from primitive approaches to craftsmanship, to MIL-STDs, to general industry standards, to international standards, and to specific industry standards.

However, none of them has been successful enough to say with a sense of certainty that quality has overcome its nemesis of timing and price. (We purposely mention price because that is the cost per unit on which most organizations focus. Organizations should focus instead on the lifecycle cost for a best estimate for the true cost.) To be sure, quality has evolved to levels of unprecedented magnitude, but we still have a long way to go.

In fact, for years we have been conditioned to think that a successful organization is one that will have a good product and a fast way to bring it to market. It may have worked in the past for some. However, in the world of today, the options of price, time, delivery accuracy, and dependability from both OEMs (customers) and suppliers may be summarized in three categories. They are

1. Good and fast delivery of a finished product: This implies quality and expensive product or service. In everyday language, we think of this option as a way to bring *all* resources together with top priority. This means we will not cut corners but rather we will encourage and use human and technological resources to ensure the product is completed on time and will be of the highest quality. The cost may be high up front, but in the end it will be more cost-effective and it will gain favorability from the ultimate customer. This option is by far the best!
2. Good and cheap delivery of a finished product: This implies slow and mass production of a product or service. In everyday language, we think of this option as a potential discounted product and as a consequence, we give it less priority and certainly less human and technology resources. The result of this option is that the quality is going to be acceptable; however, it will take a long time to accomplish based on delays in design or manufacturing, and/or delivery concerns.
3. Fast and cheap delivery of a finished product: This implies inferior product or service. In everyday language, we think of this option as *you get what you pay*. Obviously, this option is lacking in many ways, but predominantly in lower priority and fewer allocated resources. Its purpose is to deliver it at all costs on time regardless of quality. This is the worst option.

Unfortunately, many organizations follow Option 2 or 3. How else can we explain major mishaps in so many industries and in some cases fatalities because of lack of quality? The latest disasters in the automotive industry because of some fundamental mistakes (oversights?) prove the point. They are

- Power steering failures
- Accelerator problems (unintentional acceleration)
- Transmission gear jumps from park to reverse
- Ignition issues (fire incidents)
- Wiring issues (fire incidents)
- Brake issues (unusual corrosion)
- Nondeployment of air bags

These failures include automakers from United States, Germany, France, Sweden, and Japan, and involve millions of recalls in a variety of models over the last 10 years. What is interesting about these failures is the fact that all of the failures have occurred to critical or essential items in the automobile. In fact, after over 100 years of producing these cars, one would think that they would have perfected the items in question. Obviously, they have not. The question is why? There are many reasons for it, but we believe the primary reasons are the desire to cut costs and to shrink the time from development to Job 1 (the production date). Furthermore, artificial due dates end up in failures—quite often in fatalities—due to the tendency to compromise (cut corners) the standards, procedures, and practices that define the quality of the products/services.

It is interesting to note here that even compliance with international standards, recipients of the *Malcolm Baldrige National Quality Award* or strong followers of the Six Sigma methodology have not prevented catastrophic failures. The reason is that systems exist for fundamental improvement of quality, but a strong strategy for its implementation and sustainability is lacking. We must note here that even the certification to international and/or industry standards is not a guarantee that failures will not occur.

This book is focused on *doing things right the first time* including the launching requirements for a flawless implementation of a program and/or service. We have addressed some generic issues, especially from the automotive industry, which we believe transcend industries and may in fact be equally applied to service organizations with minor modifications.

In a sense, we have tried to summarize this effort in the following principles:

- More service. More service means real-time problem solving by experienced and dedicated customer service groups that are on site at each facility.

- More reliability. More reliability means strategically located production facilities that offer the same flexibility, quality, and service every time everywhere.
- More quality. More quality means quality management systems are in place that, coupled with state-of-the-art appropriate and applicable processes and unparalleled attention to detail, will ensure that a product meets the highest standards.
- More flexibility. More flexibility means a lean organization with the ability to quickly adapt to instantly changing business environments and production design requirements.

We also have recognized that the engines of change in both OEMs (customers) and suppliers are the leaders of management. It is management's role to direct and influence the organization. This leadership must be displayed on a daily basis through commitment and behaviors. Without management's commitment, there is no *do it right the first time*. The necessary behaviors for strong leadership are

- Integrity (does the right thing)
    - Exemplifies honesty and maintains trustworthiness
    - Exercises principled judgment, especially on the tough calls
    - Keeps one's word despite the consequences
- Courage (takes action in the face of challenge)
    - Offers new ideas that question the status quo
    - Takes risks in championing better ideas
    - Demonstrates judgment and self-confidence, even in stressful situations
- Teamwork (collaborates to achieve results)
    - Values team members with different ideas, points of view, and backgrounds
    - Acts to break down barriers and chimneys (silos) to innovative team ideas
    - Demands team-oriented behavior and insists on personal accountability for such behavior
- Durability (perseveres despite hardship)
    - Maintains originality and creativity in staying the course to achieve agreed upon objectives
    - Shows tenacity and boldness in securing and using resources
    - Maintains inspiration, focus, intensity, and persistence, even under adversity

- Communication (exchanges information and ideas that impact and influence others)
    - Listens completely, and then confidently speaks up on the issues
    - Provides concise, compelling, and innovative evidence to support positions
    - Demonstrates sensitivity to language and cultural communication requirements
- Systemic thinking (sees beyond the details)
    - Thinks cross-functionally about ideas that impact the business
    - Boldly pursues ways to improve processes and incorporate new ideas
    - Inspires systemic change efforts that make a difference
- People development (teaches, develops, and motivates people)
    - Values and confidently promotes a diversity of new ideas and a diverse workforce
    - Acts to enhance the creativity and professional development of self and others
    - Treats everyone with fairness and respect regardless of position or social status
- Desire to serve (demonstrates personal commitment)
    - Seeks new ways to ensure ultimate customer enthusiasm
    - Determined to achieve the organization's objectives and act in the organization's best interests.
    - Accepts willingly the challenge of different functional and geographical assignments
- Drive for results (gets the job done)
    - Develops challenging, innovative objectives and accepts personal responsibility for accomplishing them
    - Prioritizes resources, inspires performance, and measures outcomes
    - Negotiates agreements that move the business forward
    - Focuses on prevention rather than detection methods to improvement
- Innovation (applies learning for competitive advantage)
    - Demonstrates adaptability and flexibility in evaluating creative ideas
    - Applies lessons learned from successes as well as failures to inspire new ideas
    - Dares to dream about and implement new ways of doing business

- Business acumen (understands the essential requirements of our business)
  - Knows the organization's business operations and the global business context in which the organization operates
  - Knows the basic business principles used to achieve quality, customer, and profit outcomes in a global competitive environment
  - Demonstrates functional and technical proficiency
- Quality methods (understands what it takes to do quality work)
  - Shows passion for achieving quality
  - Measures and monitors quality on an ongoing basis
  - Demonstrates resolve for meeting the ultimate customer's quality requirements

Hopefully, the reader will agree with our approach.

Section I

Original Equipment Manufacturing (OEM) Responsibilities

With the acceleration of innovation in all industries, the need to ramp up to faster volume (if manufacturing) and increases in customer satisfaction (if service) while contending with more complex products/services variants is greater than ever.

To stand out, organizations need to break through with appropriate and applicable approaches with maximum and timely effectiveness. We believe these approaches are based on the principle of *predictive launching*. Maybe they will not accomplish 100% of the goal, but they are sure to be infinitely more successful if they follow the approaches we mention here. The focus is on attaining first-time quality targets, minimizing production issues, optimizing delivery times, and repeating the successes through standardizing the process systemically in the entire organization.

To be sure, original equipment manufacturers (OEMs) (customers) is faced with increasing demands for profitability, especially from global competition, finite resources, shareholder expectations, consumer demand, and rapid technological innovations. As much as the task of improvement seems a herculean one, there is a way to counterbalance these concerns. The way is through a proper and effective strategy. The strategy must be focused on setting a direction for the organization, marshalling support from within and without (external sources [suppliers]), and applicable methodologies for improvement.

In this part of the book, we will examine some critical issues that can help all OEMs (the customer) to develop the appropriate strategies for flawless launching. In Section II, we will focus on supplier issues.

1

Timing (Product Development)

Overview

Everything begins with an idea followed by a concept, development, production, and finally delivery to the ultimate customer (consumer). Whereas one can indeed verbalize the process and in fact it is a very easy process to do, the reality of implementing all of these stages is that it is very difficult to harmonize them and bring them to a conclusion on time. To put this in perspective, think of the typical automobile, which has over 12,000 individual parts, yet they all have to be planned for production on a specific automobile. An even more incredible statistic is the over one million individual parts of a typical plane that have to be planned for production on a specific plane.

Therefore, the intent of timing is to recognize this difficulty and plan accordingly all the phases involved in the generating idea/concept to the delivery of the product/service to the ultimate customer. This means that all industries have their own requisites of timing; however, depending on the product/service, this timing may vary considerably. We must emphasize here that this definition of timing must be *appropriate, applicable,* and *doable* for the specific industry and its goal of introducing a new product/service in the marketplace. Unless these requirements are met, the organization will experience delays, overruns, and missing deadlines. A good timing program is very important to both the original equipment manufacturer (OEM) (customer) and the supplier. For example, in the automotive industry, suppliers are very critical to the entire process of development as they provide (generally) 70%–80% of the value of a typical vehicle. Furthermore, a good timing plan controls the information flow in both directions (OEM↔supplier), helps in standardizing the business across divisions (nationally and internationally as applicable), and helps the OEM attain a best-in-class (BIC) level of quality at Job #1 (launching day). A typical development plan will include, at the very minimum, the following stages:

- Validation of facilities and launch ramp-up
- Validation of product and process and verification facilities
- Verification of product and process design

- Optimization of product and process design
- Confirmation of product and process design
- Identification of potential quality and capacity capability
- Validation of an established program for customer satisfaction

To be sure, any organization must define for itself what is really meant by "appropriate," "applicable," and "doable" timing for its business. In fact, the original timing may be revised on a consistent basis to adjust for lessons learned. This adjustment may be because of supplier issues or recommendations, or the OEM's redefinition of requirements or technological innovations. Furthermore, this adjustment makes the process a *continual improvement* or rather a *living practice* for everyone concerned. This means that customers' and suppliers' management, engineers, and manufacturing establish the parameters of the timing needed to complete the tasks via cross-functional teams and open communication.

The team—and this is very important—must be given specific guidance for improvement and meeting expectations, but also it must be based on consensus-based decision making. Perhaps two of the most important guidelines for this are to

- Create a single—as much as possible—process for all customer brands and locations. In other words, standardize as much as possible.
- Build upon the lessons learned and incorporate BIC approaches—whenever possible—via benchmarking from any source that offers BIC methodologies for improvement.

This team with its appropriate and applicable representatives must develop and align products, services, and suppliers with related roles and responsibilities as well as appropriate structural support to improve the competitiveness of the company and to improve the productivity of the supplier. Therefore, the scope of the team is to harmonize the process of receiving products or services from a supplier on a consistent basis, on time, as well as meeting or exceeding the requirements from their suppliers. Some of these expectations and deliverables are

1. Define a generic agreed upon process to be flexible to meet minimum changes without setting the program back or making the program have budget overruns.
2. Define a robust program to be modified based on new information, technology, and methodologies learned from BIC studies. This will facilitate
    a. Early concerns and involvement with suppliers
    b. Cross-functional meetings at the supplier's facilities

   c. Agreement on key metrics for reporting quality based on data-driven methodologies
   d. Single voice for the cross-functional and multidiscipline team
3. Define (and follow) detailed roles and responsibilities for all concerned.
4. Define the structure (meetings, decision making, information flow, reporting, and communication) to enable a successful resolution for the customer and supplier.
5. Define the reconciliation process for concerns in such a way to allow execution of the process within existing constraints. Make sure this reconciliation is based on a *win-win* resolution. Any other approach will fail!

The idea here is not to make things difficult for anyone but rather to create a culture that works for the benefit of both OEM and supplier as well as to incorporate improvements—as appropriate—during the development and implementation of the program.

Traditionally, many OEMs directed their supplier communication within functional boundaries. To be sure, the complexity of the customer/supplier business relationship creates concerns that must be resolved in most cases with cooperation that cuts across internal customer functions and organizations. That is the reason why in many parts of the world, many OEMs have adopted a fragmented, chimney approach to working with suppliers. Despite the fact that much of the work with suppliers is interrelated between the different customer functions, they often drive separate, disconnected work-streams with suppliers.

As a result, situations are created where a customer gives conflicting direction to the supply base, and, worse, the problem at hand is not resolved or help is not offered to the supplier to solve it. This could be avoided if the appropriate time and study, that is, benchmarking or experience from lessons learned, were used with a cross-functional team of the customer and supplier at key milestones within the program development cycle. This simple approach can be more effective and efficient to resolve issues, concerns, and problems, and ultimately improve launch readiness and quality.

By now, the reader should have realized that the timing for everyone (customer and supplier) has become an integral part of reducing the time from development to production. Therefore, it must follow Lean approaches—however appropriate and applicable but more importantly *realistic milestones*—so that waste and inefficiencies are eliminated or at least minimized. Of course, this depends on good interactions with suppliers and excellent communications as to what the requirements are from the customer.

Principles of Organizational Timing

In timing, there are several principles to be recognized and followed. We believe that the most important ones for a successful implementation are the following eight.

1. Supplier launch success is a cross-functional, shared responsibility of customer (product development, buyer, supplier engineer) and supplier (core team members). This principle is very important, but not always recognized within the customer culture. This of course implies correctly that supplier success is a function of a number of different team members. No single function—on its own—can deliver quality from a supplier. It is always a team undertaking. The core team is made up of individuals who have *ownership, authority,* and *responsibility* for the program. There are others to be sure, but these folks are the highest leverage team members during a launch.
2. Customer cross-functional teams will work proactively with the highest leverage (priority) suppliers on a program. The second principle acknowledges the fact that while it would be ideal to have cross-functional customer teams working with every single supplier of new-tooled end items on a program, it is not possible or practical to execute that way. Therefore, within the process, the team will target its cross-functional efforts on the most important subset of suppliers. It should also be pointed out that this process is designed for those major programs that have the possibility of the highest degree of engineering change or historical problems. (Generally, a program's suppliers are defined as "priority" and "nonpriority" or "critical" and "noncritical." Priority suppliers have new-tooled end items that present risks to the launch [site, part, or program risk—no historical background, financial stability]. Therefore, the timing supplier engagement process usually focuses the cross-functional customer resources on the priority or critical suppliers. Nonpriority or noncritical suppliers should be required to conduct self-assessment and submit to the customer any deficiencies or problems. Generally, on-site visits by the supplier engineer are focused on the priority and most critical suppliers.)
3. Supplier engagement will begin early in a program—immediately after sourcing. The third principle is very important as well because generally it is a departure from past practice for both customer and supplier. In this principle, we emphasize the need to begin working with these critical suppliers as soon as they are sourced. We do that because we want to identify concerns early and fix them before it becomes prohibitively expensive to do so. Fixing issues or concerns

may help us avoid problems later in the process. It also allows the customer's manufacturing plants to focus on their internal concerns instead of supplier concerns during preproduction builds.

4. Customer cross-functional teams will visit the key supplier manufacturing facility a minimum of four times. Perhaps the most critical of all the principles is the customer team's responsibility to visit the supplier's manufacturing facility at least four key times during the development cycle. (Obviously, the frequency of the visit depends on the customer, the supplier's history, and the product. Here we use four *only* as a guideline.) To help the supplier launch parts, the team has to have a common understanding of the parts and process. These visits are synchronized to *key program milestones* to ensure that supplier deliverables support customer deliverables and are on time.
5. The team will provide the supplier with a single customer voice on key launch matters. As we have discussed before, a key goal of timing supplier engagement is to provide a single voice to the suppliers. *No* customers can afford to waste time anywhere internally in the organization. If they do, there is a high price to be paid from both competitor and cost perspectives. Therefore, ensuring appropriate and applicable interface with suppliers should be as efficient as possible. Underlying advanced product quality planning (APQP) and production part approval process (PPAP) processes and roles and responsibilities will remain the same. It is important to understand that while the timing supplier engagement process defines a new way to work with the supply base, it does not change any of the underlying processes that suppliers and customers have used in the past. The elements of APQP and PPAP are still the same and still reflect industry standards. We must emphasize here that we are not proposing any changes in *what* we do, but we do strongly suggest that the change has to be in *how* we do it. (Remember, all customer roles and responsibilities within this process are structured around ownership of the key elements of the industry standard APQP and PPAP. APQP and PPAP are existing industry processes/standards and are both fundamental building blocks of any timing scheme.)
6. Team engagement will continue until the supplier has successfully completed all PPAP requirements and fully met all program ramp-up volumes. This means the part submission warrant (PSW) has been approved by the customer/supplier engineer and all functional tests have been completed and approved by the customer's plant. It probably sounds obvious, but the customer team (especially the customer engineer) needs to stay working with the supplier until the parts are fully PPAPed.

7. The customer will implement the supplier engagement process consistently on a global (if applicable) basis in alignment with the global makeup (if applicable) of the supply base.
8. Finally, since the timing is a process that is used to optimize the overall timing for the product/service, it must be applied equally to every supplier. In other words, it must be the customer's standard for any supplier no matter where they are.

Special note: The reader may recognize that some of the points mentioned here are specifically for the automotive industry. However, the principles of APQP may be applied to any organization and to any industry. The discussion of the PPAP is only appropriate for product. However, in service industries the principles of the PPAP may also be applied as appropriate especially during formative and summative evaluation.

APQP (Based on the Automotive Industry)

In the most generic form, APQP is nothing more than a planning methodology to make sure that a supplier fulfills all the customer's requirements and delivers the product/service on time, within cost, and with all quality requirements completed without any difficulty or delay. Table 1.1 shows the key characteristics of an APQP, the relationship to the PPAP, and some expected deliverables.

For more detailed information, the reader is encouraged to see AIAG (2008a,b) and Stamatis (1998).

Typical Example of Timing in the Automotive Industry

The following example is a typical implementation of a timing process for a critical part. The items identified here are not complete but rather an example of key milestones that an organization may develop for their product. In addition, the months are not real. They are an example of timing that may be considered. They are used here to show the flow and the importance of milestones.

- KO, kickoff: About 40 months before Job 1
- PTC, program target compatibility: About 35 months before Job 1
- PA, product approval (VP level): About 30 months before Job 1
- M1DJ, data judgment: About 27 months before Job 1
- FDJ, final data judgment: About 20 months before Job 1

TABLE 1.1

Typical Example of APQP/PPAP and Expectations

Item	APQP	APQP/PPAP	Examples of Deliverables
1	Sourcing decisions	Sourcing decision	1. Required sourcing agreement signed—site identified 2. Required supplier commercial and program agreements signed 3. All production tool orders issued to suppliers 4. Final mix, maximum weekly volume communicated and agreed on
2	Customer input requirements	Customer input requirements	5. Program expectations cascaded to suppliers
3	Craftsmanship/appearance approval report (AAR)	Craftsmanship/AAR	6. AAR approved 7. Color changes completed supporting color harmony
4	Design failure mode and effect analysis (DFMEA)	DFMEA	8. DFMEA completed 9. Special characteristics cascaded to suppliers
5	Design/manufacturing reviews	Design/manufacturing reviews	10. Quality issues closed and effective
6	Design verification plan and report (DVP&R) material, performance test results	DVP&R material, performance test results	11. Design verification (DV) testing complete
7	Subcontractor APQP status	Subcontractor APQP status	12. Subsupplier PPAP timing plan completed 13. Subsupplier PPAP run-at-rate (or equivalent) completed 14. Subsupplier PPAP capacity (or equivalent) completed
8	Facilities, tools, and gauges	Facilities, tools, and gauges	15. Facility and tooling timing plan completed 16. Facilities/tools/gauges are at the final production 17. Supplier's OEE plan is confirmed by surrogate 18. Gauge plan (including released gauge drawings) completed 19. Supplier's demonstrated OEE (run-at-rate) supports capacity requirements
9	Prototype build control plan	Prototype build control plan	20. Prototype build control plan completed

(Continued)

TABLE 1.1 (Continued)

Typical Example of APQP/PPAP and Expectations

Item	APQP	APQP/PPAP	Examples of Deliverables
10	Prototype builds	Prototype builds	21. Plan to complete tool orders release to suppliers in support of system build 22. Plan to complete parts orders to suppliers in support of product build 23. Plan to complete parts orders to suppliers in support of product build
11	Drawing and specification design records	Drawings and specifications	24. Design styling confirmation 25. Plan to complete parts at a level that can be tooled for production release by final data judgment
12	Engineering change document	Engineering change	26. All customer engineering changes approved and recognized by the supplier
13	Team feasibility commitment/ customer engineering approval	Team feasibility	27. Design, process, and timing feasibility confirmation
14	Manufacturing process flowchart/ process flow diagram	Manufacturing process flow	28. Final process flowchart supports PPAP phase
15	Process failure mode and effect analysis (PFMEA)	PFMEA	29. Final PFMEA completed with linkages
16	Measurement systems evaluation/ measurement system analysis studies	Measurement system analysis	30. Gauge repeatability and reproducibility (R&R) results $\leq$10% per PPAP customer specifics
17	Qualified laboratory documentation	Qualified laboratory documents	31. Supplier internal and external laboratory compliance
18	Checking aids	Checking aids	32. Checking aids compliant with part specifications
19	Pre launch control plan	Pre launch control plan	33. Pre launch control plan completed with linkages
20	Operator process instructions	Operator instructions	34. Operator process instructions completed
21	Packaging specifications	Packaging specifications	35. Packaging approval process completed
22	Production trial run	Production trial run	36. PPAP run-at-rate (including sub-suppliers) completed

TABLE 1.1 (Continued)

Typical Example of APQP/PPAP and Expectations

Item	APQP	APQP/PPAP	Examples of Deliverables
23	Production control plan/control plan	Production control plan	37. Production control plan complete with linkages
24	Dimensional results	Initial process capability	38. Initial process capability results (Ppk $\geq$ 1.67)
25	Initial process capability study	Dimensional results	39. 100% of required measurement points within tolerance
26	Production validation (PV) testing	PV testing	40. PV testing complete
27	Part submission warrant (PSW)	PSW	41. PPAP quality characteristics completed 42. PPAP all streams for quality characteristics completed 43. PPAP capacity completed 44. Supplier's demonstrated OEE (final capacity) support capacity requirements
28	Bulk material requirements	Bulk material requirements	45. Bulk materials checklist included in PPAP submission
29	Sample product	Sample product	46. Sample product produced with customer's identification
30	Master sample	Master sample	47. Master sample approved
31	Record of compliance	Record of compliance	48. Customer-specific requirements documented

- VP, verification prototype: About 17 months before Job 1
- PEC, preliminary engineering completion: About 10 months before Job 1
- LR, launching readiness: About 6 months before Job 1
- TT, tooling trial: About 5 months before Job 1
- PP, pilot production: About 4 months before Job 1
- MP, mass production: Zero months before Job 1
- J1, job one: Full production

References

AIAG (2008a). *Advanced Product Quality Planning and Control Plan*. 2nd ed. Southfield, MI: Chrysler Corporation, Ford Motor Company, and General Motors. Automotive Industry Action Group.

AIAG (2008b). *Production Part Approval Process*. 4th ed. Southfield, MI: Chrysler Corporation, Ford Motor Company, and General Motors. Automotive Industry Action Group.
Stamatis, D. (1998). *Advanced Quality Planning*. New York: Quality Resources.

Selected Bibliography

Bhote, K. (1991). *World Class Quality*. New York: American Management Association.
Griffiths, D. (1990). *Implementing Quality with a Customer Focus*. Milwaukee, WI: Quality Press. White Plains, NY: Quality Resources.

2

Basic OEM Quality System

Overview

In today's world of globalization, OEMs must have some criteria for quality, capacity, delivery, and performance. Those criteria, depending on the organization and product, do vary; however, all of them have four basic characteristics and requirements. They are

1. Certification to ISO 9001 standard (ISO 2008)
2. Certification to ISO 14001 standard
3. Certification to industry standards, that is, ISO/TS 16949–automotive; AS 9100–aerospace; and others
4. The organization's own standards, that is, Ford Motor Company's Q1 (Ford Motor Company 2013), Chrysler's Pentastar, and GM's Global Quality

The first three are generally the basic prerequisites of a supplier being certified in order for a customer to pursue doing business with a particular supplier. In fact, for Ford Motor Company, unless the supplier is certified to ISO/TS, it will not be qualified for the Q1 certification. (In special cases, this requirement is waived; however, this is not common.) Let us examine these four certifications in a cursory approach.

International Industry- and Customer-Specific Standards

Certification to ISO 9001

This is a general and very basic standard dealing with overall quality in any organization. The effectiveness of the standard in any organization is based on third-party certification. The customer's supplier engineer is responsible for making sure that the third-party auditor is authorized by the International Automotive Task Force (IATF) and validating the certificate of the certification (date and scope). Specific follow-up may be required if during the customer audit there are significant deficiencies in the supplier's quality system. The standard is made up of the following elements (clauses):

1. Scope
2. Normative reference
3. Terms and definitions
4. Quality management system
5. Management responsibility
6. Resource management
7. Product realization
8. Measurement, analysis, and improvement

Annex A: Control Plan Requirements

What is important about these elements is that (1) items a–c are not certifiable, (2) items d–h are certifiable, (3) the annex is a guide to a control plan, and (4) the certifiable items include four specific words with specific meaning. They are

- "Shall" indicates a must requirement.
- "Should" indicates a recommendation.
- "Note" is a guideline for understanding.
- "Such as" are examples given for guidance only.

In addition to these special words, the standard is revised (updated) every 5 years and it is up to the user to keep up to date. The current standard is the ISO 9001:9008; however, the draft for the ISO 9001:2015 is available and there are some pending changes.

Part of this certification process to the ISO is the need for most organizations to follow the ISO 14001 environmental standard as well. Both standards (ISO 9001 and ISO 14001)

- Focus on the achievement of results.
- Satisfy the needs, expectations, and requirements of the customer.
- Facilitate planning and allocation of resources.
- Evaluate the overall effectiveness of the organization (system, *not* product). This is very important to differentiate and part of the reason the industry standards were developed was to fill that void.

Certification to ISO 14001

Specifically, the ISO 14001 mandates that organizations need to achieve and demonstrate sound environmental performance by controlling the impact of their activities, products, or services on the environment. A summary of the elements is shown in Table 2.1.

Certification to Industry Standards

Generally, the industry standard is a combination of the ISO 9001 standard plus generic industry definitions and additional quality requirements common for the industry. A typical one is the automotive ISO/TS 16949; another one is the aerospace AS 9100, and there are many more. The format for the industry standard is that it includes all requirements of the ISO generally within a box. The information outside the box is the additional requirements of the specific industry. The specific words mentioned previously are the same and have the same meaning as the standard proper.

Generally, the goal of the industry standard is the development of a quality management system that provides for continual improvement, emphasizing defect prevention and the reduction of variation and waste in the supply chain. To do that, it covers the same exact elements of the ISO and goes beyond them. For purposes of an example, we have chosen the automotive ISO/TS 16949. For discussion that is more extensive, the reader is

TABLE 2.1

Summary of ISO 14001 Key Elements

Element or Clause	Title	Key Components
4.2	Environmental policy	Policy scale, continual improvement, environmental regulations, objectives, and targets
4.3	Planning	Aspects, legal, objectives and targets, environmental management programs
4.4	Implementation and operation	Structure and responsibility, training awareness and competency, communication, documentation, operational control, emergency preparedness
4.5	Checking and corrective action	Monitoring and measurement, corrective and preventive action, records, audits
4.6	Management review	Adequacy and effectiveness

In essence, the ISO 14001 requires insurance that management commitment and support for continual improvement and environmental sustainability are always present and communicated throughout the organization via

- Developing an environment policy
- Ensuring availability of resources
- Planning and implementing this policy in the organization
- Auditing and reviewing this policy for compliance and continual improvement

encouraged to see Stamatis (2004). The elements are presented here with a minimal explanation:

1. *Scope*: Can be applied to (only) manufacturing sites throughout the supply chain (including customer or affiliate assembly and manufacturing centers). However, it does not cover warehousing.
   a. Manufacturing
      i. Production materials
      ii. Production or service parts (if applicable)
      iii. Assemblies
      iv. Heat-treating, welding, painting, plating, or other finishing services
   b. Supporting functions: On-site or remote (they are part of the site audit). The only permitted exclusion is for product design and development (manufacturing process design is not excluded).
2. *Normative reference*: Normative references are required. In this case, the normative standard is the ISO 9000:2008, Quality management systems–Fundamentals and vocabulary: for definitions or clarification of terms and words.
3. *Terms and definitions*: These are used by third-party certification bodies (CBs) to identify supply chain hierarchy, defined in Technical Specification (TS) 16949. When the company being certified is Tier 1, the following definitions apply: OEM/affiliates/receiving plants are referred to as "Customer." On the other hand, Tier 1 is referred to as "Organization" (the Organization is the entity being certified). Tier 2 is referred to as "Supplier," and any Tier 2+ is recognized as "Subcontractor." In addition to these generic terms, there are additional definitions with which the reader should be familiar. They are
   a. Control Plan: Specified in Annex A
   b. Design Responsible Organization: Authority to establish or change product specification
   c. Error Proofing: Prevent manufacturing nonconforming products through product design and manufacturing process design
   d. Laboratory: Facility for inspection, test, or calibration
   e. Laboratory Scope: A controlled document of methods, operations, and equipment
   f. Manufacturing: Process of making or fabricating, production materials, service parts, assemblies, special processes (heat-treating, welding, painting, plating, finishing, etc.)
   g. Predictive Maintenance: Use of data to predict when maintenance is required, rather than using a fixed schedule, and to

avoid failure modes and problems while eliminating unnecessary maintenance

h. Preventive Maintenance: Process design element action to eliminate causes of equipment failure
i. Premium Freight: Costs or charges that exceed contract deliver
j. Remote Location: Location that supports a site and where non-production processes occur
k. Site: Location where value-added manufacturing occurs
l. Special Characteristic: Product or manufacturing parameter that affects safety, fit, function, performance, or subsequent processing

Special note: We must emphasize here that Scope, Normative References, Terms, and Definitions are *not* auditable or certifiable. Certification is only for Elements 4–8.

4. *Quality management system*: This element requires that a documented, implemented, and maintained system exists and is continuously improved. Furthermore, it is process-based and it holds the Tier 1 (organization) accountable for outsourced processes. A typical quality management system shall include the following documentation:
   a. Quality policy
   b. Quality objectives
   c. Quality manual
   d. Documented procedures
   e. Planning, operation, and process control documents
   f. Customer engineering standards/specifications (customer driven)
   g. Evidence of conformity to requirements in process and operations
   h. Change control for supplier-initiated changes (customer driven)
5. *Management responsibility*: Top management is the people who control the strategic decision making for the organization. (Who spends the money or authorizes such? Who resources the people, equipment, and other? Who approves and instigates policies and procedures?) Top management must take an active part in knowing the inputs and outputs of the Quality Management System, and be accountable for results. This means top management, through management reviews, shall provide evidence of its commitment to develop, implement, and improve the quality system. Specifically, top management should review and evaluate
   a. Results of audits
   b. Customer feedback

c. Process performance and product conformity
d. Status of preventive and corrective actions
e. Follow-up actions from previous management reviews
f. Changes that could affect the quality management system
g. Recommendations for improvement

6. *Resource management*: This element focuses on *employees affecting quality*. This means that employees affecting quality shall have appropriate education, training, skills, experience, on the job training, or any combination of these. As a consequence, the scope, that is, who is included to do their tasks and an organizational chart to see their authority and responsibility in the organization, should be available. This information should be asked by the customer and provided by the supplier. Another issue of evaluating here is cleanliness. However, the evaluator must be cognizant of the different commodities produced because they may have different criteria. For example, a sand casting will differ considerably from a spark plug production. Some specific items to audit and evaluate the existence of policies and procedures or systems are
   a. The maintenance of appropriate training records for employees as evidence of keeping or improving skills for quality
   b. The maintenance of a process to motivate and empower employees
   c. The presence of a program to promote quality and technological awareness throughout the organization
   d. The presence of an opportunity to utilize a multidisciplinary approach (Cross-Functional Work Team) to establish work plans that consider all aspects of Product Realization
   e. The maintenance of emergency contingency plans
   f. The presence of a system to minimize potential risks to employees
   g. The existence of procedures to provide a clean environment consistent with product and manufacturing process needs (as required)

7. *Product Realization*: By definition, product realization is to produce a product from conception to delivery, to the end customer using a disciplined approach. For most customers and suppliers, the most common and frequent approach to product realization is APQP. In some organizations, the APQP is combined with the timing and milestones of a given organization. In either case, the supplier must have documented evidence that it can effectively deliver a product (cradle to grave, as applicable). Product realization generically must include
   a. Quality objectives and requirements for the product
   b. Established processes, documents, and resources

c. Verification, validation, monitoring, inspection, and test activities, with criteria for acceptance
d. Recorded evidence the product realization process is working

On the other hand, a good product realization approach must specifically address the following:

- Requirements
  - Acceptance level defined (set at zero defects for attribute measurements)
  - Customer confidentiality protected
  - Government regulations, safety, recycling, Restricted Substance Management Standard (RSMS) understood and met
  - Tier 1 must follow PPAP, subtiers must comply with PPAP
  - Care for customer-owned property (e.g., tooling, semi-finished parts)
- Prevention
  - Supplier request for engineering approval (SREA) used for any zero cost manufacturing process changes by the supplier
  - Error prevention techniques are to be used as the preferred process control methods, especially when human inspection is the major control
  - Control plans must have specified reaction plans to address unstable or out of control conditions
  - A process must exist to prevent occurrence of customer concerns
  - The organization must have a system to maintain preventative-predictive maintenance; record and monitor annual or at least quarterly findings
- Communication
  - Tier 1 must be able to communicate in the language and format of the *customer* (e.g., CAD data and data exchange, problem-solving technique, compatible software)
  - Work instructions at workstations
- Monitoring/measuring
  - Tier 1 suppliers use quality operating system (QOS) to monitor, analyze, and report measurements and summarize the status of the organization to management
  - Validate first piece and job setup
  - Statistical process control (SPC) and gauges referenced in the control plan
  - Gauge repeatability and reproducibility (R&R) control and calibration

  - Define and document internal/external laboratory scope (for PPAP and control plan measurements)
  - Capacity forecast and monitored according to agreement with customer
- Subtier supplier management
  - Tier 1 shall maintain a quality management system with subtiers
  - Unless waived by the customer, Tier 2 suppliers shall be third-party registered to ISO 9001 latest revision by an accredited CB (appropriate certification letter or note must be available if a waver is issued)
  - Tier 1 must monitor Tier 2 on rejects, parts per million (PPM), field returns, and delivery
  - The organization must have a strategy for controlling incoming product quality

8. *Measurement, Analysis, and Improvement*: The focus of this element is to have the supplier demonstrate that an effective measurement system exists, appropriate analysis is conducted with applicable data, and a commitment to continual improvement exists throughout the organization. This is verified and validated with annual internal audits of the quality management system. Frequency can be adjusted if concerns exist. One of the primary concerns is the use and understanding of statistical concepts throughout the organization (stability, variation, C_{pk}, C_p, P_{pk}, P_p, over-adjustment, etc.). Note that for initial process capability demonstration, only the P_{pk} should be used. Some customers use computer software systems to monitor and react to items that affect the customer in any way. Other areas of importance are
   - Appropriate resources for appearance items (lighting, masters, equipment) (Note: Special skills are considered a resource)
   - 8Ds and error proofing for problem resolution
   - Control of nonconforming product
   - Tier 2 monitoring, especially of pass-through characteristics
   - Unidentified/untagged parts shall be classified nonconforming
   - Rejected parts from a customer are analyzed (plants, dealerships, engineering)
   - Must follow Annex A requirements for control plan (especially FMEA linkage)

Annex A–Control plan requirements: Annex A contains the contents of the control plan, in detail. It is not unusual to find individual customers that add or clarify the control plan requirements with their own specific requirements as part of their quality system. Their specific requirements also become the roadmap for their supplier to meet both the requirements and the expectations of the customer.

Typical specific customer requirements that the standard refers to may include

- Requirements for special processes, that is, heat-treating, welding, and so on
- Requirements for DFMEA, PFMEA, critical characteristics, and control plan linkages
- Lean techniques
- Requirements for subcontractor development
- Requirements for QOS
- Requirements for additional expectations for APQP
- Requirements for contingency plans
- Requirements for logistics and laboratory
- Requirements for environment such as: ISO 14000
- Requirements for dealing with customer concerns

Customer-Specific Standards

Every organization has a culture and a specific way of doing business. This specificity carries itself to the quality requirements and expectations within the organization itself but more importantly to its supplier base. To be sure, the internal requirements may be different and the implementation process may be less structured for many reasons. However, when an organization depends on suppliers for their quality, it is imperative that it is structured, systematic, and standardized. The reason for this is that the intent of the organization's perception of quality must be translated to the supplier in a way that the supplier will find value and want to actively participate in a system that is considered fair and excellent. It is a way for the customer to identify a superb supplier who is willing and able to perform on a consistent basis to the highest quality level and the greatest reliability.

In addition, this customer standard assures that the supplier has a way of demonstrating the discipline to ensure success through continual improvement. This demonstration is evaluated generally with a point system and critical metrics by which the supplier is measured and collected by the customer at least once a year unless there are problems in the interim. In essence, the certification to the organization's standard facilitates the sourcing process without any concerns or problems.

On the other hand, noncertified suppliers to a customer's standard implies they have not demonstrated consistent and superior performance. Therefore, they cannot be sourced without proper authorization and perhaps more frequent audits as well as stricter control of the specific metrics requirements under question. The reason for the additional attention is that in this case, the supplier represents a potential risk to the customer and as such, the customer must be protected.

So, what precisely are the requirements for attaining certification from the customer? Generally, three requirements define conformance to the customer needs. They are

1. Capable systems: A fundamental requirement that is based on a third-party certification to
    a. ISO/TS 16949 Quality Management System
    b. ISO 14001 Environmental Management System
2. Self-assessment: A self-evaluation of the quality system based on internal audits to identify any nonconformities. Generally, they are conducted once every quarter (more frequently, if needed).
3. Ongoing performance: A method of evaluation to track key metrics such as overall site PPM, PPM commodity performance–production or service (if applicable), delivery performance–production or service, warranty spikes, field service actions, stop ship, and violation of trust.

When these three requirements have been completed (sometimes they may not be; however, there is a major improvement toward their completion), the customer will audit the supplier. This audit is called a manufacturing site assessment (MSA) or facility assessment. It must not be confused with measurement system analysis (MSA). Whereas the measurement system analysis measures the integrity of the measurement systems, the site assessment is not a one-time event, but rather an enabler for continual improvement. Generally, this audit covers, but is not limited to, the following primary measures:

- Implementation of fundamental quality management system processes
- Change management
- Corrective/preventive action
- Failure mode avoidance

The evaluation process is a numerical value and it depends on the customer. No matter what the numerical values are, there is always a baseline the supplier must reach. It is also important to recognize that a positive score may not be transferred to a deficient metric that may be below the minimum. All scoring adjustments—if they do occur—are based on three items: capable systems, ongoing performance, and the manufacturing site assessment. A minimum score must be reached to either gain or maintain the certification.

Generally, this minimum score is 80% of the total. Less than that means the certification will not occur or it may be revoked if the shortages of points are in critical areas.

Yet another important issue with the MSA is that the supplier must have data for a considerable period, usually at least a year. It is also important for the supplier to update the customer on a monthly basis as to the improvements or corrective actions that are necessary for that improvement.

If the issues or problems are serious, the customer supplier engineer in conjunction with the appropriate supplier personnel will work together to identify improvements and appropriate corrective actions.

The evaluation for capable systems begins with the verification and validation of the ISO/TS 16949 and ISO 14001 as well as the MSA. The intent of the ISO and ISO/TS here is for the supplier to demonstrate that the basic characteristics of a quality system have been attained. The intent of the MSA is to show continual improvement and the process of correcting deficiencies. Key metrics in which the customer is interested are

- A track record of reduction for PPM (production and or service)
- Specific PPM commodity requirements, such as
    - Delivery performance, generally above the 8%–10% rating of the total rating scheme
    - Weighted 6-month average
    - Results of two complete months of production or service
    - Results of current running month even though it may be a partial month

The evaluation for warranty spikes deals with problems and metrics that are the result of field service actions (failures that have been experienced in the field) as well as stop shipments to the customer. In addition, a close evaluation is for historical issues or problems within the most recent 6 months (infant mortality) that are the supplier's responsibility. The expectations are zero for this period.

Disposition

To start the discussion of any disposition, we must make sure that both the customer and the supplier understand the requirements. Therefore, it is essential for the customer to have written procedures (and reference them when dealing with the supplier) that deal with the disposition of the MSA results so that there is no confusion if there is a conflict in understanding or implementing the requirements.

With that in mind, let us now describe a typical, yet very fundamental, approach of any customer's quality system and their expectations of their supplier. Of course, a customer may add others or modify what we have identified here depending on the product, service, and organizational relationship with the supplier.

Disposition Required

This is the first step of the process. It is generally assigned to a supplier site upon the assignment of a customer supplier engineer. This act is quite often a placeholder until and when the site is assessed, usually no more than 6 months. Under this condition, no sourcing is permitted. However, an MSA may change the disposition as it will determine the actual status code.

Working toward the Quality System

Perhaps this is the second most important disposition that needs to be understood by both customer and supplier. One of the significant factors is to determine whether the site is a green field—a planned facility but not build yet—or a brown field—an existing plan with a history of producing parts either for the customer or for some other similar customers. In both cases, if the supplier has the potential to obtain the certification of the customer's expectations, then the appropriate customer representative will give an approval to proceed based on the reputation of the supplier and surrogate data. The implication of both the surrogate data and the reputation of the supplier is of profound importance especially because major decisions about specific parts or services must be made in advance. If the information at this stage is not correct, problems of both quality and delivery may appear later in the planning stages. Finally, in both cases—green field or brown field—once production begins, the supplier's site must meet all requirements of the MSA within a year. Specific requirements and dates are determined by the individual customers.

Approval Denied

This disposition is the least wanted and discouraged because (1) the supplier was given permission to pursue the certification but failed to meet the requirements about a year after the production-established milestone, (2) the supplier was legitimately working toward certification but violated performance metrics such as PPM, delivery, capable systems, field service actions, stop shipments, violation of trust, etc., and (3) after an initial assessment of positive indicators, it is determined that the key requirements of the MSA will not be met within the year after the production-established milestone. At this point, the customer may still source the supplier with deviations, in the hope that as time proceeds the key metrics will be improved and monitored.

Approval Achieved

The goal of MSA is to make sure that a supplier site has met all requirements. This implies that the supplier is ready to supply parts or services to the customer. Upon satisfactory completion, proper and formal recognition by the customer is conferred and a plan for developing strategies for continual improvement is put in place.

Approval Revoked

By all measures, this classification is the worst. It signifies that the supplier site has failed to maintain the MSA requirements. When this happens, the supplier cannot regain the approved status for at least six months. However, the most common timeline for reinstatement is about 18 months. The revocation is a major event for both parties and is the beginning of a possible resourcing of the business elsewhere. The revocation is also communicated to the certification body of the ISO/TS registration with the possibility of revoking that certification as well. It must be noted here that in rare occasions (such as sole supplier) the supplier may continue to be sourced with deviation and greater controls to assure delivery of good product.

Tier 2

This is perhaps the most confused disposition in any customer's quality system. In our opinion, it is confusing because it is not communicated properly to the supplier. A customer may pursue two distinct situations. The first is to help directly the Tier 2, if the Tier 2 is a direct subsupplier, to make sure that quality, cost, performance, and delivery are acceptable. The second is to wait for the Tier 2 to ask for help whether the Tier 2 is a direct or subsupplier. In either case, the responsibility to make sure that the Tier 2 subsupplier is delivering good parts/service on time always lies with the supplier. That is an APQP as well as a customer requirement. The supplier is responsible. If the Tier 2 subsupplier is a directed supplier (the customer has directed the supplier to use a particular subsupplier), then the MSA is an acceptable method for investigating the site. Even though the MSA takes place in the subsupplier's site, all performance tracking must be part of the Tier 1 quality system.

Waived

This kind of disposition is for a supplier site that meets the MSA ongoing performance requirements and expectations. It is very limited in its application and very specific on its scope. A waived disposition implies that the site demonstrates an acceptable per the MSA ongoing performance metrics but for other reasons, for example, commercial reasons, the supplier site does not qualify for full certification. Another reason is the lack of third-party certification to an industry standard (i.e., ISO/TS 16949), but compliancy to an equivalent quality management system. It is also important to note here that a waived disposition may develop into disapproval status if the nonconformances are, in part, violations of performance metrics agreed on by both the customer and supplier. Typical issues in this category are PPM, delivery, field service actions, stop shipments, and violation of trust.

Nonmanufacturing Site

This disposition is reserved for sites such as warehouse, sequencer, sales office, technical center, and P.O. Box. Obviously, these sites cannot be sourced, and if they are currently sourced, the buyer must be notified and removed from the list of pending suppliers. No nonmanufacturing suppliers are qualified to be part of the MSA by definition.

References

Ford Motor Company. (2013). *Q1*. Dearborn, MI: Ford Motor Company.

ISO 9001. (2008). *Quality Management System*. Geneva, Switzerland: International Organization for Standardization.

Stamatis, D. (2004). *Integrating ISO 9001:2000 with ISO/TS 16949 and AS9100*. Milwaukee, WI: Quality Press.

Selected Bibliography

Ruxton, W. (1984). *Doing Things Right*. Fort Washington, MD: National Tooling and Machining Foundation.

3

Manufacturing (Facility) Site Assessment (MSA)

Overview

In Chapter 1, we discussed the issue and significance of timing. We focused on the predominant reasons for which timing is essential and it is a required tool to be defined by the organization (customer). In summary, we identified four key characteristics of a good timing schedule:

1. Early involvement. Work with suppliers more consistently, earlier in the product development process. Early involvement will allow the customer to work with its suppliers more effectively, earlier in the product development process.
2. Teams. Utilize cross-functional customer teams to work together with suppliers at their facilities. The creation of cross-functional teams working together to resolve issues at the supplier sites will help build teamwork and trust from all involved.
3. Standardization. Adopt a single, global, disciplined process to minimize rework and redundancy. Standardizing process tools and forms will build effectiveness and consistency as well as eliminate redundancies.
4. Governance and reporting. Use common, clear deliverables and a structured management review cadence. This means the governing approach to supplier engagement and measuring the results the same way each time, across all regions (nationally and internationally, as applicable) will drive consistency for all involved, both from within the customer and with its supplier partners.

The intent of these four characteristics is to make sure that the supplier delivers products and services on time with correct designs, proven performance, best-in-class (BIC) quality, competitive cost, and required capacity.

To optimize the implementation process and validate the results, both the customer and the supplier play major roles by making sure the plan is appropriate and realistic, the execution of the plan is on time, and the results are indeed what was planned. The tools for following these three requirements are

- Plan: Advanced Product Quality Planning (APQP 31 Tasks)
- Execution: Production Part Approval Process (PPAP 18 Tasks)
- Results: Part Submission Warrant (PSW 31 Unique Tasks/48 Deliverables)*

This means that the scope of activities for the supplier engagement process ranges from sourcing and design to quality and PPAP. These activities start early on from initial sourcing and continue through to Job 1. In addition, these activities encompass many events that require everyone to be on the same page and working to one goal. The goal is "flawless" launching and consistent production. Therefore, the key elements are

1. Earlier involvement with the supply base initiated at program target compatibility (PTC) or 27 months prior to Mass Production–Build Start (MP1) and continuing through to PPAP
2. Cross-functional engagement with members from engineering, purchasing, customer quality engineer, and the supply base for high-priority supplier components
3. Apply appropriate and applicable time requirements to particular products and services
4. "Priority" suppliers and supplier component selection based on a fact-based model consistently applied and including approximately 30% of all new-tooled end-item (NTEI) suppliers
5. Data-driven quality of event criteria consisting generally of 48 cross-functional deliverables for each NTEI with quantitative expectations through the use of the external supplier APQP/PPAP readiness assessment
6. Detailed roles and responsibilities for each of the cross-functional team members and supplier
7. Detailed governance and reporting requirements throughout the program development process
8. Clear linkage to the corporate program management process through the quality panel process

* The tasks and deliverables generally depend on the customer. However, the numbers here represent the most common ones used in the automotive industry.

Typical MSA Form

There are many variations of the form for collecting and evaluating data from a facility assessment. Figure 3.1 is one such form.

The form is simple. However, some items need to be explained. They are as follows:

Item: The item of the APQP.

Change from last evaluation: The status from the last evaluation.

Rating: Current rating of the item.

Expectations: Obviously, the expectations will depend on the customer, product, service, and supplier. Here we give some very basic categories from a manufacturing perspective; however, this list may also be modified to address issues in service, pharmaceuticals, chemical, financial, and many other organizations. They are

- Planning for manufacturing process capability
- FMEAs/control plans
- Employee readiness/training review
- APQP/launch/PPAP
- Managing the change
- Subsupplier quality management
- Control of incoming quality
- Control plans and operator instructions
- Process variability monitoring/reduction
- Measurement system capability, calibration, and use
- Control of parts, part identification/packaging/shipping
- Testing/engineering specifications
- Preventive maintenance/housekeeping
- Manufacturing flow/process engineering/Six Sigma and Lean manufacturing metrics
- Problem solving/corrective actions

Suppliers must list evidence to support each expectation. Generally, this column will identify both element and subelement requirements with the genetic color code of G, Y, and R. For green (G), the expectation element must have zero red subelements and zero yellow subelements. All subelements are green. For yellow (Y), the expectation element must have zero red subelements and one or more yellow subelements. For red (R), the expectation element can have one or more red subelements. In some cases, the element may not be complete, so

<table>
<tr><th rowspan="2">Item</th><th rowspan="2">Change from Last Evaluation</th><th rowspan="2">Rating</th><th colspan="7">Date:
Supplier name:
Supplier code:
Contact person:
Contact information:
Engineer's name:
Item:
Approved by:</th></tr>
<tr><th>Expectations</th><th>Comments or Evidence</th><th>Action Plan</th><th>Resp.</th><th>Due Date</th><th>Actual Date</th><th>Expectation Guidelines</th></tr>
<tr><td></td><td></td><td></td><td></td><td></td><td></td><td></td><td></td><td></td><td></td></tr>
<tr><td></td><td></td><td></td><td></td><td></td><td></td><td></td><td></td><td></td><td></td></tr>
<tr><td></td><td></td><td></td><td></td><td></td><td></td><td></td><td></td><td></td><td></td></tr>
<tr><td></td><td></td><td></td><td></td><td></td><td></td><td></td><td></td><td></td><td></td></tr>
</table>

FIGURE 3.1
Typical MSA form.

the appropriate designation is incomplete (INC). INC is appropriate only if one or more subelements are not ranked red, yellow, or green.

For the subelement expectations, it is important to remember that a G implies a High Quality of Event/Meets the Expectation and the Intent of the Expectation Guidelines/Item is Complete. A Y implies a Marginal Quality of Event/Marginally Meets the Expectation and the Intent of the Expectation Guidelines/Concurred and timed recovery plan is in place and on schedule. An R implies an Unacceptable Quality of Event/Does not Meet the Expectation and the Intent of the Expectation Guidelines and there is no plan or it is inadequate to meet the schedule.

Comments or evidence: All expectations regardless of color rating must have documented evidence listed including examples and part number viewed.

Action plan: Action plan, responsibility, and due date are required entries for all yellow- and red-rated expectations.

Responsibility (Resp.): The person who is responsible for making sure that the element requirements will be completed.

Due date: The date that the requirements are due to be completed.

Actual date: The date that all requirements for the element are completed.

Expectation guidelines: Generally, they are given by the customer as a springboard of ideas. They are not meant to be all-inclusive. It is very important to mention here that whereas the expectations were given in a list of 15 categories (see previous list), these guidelines provide detailed insights as to what exactly the customer is looking for from the supplier. Obviously, they are not exhaustive. Each customer–supplier relationship should have a good understanding of what is needed and wanted, and the guidelines should support that goal. Here we provide a sample of what may be considered a guideline for each of the 15 categories.

- Planning for manufacturing process capability. Minimum requirements should include all ISO 9001 and industry-specific items with a focus on the quality manual, procedures, instructions, records, internal audits, and corrective actions, as well as specific metrics for input and output variables, management responsibility, and so on.
- FMEAs/control plans. Minimum requirements should include current FMEAs, process flow charts, and control plans. Key inputs for FMEA should include robustness and appropriate definitions for critical and significant characteristics. Emphasis on prevention should be the prevalent approach as opposed to detection and control. Control plans should include the characteristics that

are critical to quality and must be linked to the PFMEA as well as identify appropriate and statistically sound sampling and frequency of sampling.

- Employee readiness/training review. Minimum requirements should include criteria for selecting appropriate operators for individual tasks and provide the applicable training. How the supplier identifies the training needs and how training is implemented throughout the organization is fundamentally important. Management reviews are to be monitored and appropriate actions should be taken in order to make sure that all employees are equipped with the knowledge needed to do their assigned task (both operators and management personnel).
- APQP/launch/PPAP. Minimum requirements should include monitoring and reviewing the APQP/launch/PPAP requirements on a regular basis. Appropriate actions should be planned for any known delays in any missed milestones. The supplier should have appropriate links with the customer to communicate any changes or delays to the program involved. This means that the supplier must (1) have regular "fresh eye" reviews to make sure that all action plans are being followed and implemented for the items that need improvement, (2) have superb launch plans in place to ensure a safe launch of the new product, (3) have inspection data to validate unusual events, and (4) have all the appropriate data that demonstrated completion of the PPAP. (Note: Some customers may require additional information for capacity and what happens if interrupted production occurs.)
- Managing the change. Minimum requirements should include appropriate procedures to implement necessary changes that may involve the customer. These procedures must be validated with the most recent change.
- Subsupplier quality management. Minimum requirements should include a defined process by the supplier for managing all subsuppliers. Subsuppliers must be certified to the appropriate international standards and—if they exist—industry standards. The subsupplier must have a documented problem-solving methodology. On the other hand, the supplier must (1) have evidence of conformance for all products that the subsupplier provides including PPAP—if required, (2) visit the facility of the subsupplier at predetermined intervals and conduct assessments on their quality system; the findings of the assessment must be documented and communicated to the subsupplier, and (3) have a procedure to conduct random incoming quality audits to verify incoming quality as well as the corrective actions implemented by a subsupplier.

- Control of incoming quality. Minimum requirements should include a process to review subsupplier quality systems including a resolution process of incoming inspection rejects. In case of incoming nonconformance material, there should be a specific containment area where the rejects are contained until final disposition.
- Control plans and operator instructions. Minimum requirements should include appropriate and applicable work instructions with all necessary detailed information including visual aids. In case of multilingual personnel, the supplier must demonstrate that all personnel understand work instructions. Work instructions should include a reaction plan if a suspect item gets through the system. All control plans must be accurate and current.
- Process variability monitoring/reduction. Minimum requirements should include operator-based statistical process control (SPC) or similar methods of monitoring processes. The usage of SPC should be focused on variable charts whenever possible and encourage the transfer of attribute charts to variables. The operators must demonstrate whether they know how to use the information from the SPC and that information is utilized appropriately to either reduce or maintain variability. C_{pk} or P_{pk} must be calculated, monitored, and understood especially for special characteristics.
- Measurement system capability, calibration, and use. Minimum requirements should include validation of gauge availability and appropriate capability. Gauge R&R is performed for both variable and attribute data. Preference is given to the ANOVA approach with more than five distinct categories. Records of damaged gages should be kept and make sure they are in compliance. Measurement instruments should be 10 times more accurate than the measurement result. Mistake-proofing devices must be tested on a regular basis for possible failures.
- Control of parts, part identification/packaging/shipping. Minimum requirements should include appropriate and applicable controls and processes in place for handling rework, repair, and scrap including clear documentation for evaluation, quarantine, segregation, and disposition. In addition, all handling should be monitored for damage with appropriate records.
- Testing/engineering specifications. Minimum requirements should include a documented process to secure requirements in engineering specifications or print as well as specific customer requirements. In addition, a supplier must have a process to

record and report specification failure as well as a clear reaction plan to identify who is to act and what has to be rejected, recalled, or stopped from shipment to the customer if necessary.

- Preventive maintenance/housekeeping. Minimum requirements should include appropriate and applicable records that demonstrate the maintenance and the life cycle of equipment and tooling. One such metric should be the overall equipment effectiveness (OEE). Records should also be available for documenting improvement in the effectiveness of preventive maintenance and applicable housekeeping procedures.
- Manufacturing flow/process engineering/Six Sigma and Lean manufacturing metrics. Minimum requirements should include a documented system that drives improvement in the organization. Examples may be just-in-time (JIT), Kanban, Takt time, change over time, Lean tools, robust designs and processes, Six Sigma metrics, and many more.
- Problem solving/corrective actions. Minimum requirements should include a documented problem-solving methodology that identifies the escape point and the root cause of a problem. Linkages must be documented in the FMEA and control plan as well as capturing lessons learned for future prevention actions.

Selected Bibliography

http://www.docstoc.com/docs/39317872/Q1-Assessment-Evaluation-Matrix. Q1_Site_Assessment_Evaluation_Matrix.xls.

http://www.iatfglobaloversight.org/docs/FordspecTSJune2013.pdf.

http://www.scribd.com/doc/60087567/FordQ1Requirements.

http://www.stpaul.gov/DocumentCenter/Home/View/15297.

Chrysler Motor Company. (2008). *Quality Verification and Validation System at Chrysler*. Auburn Hills, MI: Chrysler Manufacturing Quality.

General Motors Company. (2011). *Quality System Basics*. Detroit, MI: GM Purchasing and Supply Chain.

4

Sourcing

Overview

In the last chapter, we discussed the value of manufacturing site assessment (MSA) and the actual visit of the customer to the supplier facility. We discussed the intent of the MSA as a verification of the ongoing improvement and any appropriate adjustments that may have taken place since the last evaluation. In this chapter, we focus on the selection process of a particular supplier based on quality requirements that the OEM has defined for flawless launching. We have chosen an automotive situation to demonstrate the steps a customer has to follow. The reason for the automotive example is that it provides the reader with a comprehensive approach. We hope that the steps in modified fashion can transfer to other industries as well.

Every beginning is difficult. When an organization is about to undertake the implementation of its quality system in its supplier culture, it is imperative that the customer do two things. The first is to communicate why the supplier's quality is important to the customer and "what is in it" for them. The second is to communicate the need of the customer to have suppliers of the highest quality. This means that the customer must identify candidate suppliers with strong quality track records. In essence, sourcing for quality (SfQ) is a comprehensive quality evaluation of a supplier site that is being considered for new business or resourcing existing business. In other words, SfQ is a process that (1) ensures quality as a prerequisite to awarding business to qualified suppliers by evaluating direct performance history or surrogate and (2) evaluates the risk of the supplier doing business with the customer based on two categories. The first category is based on an evaluation of compliance to the customer's quality requirements for a particular product as well as the capability to manage the new product effectively. The second category is generally broader on quality and evaluates timing to ensure that a PSW can be achieved on time, and accommodate new design complexity and new manufacturing technology readiness.

Supplier Sourcing Begins

The timing supplier engagement process will begin once the suppliers are sourced. In the automotive industry, generally the timing breaks the vehicle into Powertrain (PT), Underbody (UN), and Upperbody (UP). Sourcing completion is generically required by program target compatibility (PTC), which is about 40 months before Job 1.

Priority Suppliers Selected

Once sourcing is complete, the supplier program lead engineer and the product development program lead will work with the process manufacturing team (PMT) leaders and supplier engineer site managers to complete the selection of "priority" suppliers for the program. A specific risk model is used to determine the priority suppliers. A typical model may look at

- Site risk: Is the supplier's quality performance acceptable?
- Part risk: Is the part/system complex such that it requires close cooperation to launch?
- Program risk: Are we kicking off parts late in the process, or are we launching brand new technology on the vehicle?

Note here that after the PTC and M1DJ milestones, the suppliers ought to be selected on an incremental priority based on the three categories just mentioned.

Cross-Functional Teams Formed

The supplier engineer and product development program leads will notify the appropriate design engineers, buyers, site engineers, and other analysts (as required) that their suppliers are designated as priority for a given program. This notification will include specific timing and reporting requirements for the program. One of the paramount objectives here is to notify the supplier of the physical audit to the supplier's facility. Often, these audits are spread throughout the overall timing and generally, there are four. The first one is about at PTC and the second one is at about final data judgment (FDJ). At these audits, the evaluation of whether to proceed with this particular supplier is made. This determination generally allows the supplier to get prepared for the initial PPAP (run-at-rate). The third audit is generally at the launch readiness (LR) timing and the fourth is at pilot production–built start (PP). This is where the PPAP is completed and the PSW is signed off.

The team responsibilities vary depending on the supplier, the product, and the relationship of OEM and supplier. To be sure, there is a team lead at all times; however, each team member (under the tutelage of the lead) still has the responsibility between meetings to

- Resolve open concerns
- Communicate with other team members as needed
- Bring together other team members to close out concerns as necessary

It is recognized that all team members will interface with the supplier regularly between on-site evaluations and, in many cases, this interaction will occur at the supplier's site.

Team Kickoff Meetings Begin

Prior to the on-site evaluations, the cross-functional OEM team members will meet with their supplier team to review program expectations and begin the preparations for site visits. (*Note*: Actual review timing will depend on the specific needs of the component and the program.) This is called the *kickoff* meeting. The actual location is determined and generally is based on the team's preference. It may be held at the OEM's facility or supplier's facility, or it may be held via teleconference. No matter where or how it is held, the duration is approximately 2–3 hours and the participants are the design engineer (lead) with the site engineer, buyer, and supplier representatives. The purpose of this is to introduce everyone and to

- Provide them with an overview of the timing process as well as the expectations of the engagement process
- Communicate team expectations prior to on-site audit (evaluation) #1, which is a key deliverable, as well as reaffirm the follow-up audits and confirm the appropriate timing for the required completion of the program on time and on budget with all the quality requirements

The cross-functional team's first visit is called On-Site Evaluation #1, which is a Verification of Supplier Failure Mode Avoidance Strategy and Manufacturing Plan. The specific timing for this depends on the section of the vehicle, that is, PT, UN, or UP. In either case, the on-site evaluations are timed to align with the specific milestones. The intent here is to make sure the supplier's work is completed and in step with the OEM's work. Therefore, the first on-site evaluation is intended to be completed by M1DJ for UN parts and FDJ for UP parts to ensure that the up-front failure mode avoidance work is completed. Getting the DFMEA and control plan finished along with the critical and significant characteristics is vital for the supplier to complete its planning activities.

Furthermore, the first evaluation is also targeted at getting the supplier's capacity planning (facilities and tooling) in line with the OEM's expectations.

The buyer plays a key role in this meeting, as important funding decisions need to be made. A typical summary of the #1 on-site evaluation is shown in Table 4.1.

The cross-functional team's second visit is called On-Site Evaluation #2, which is a Verification of Supplier Launch Preparation. This visit will always occur prior to the supplier's shipment of parts to system parts builds. This evaluation is generally completed before the system builds and the intent is straightforward, that is, ensure that the parts the supplier will ship for the system builds meet both the design intent and the quality intent required to have a successful prototype build. The rationale for this is to gain confidence in the supplier's performance. The more confident the customer is in the robustness of the supplier's system parts, the higher quality of event we can expect for the overall program. By gaining confidence, it is possible to accelerate the timing and expect a smaller margin for error—which of course does reduce the risk.

The second evaluation is also important from a manufacturing planning standpoint at the supplier's facility. In essence, this evaluation focuses on

- Facilities and tooling planning/installation
- PFMEA and control plans

TABLE 4.1

Summary of On-Site Evaluation #1

APQP Item (see Table 1.1)	Deliverables	Outcomes
1	Sourcing decision • Required sourcing agreement signed, site identified • Required supplier commercial and program agreements signed • Final mix, maximum weekly volume communicated and agreed	Updated supplier APQP/PPAP Readiness assessment and confirmation of the required deliverable status (if applicable) Agreement on failure mode avoidance strategy (completion of DFMEA and agreement on special characteristics) Agreement that capacity plan is achievable and consistent with customer's requirements
2	Customer input requirements • Program expectations cascaded to suppliers	Plan developed to support prototype builds (prototype control plan and build plan)
4	DFMEA • DFMEA completed • Special characteristics cascaded to supplier	Updated Manufacturing Site Assessment (MSA): As required Agreement of timing for next on-site evaluations
8	Facilities, tools, and gauges • Supplier's OEE plan is confirmed by surrogate data	
11	Drawing and specification/ design records • Design styling confirmation	

- Gauging
- Staffing and training planning
- VP prototype build readiness
- Run-at-rate preparation

A summary of the #2 on-site evaluation is shown in Table 4.2.

TABLE 4.2

Summary of On-Site Evaluation #2

APQP Item (see Table 1.1)	Deliverables	Outcomes
1	Sourcing decision • All production tool orders issued to suppliers	Updated supplier APQP/PPAP readiness assessment and confirmation of the required deliverable status (if applicable) Acceptance of parts to support VP prototype build (prior to shipment) Concurrence on manufacturing plans in support of run-at-rate event Agreement on PFMEA Updated manufacturing site assessment (MSA) as required Agreement of timing for next on-site evaluations
7	Subcontractor APQP status • Subsupplier PPAP timing plan completed	
8	Facilities, tools and gauges • Facility and tooling timing plan completed • Gauge plan (including released gauge drawings) completed	
9	Prototype build control plan • Prototype build control plan completed	
10	Prototype builds • Plan to complete tool orders release in support of VP build • Plan to complete parts orders to suppliers in support of VP build • Prototype parts achieve 100% of required print specifications	
11	Drawing and specification design records • Plan to complete parts at a level that can be tooled for production Release by <FDJ> + 4 weeks	
13	Team feasibility commitment/ customer engineering approval • Design, process, and timing feasibility confirmation	
15	PFMEA • Final PFMEA completed with linkages	

The cross-functional team's third visit is called On-Site Evaluation #3, which is a Verification of Supplier Capability. This visit will always occur prior to the supplier's shipment of parts for production tool tryout (TT) builds. This visit is important because the goal is to achieve a completed PSW for the builds. Again, the customer team has to be confident in the design level and quality level of the parts.

To be sure, not all parts will achieve PSW by TT, so the team must also ensure that the containment plans for the builds are clearly communicated to the supplier so that all the goals of the TT vehicles can be met. It is also important to point out that some customer teams will go to the supplier site more than four times. It will depend on the needs of the program or the supplier. This means that individual team members will also visit the supplier as required; this process ensures they come together at the supplier a minimum of four times. In essence, the purpose of this evaluation is to make sure the supplier is ready for

- Completion of PPAP requirements for run-at-rate
- Dimensional results
- Production validation testing
- Process capability
- Appearance approval

A summary of the #3 on-site evaluation is shown in Table 4.3.

The cross-functional team's fourth visit is called On-Site Evaluation #4, which is a Verification of Supplier Capacity. This visit should always occur prior to the supplier's shipment of parts for Mass Production 1 (MP1) builds or Job 1 milestone. The intent is to ensure the supplier fully meets the capacity requirements for the program. This visit provides closed-loop feedback on how well the planning activities, which started at the first on-site evaluation, have been realized in production. From a team perspective, the job is not complete until the supplier is at the fully required production rate. In essence, the purpose of the final evaluation is to ensure the following:

- Completion of PPAP including capacity
- Operators are staffed and trained
- Ability to meet launch ramp-up curve
- Ability to meet ongoing production
- OEE confirmation

A summary of the #4 on-site evaluation is shown in Table 4.4.

TABLE 4.3

Summary of On-Site Evaluation #3

APQP Item (see Table 1.1)	Deliverables	Outcomes
3	Craftsmanship/appearance approval report • AAR approved	Updated supplier APQP/PPAP readiness assessment and confirmation of the required deliverable status (if applicable) Signed run-at-rate PPAP warrant (quality requirements for single production work stream) Updated manufacturing site assessment (MSA) as required Agreement of timing for next on-site evaluations
6	DVP&R material/performance test results • DV testing complete	
7	Subcontractor APQP status • Subsupplier PPAP run-at-rate and quality requirements (or equivalent) completed	
8	Facilities, tools, and gauges • Facilities, tools, and gauges are at the final production location • Suppliers demonstrated OEE (for run-at-rate) supports capacity requirements	
12	Engineering change documents • All Customer engineering changes approved and recognized by the supplier	
14	Manufacturing process flowchart/process flow diagram • Final process flowchart supports PPAP run-at-rate event	
16	• Measurement system evaluation/measurement systems analysis studies • Gauge R&R results ≤10% per PPAP customer specifics	
17	Qualified laboratory documentation • Supplier internal and external laboratory compliance	
18	Checking aids • Checking aids compliant with part specifications	
19	Prelaunch control plan • Prelaunch control plan completed with linkages	
20	Operator process instructions • Operator process instructions completed	
21	Packaging specifications • Packaging approval process completed	
22	Production trial run • PPAP run-at-rate (Including subsuppliers) completed	
23	Production control plan • Production control plan completed with linkages	

(*Continued*)

TABLE 4.3 (Continued)

Summary of On-Site Evaluation #3

APQP Item (see Table 1.1)	Deliverables	Outcomes
24	Initial process capability • Initial process capability results (Ppk ≥1.67)	
25	Dimensional results • 100% of Required measurement points within tolerance	
26	Production validation (PV) testing • PV testing complete	
27	PSW • PPAP quality requirements for one stream complete • PPAP quality requirements for multiple streams complete	
28	Bulk materials requirement • Bulk materials checklist included in PPAP submission	
29	Sample product • Sample product produced with customer identification	
30	Master sample • Master sample approved	
31	Record of compliance • Customer-specific requirements documented	

TABLE 4.4

Summary of On-Site Evaluation #4

APQP Item (see Table 1.1)	Deliverables	Outcomes
3	Craftsmanship • Color changes completed supporting color harmony	Updated supplier APQP/PPAP readiness assessment and confirm required Deliverable status (final status) Signed final PPAP warrant and capacity analysis report Updated manufacturing site assessment (MSA), as required
5	Design/manufacturing reviews • Quality issues closed and effective	
7	Subcontractor APQP status • Subsupplier PPAP including capacity (or equivalent) completed	
27	PSW • PPAP with all quality requirements completed (including multiple work streams) • PPAP including capacity completed • Supplier's demonstrated OEE completed PPAP supports capacity requirements	

General Procedures for Sourcing

As previously mentioned, there are two categories for sourcing. The first is the new or nonemergency case and the second is an existing case. In both cases, a supplier's quality system may or may not be prequalified. Let us start our discussion with the first category.

New/Initial Sourcing and Nonemergency Resourcing

It must be emphasized here that the supplier in this case must be prequalified to the customer's quality requirements as well as being a manufacturing site. If these two conditions are not met, then sourcing may be delayed, waived, or not issued. For the actual site to be officially sourced, a customer technical or quality specialist or a buyer must have approved the process.

On the other hand, obtaining approval for a non-prequalified supplier, the sourcing must also be approved with the appropriate authority signatures and documented consistent with the buyer. The sourcing approval for this situation must be obtained prior to awarding business to the supplier manufacturing location. If and when the approval is secured, a new manufacturing code is issued by the customer's purchasing department. The process is customer dependent and there are no general rules to be followed other than making sure that all appropriate and applicable forms, contracts, and procedures have been agreed to and signed off. Special attention should be given to significant changes that require strategy changes.

Whereas the requirements for the pre-approved supplier are fundamentally based on historical performance and the facility assessment, all requests to approve non-pre-approved sourcing must include an approval package to make sure that essential quality, safety, and production issues are well covered:

- Parts list, including volumes, and using plants
- Safety/critical designations
- Estimated dates of the first production for new sourcing (as applicable)
- Customer's assessment of the supplier site's potential to achieve or regain approval status with the planned date
- Sourcing rationale
- MSA

Sourcing to Sites with a Waiver

A waver should not be the *modus operandi* of doing regular business. However, it does happen for several reasons, including lateness of sourcing or even unique supplier availability. A waiver allows the supplier to do business with the customer under strict rules and requirements. In some cases, multiple site waivers may be approved through one waiver form for a group of supplier sites where the acceptable, alternative quality management plan is identical, for example, several parts from multiple sites under a single contract and single supplier management. It is imperative that the list of supplier sites for the multiple site waiver is documented on a standardized form or quite often be part of the production sourcing approval package and referenced on the request form.

On rare occasions, "multiple year waivers" may be permitted under very strict guidelines of the customer, including a written contract, and documentation that the supplier has its own qualified quality system and it is effective.

Emergency Resourcing to a Nonqualified Facility

If and when an emergency resourcing action to a nonqualified supplier is required, the resourcing buyer, in conjunction with the appropriate supplier engineer, is required to initiate specific documentation for the emergency and appropriate deviation records. Generally, emergencies are limited to labor strikes, acts of God, bankruptcy, and imminent capacity issues. Here it is very important to mention that this emergency is of finite time, usually no more than 30 days, at which time the customer should be able to find a replacement.

5

Segmentation

Overview

In the last chapter, we discussed the function and importance of having selected the right suppliers for a given program or service. In this chapter, we will discuss the evaluation of those suppliers that qualify to do business with the organization and provide some critical analysis as to why the suppliers must meet the requirements of the customer. In essence, we will focus on the methods and metrics necessary to manage the supply base effectively.

The goal of the segmentation process is to help the original equipment manufacturer (OEM) personnel understand the quality segmentation process and how to manage a supplier site appropriately. To do this, a process that focuses technical resources toward specific suppliers and a method to align quality improvement efforts toward customer satisfaction and safety must be developed and communicated throughout the organization as well as to the individual supplier.

A good way to do that is for the OEM to have a good data model based on some form of quality management system and commodity prioritization. Obviously, this method must be robust and able to highlight and trigger awareness where resources are needed to avoid problems of capacity or delivery.

Segmentation Definition

A good segmentation analysis is a way to

- Define a method of determining the level at which a supplier's quality performance and continuous improvement efforts should be monitored compared with the OEM's basic quality system.
- Monitor suppliers who provide low impact on customer satisfaction and safety parts and have a history of good performance. This form

of monitoring requires very different management techniques from suppliers who provide high impact on customer satisfaction and safety parts and have a history of issues. This history may be direct with the OEM or from surrogate OEMs.

- Examine each supplier site's product and production history to determine the risk posed by that supplier site to the overall success of a program. Quite often—especially if it is a new supplier—historical surrogate data will be used to establish capacity and quality performance.

The reader must understand that supplier segmentation is a process that focuses technical resources toward specific suppliers (current or new). The reason for this undertaking is to align quality improvement efforts toward customer satisfaction and safety. To do this alignment, appropriate and applicable data for basic quality adherence and commodity prioritization must be in place to properly identify potential issues and commit the necessary resources to circumvent delays of any kind.

Segmentation from the supplier's perspective will help the company reach three important goals:

1. Improving quality
2. Assuring safety
3. Improving customer satisfaction

From the OEM's perspective, segmentation helps to achieve at least the following:

- Defines the process for focusing the technical quality representative resources by standardizing OEM resources based on basic quality system and customer satisfaction
- Prioritizes improvement efforts on commodities that have a larger impact on customer satisfaction and safety
- Aligns purchasing priorities with corporate objectives for quality, safety, and customer satisfaction
- Promotes the development of self-reliant suppliers
- Provides new quality representatives with an environment where it is possible to practice supplier improvement rather than crisis management
- Allows purchasing and all appropriate personnel to focus on higher risk suppliers

Therefore, to optimize segmentation, the OEM must group suppliers into segments that allow purchasing to differentiate between suppliers to apply the appropriate level of technical resources.

Segmentation Model

A typical basic segmentation model may be separated into four segments:

1. Critical: High impact on customer satisfaction/safety with no documented basic quality system approved by the OEM. Therefore, it is a high-risk supplier.
2. Significant: Low impact on customer satisfaction/safety (CS/S) with no documented basic quality system approved by the OEM. Therefore, it is a high-risk supplier.
3. Experienced: High impact on CS/S, with documented basic quality system approved by the OEM. Therefore, it is low risk.
4. Autonomous: Low impact on CS/S, with documented basic quality system approved by the OEM. Therefore, it is low risk.

This basic model is shown in Figure 5.1. The reader will notice that Segments 1 and 2 are not good contenders for full participation. One may even say that they operate under a "probation" status, so to speak. It is up to them to improve and operate in Segment 3 or 4. On the other hand, if they do not improve, they will be denied participation as a full supplier. (A site is categorized as having a high impact on CS/S if it has at least one commodity designated as high impact on CS/S.)

Segment 3 or 4 should be the most common category with which an OEM has to be concerned. The reason is that a supplier under either category has established (with historical data) that it has a good and sound basic quality system and it follows it.

In addition, if a supplier with a previous history is beginning to produce a new product, it will be placed in the left segment. This is because the new product will demand different resources and it has to be documented as to

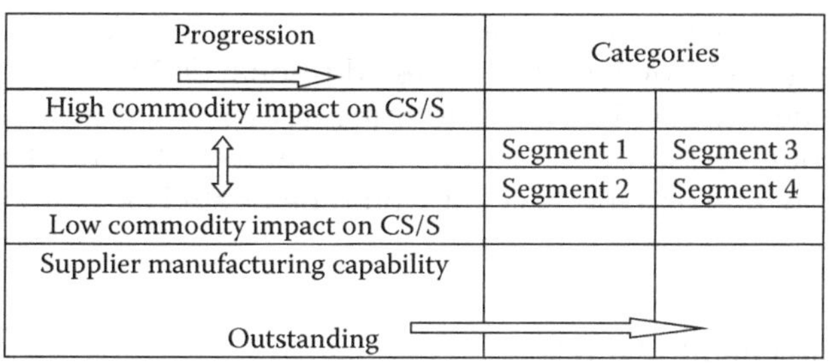

FIGURE 5.1
Basic model of segmentation.

whether the supplier can produce it. On the other hand, if a supplier is awaiting approval, it is to be segmented along the x-axis (the left or right side of the model) depending on the discretion of the OEM's representative.

Overlays

In the segmentation process, sometimes the requirements just mentioned do not fit a particular supplier. In that case, the customer should have a contingency plan. Generally, this plan is addressed with what is called an *overlay*. An overlay is a label given to supplier sites to identify a special case where a site does not fit into the normal rules of segmentation. The purpose of an overlay is to either add or remove resources based on the impact that a site has on the OEM. The three categories are (1) quality overlay, which requires a segment change; (2) launch overlay, which does not require a segment change; and (3) limited involvement, which does not require a segment change.

Quality Overlay

Quality overlay is a label given to a supplier with a basic quality system defined by the OEM (an example is the Q1 program from Ford Motor Company) if the status of this basic quality is segment 3 or 4, that would be allowed to be segmented in segment 1 or 2 based on the manager's discretion to apply more resources to correct issues at the supplier. For example, at the site in question, segment 3 or 4 would move to segment 1 or 2 because the requirements of the basic quality system are not fulfilled. Furthermore, the OEM representative re-evaluates the site at some future time interval (e.g., 6 months for potential resegmenting) or status change.

Launch Overlay

Launch overlay is a label given to a given and approved supplier, with a positive status (segment 3 or 4) that requires additional OEM resources due to the support of a new model launch or to mitigate risk at a site due to increased business through re-sourcing actions. For example, the supplier site would remain a segment 3 or 4; however, it would be treated as a nonqualified segment 1 or 2.

Limited Involvement Overlay

Limited involvement overlay is a label given to a supplier without a basic quality system, or a supplier that had an approved quality system in place but it has been revoked or denied because of noncompliance to minimum quality requirements. It is important to note here that the limited involvement usually requires a reduction of OEM resources because more often than not, the supplier has already demonstrated knowledge and performance in the past. For example, if the supplier site has lost the approval of the OEM, that supplier would remain in Segment 1 or 2. However, the supplier would be treated as an approved supplier under Segment 3 or 4. Obviously, this is a special case and it will take effect under these conditions:

- An organization outside the purchasing department has accepted responsibility and applied resources to manage nonquality improvement actions (e.g., delivery or environmental).
- The site is classified short-term eliminate (STE) or equivalent and has an acceptable site assessment action plan that creates limited risk to the OEM.
- The supplier has acceptable quality performance and an approval of the basic quality system is pending and will be awarded on receipt of at least the certification of ISO/TS16949 and ISO14001. Other certifications may be relevant depending on the OEM.

Applications

The following three examples show the flow of segmentation process for different categories.

Example 1

- J&J has recently started producing odometers, *a low impact* on customer satisfaction and safety item, for one of the SUVs.
- J&J is a relatively new supplier to a particular OEM.
- Recently lost its basic quality system status due to poor quality.

Answer: Segment 4 to 2

Example 2

- LMS produces retaining clips for door panels, a *low impact* on customer satisfaction and safety item.
- Was high risk—nonapproved supplier with no verified minimum quality system. However, recently regained approval status.
- They are coming up for their monthly review.

Answer: Segment 2 to 4

Example 3

- A supplier with approved quality system and a provider of door latches and body side molding.
- Door latches are considered *high impact* on customer satisfaction and safety items—contract expired.
- Body side moldings are considered *low impact* on customer satisfaction and safety items.

Answer: Segment 3 to 4

One can see that using a segmentation strategy is very beneficial, especially if the appropriate evaluation is conducted with appropriate and applicable tools (FMEAs, control plans, PPAP, 8D, SPC, design of experiments (DOE), Six Sigma, or Lean manufacturing). This evaluation will definitely lead to customer satisfaction, stability of production, and good basic quality performance. Therefore, the particular segment (1, 2, 3, or 4) depends on the impact the product may have on the OEM, especially on customer satisfaction and safety plus approval quality status code (this is generated by the OEM's metrics).

6

Supplier Development

Overview

In the last chapter, we addressed the issue of segmentation and we offered some benefits for both the customer and supplier. In this chapter, we focus on supplier development.

ISO 9001:2008 contains two subclauses—7.4.1, Purchasing Process, and 7.4.2, Purchasing Information—that make an organization responsible for ensuring that its suppliers are capable of meeting the organization's (and its customer's) specifications, but they do not require the organization to develop its suppliers so products and services purchased from those suppliers are improving. ISO 9001:2008 also requires the organization to continually improve the effectiveness of its quality management system (QMS), so any competitive company should include supplier efficiency and effectiveness as improvement goals. (Note: By the end of 2015, a new revision of the ISO 9001 is expected and the anticipated changes will follow a structure that makes it easier to use in conjunction with other management system standards, with increased importance given to risk.)

ISO/TS 16949:2009 (ISO/TS 2009) goes farther than ISO 9001:2008 and specifically requires your organization to pursue supplier development, which means ensuring any supplier that affects your product has an effective QMS and is improving its processes to meet your customers' needs. Of course, the ISO/TS 16949 is an automotive-specific requirement but similar industry international standards have the same or very similar requirements such as medical device ISO 13485, aerospace AS 9100, and many others.

Therefore, we see that the international standards suggest and the industry-specific requirements require supplier development. However, what is supplier development? At a minimum, a supplier development program should be aimed at achieving the following four goals:

1. Make sure capacity requirements are met.
2. Make sure delivery requirements are met.

3. Make sure any risk is minimized by anticipation, monitoring, or mitigation.
4. Make sure quality requirements are met or exceeded. This, in turn, will cause
   a. Lower supply chain *total* cost
   b. Increased profitability for all supply chain participants
   c. Increased product quality
   d. Near-perfect on time delivery at each point in the supply chain

Understanding the Fundamentals of Supplier Development

Perhaps the most important fundamental issues that any formal supplier development program must take into consideration as part of its design to be successful (effective) are the following items:

- Establish, define, and govern the relationship between customer and supplier.
- Observe, monitor, improve, and sustain operations. Unless the operations are studied and understood, they will not be improved.
- Transfer technology and lend financial support. We are in an age where technology is changing too fast. The organization must be prepared to support innovations in production, quality, and efficiency methods for improvement.
- Train management and the workforce. One cannot perform excellence unless one is trained appropriately and timely. Everyone in the organization must be trained to do their job in an optimum way and to look for improvement opportunities.
- Enforce environmental and social compliance. In the last 10–15 years, all of us have been sensitized to both environmental and safety concerns. The ISO 14000 and the application of failure mode and effect analysis as well as hazard analysis should be part of the culture of the organization.
- Identify and mitigate all sources of risk. It is the responsibility of the supplier to make sure that all risks have been identified and appropriate and applicable corrective actions are in place to either minimize or eliminate those risks.

On the other hand, critical success factors for an ongoing successful supplier development program from our experience have been the following:

- Providing information about products, expected sales growth, etc. Poor communication is one of the biggest wastes with a Lean supply chain. Lack of information translates into additional costs (usually in the form of just-in-case inventory). Suppliers need to become extensions of their customers. However, the customers must also be sensitive to providing timely information for applicable adjustments whether it may be cost, delivery, capacity, performance, and so on.
- Training in the application of Lean and quality tools. Asking suppliers to drop their price without giving them the knowledge to lower their costs through Lean implementation is not sustainable long term. In other words, this will drive suppliers out of business, which goes against the purpose of supplier development. This actually happened under the directorship of Mr. Lopez while at GM in the early 1980s. It is profoundly important for the OEMs not to bully their suppliers—just because they can—to suppress quality at the benefit of price. Both suppliers and customers must understand that value (total cost) is more important than just plain price. It is the responsibility of the customers to educate what they really want and how the supplier can deliver it.

Therefore, to succeed, a supplier development program requires participation and cooperation from both internal and external stakeholders. A cross-functional team representing internal stakeholders and with an executive sponsor needs to be created. This team gets the internal stakeholders on board and then ensures alignment of the external stakeholders to successfully accomplish the initiative. To this end, the customer must designate someone (most OEMs actually have such a person) to facilitate both the needs of the customer and help in resolving issues and problems with suppliers. Many organizations call that individual supplier technical engineer, supplier quality engineer, supplier technical assistant, or some other name.

To be sure, not all supplier development initiatives are successful. From my experience working with the automotive and many other industries over the last 35 years, I can say 40%–60% of all initiatives have failed primarily due to poor implementation, artificial commitment from both customer and supplier, and very poor follow-up. Some specific issues that I have observed over the years are

1. Most firms engage in reactive supplier development approaches (which addresses sporadic problems), as opposed to strategic supplier development approaches (which addresses continuous improvement of the entire supply base). In other words, suppliers react to fires, not to preventing fires.
2. Most organizations focus on convenient approaches to supplier development, which include rewarding performance (*the carrot*), penalizing poor performance (*the stick*), ongoing detailed assessment

and feedback (*measurement*), and direct involvement in suppliers' operations (*hands-on approach*). In other words, customers play a game of *got you* rather than working with the supplier and demonstrating the benefits of a good long-term relationship for both.

3. Most organizations lack real data to support decisions; that is, lack of research on which approach is most effective for the best outcome or data to suggest that a combination of approaches may be appropriate under different circumstances, depending on the nature of the supplier, the type of commodity, and the management team at the supplier. In other words, most OEMs push for resolutions that are fast and temporary rather than focusing on long-term effective solutions. On the other hand, suppliers are eager to satisfy the customer as soon as possible with band-aids based on temporary fixes of the past rather than systematic problem resolution and prevention.

If the supplier is new to quality thinking, it is strongly suggested to follow the guidelines of the quality system assessment checklist, which is accompanied by all industry-specific requirements. In the case of the automotive, it is the AIAG edition 2009 (AIAG 2013).

Benefits of Supplier Development

To be sure, any endeavor to improve supplier relationships comes with effort and commitment in the form of strategy for volume, location (proximity), historical performance, and time of existence. All these must be verified and substantiated, usually with an on-site audit and paper historical evaluation.

It is of paramount importance to realize that this strategy in the past quite often never went beyond fulfilling orders and exhibited only superficial levels of collaboration—just to get by. Additionally, there was no effort to tap into the hidden potential of these relationships. In the future world of improvement, this strategy must be active and open to consider all issues and problems. If that occurs, then the extra effort typically accumulates into at least the following:

- Improving overall communication internally and externally
- Reducing product defects
- Improving on-time delivery
- Reducing cycle time
- Improving overall performance and customer satisfaction
- Reducing nonvalue activities
- Improving capacity

How do we make sure that this strategy is being followed? By an on-site audit in which both customer and OEM representatives participate. A successful audit must cover at least the following areas:

- Organization: management, people, quality, innovation
- Resources: technology, process
- Health: finance, supply risk
- Responsibility: environment, certifications

The on-site supplier audit must be carried out by a cross-functional team, and must cover various main areas as well subsections of these areas. In other words, the audits must have appropriate and applicable breadth and depth in their evaluation process. Obviously, for each main area and subarea a checklist and questions must be prepared to be investigated and answered. The results should be tabulated in a scoring system developed by the OEM and should indicate excellence as well as deficient points for improvement.

In the final analysis, a Lean supply chain may be created by commitment of both OEM and supplier to improve the overall performance and safety by doing the following:

- Redefine organization/supplier relationships.
- Develop and implement an effective e-business strategy that will enhance communication across the supply chain—especially the advanced product quality process (APQP).
- Use appropriate and applicable metrics (performance measures) that will provide feedback on the supply chain.
- Increase collaboration across the organization/supplier interface.
- Employ a *cost out* strategy, not merely reducing *price.*
- Align the customer with its suppliers so that they form a single value-generating entity.
- Improve process stability, process capability, and capacity.

References

AIAG. (2013). *Checklist to ISO/TS 16949*, 4th ed. Southfield, MI: AIAG, Daimler Chrysler, Ford Motor Company, and General Motors Corporation.

ISO/TS 16949. (2009). *Technical Specification ISO/TS 16949*, 2nd ed. Corrected copy. Southfield, MI: AIAG, Daimler Chrysler, Ford Motor Company, and General Motors Corporation.

7

Role of Supplier Technical Engineer

Overview

Perhaps the most visible customer representative in the customer–supplier relationship is the supplier technical engineer. Although different companies use different names for this representative, their function is always the same. That function is to help communicate the needs and wants of the customer and to make sure that the supplier has the ability to fulfill these needs and wants in the form of requirements. In case the supplier falls short of these expectations, it is the responsibility of the technical engineer to help the supplier catch up in its deficiencies. Fundamentally, the role of the technical engineer is to be responsible for

- Plant support
- Technical support
- Supplier site management
- Warranty reduction
- Process development
- New model programs

These responsibilities are fulfilled by encouraging and motivating the supplier to

- Act like a partner so that they can be a partner
- Trust and be trustworthy
- Communicate with consistency
- Be fact based
- Think value, not just price

Specific Responsibilities of the Stakeholders

In order for the supplier technical engineers to perform their duties, it is imperative that all stakeholders understand their roles. These roles are the basic functions that the supplier engineer has to live up to and optimize daily. Minimum requirements for a supplier are

- Deliver the right product, at the right time, in the right quantity
- Meet the technical, financial, and process capability objectives
- Staff the program with sufficient resources
- Complete product assurance and timing plans
- Challenge seemingly redundant customer requirements

Minimum requirements for purchasing are

- Identify suppliers who deliver the required product development capability
- Ensure technical resources are available to close deficiency gaps
- Verify supplier quality capabilities and utilize a metrics-driven assessment
- Facilitate a cross-functional sourcing consensus
- Issue target agreements
- Develop forward-model-sourcing strategies
- Ensure supplier responsiveness
- Drive adherence to the purchasing supplier relationship values (this is an issue of trust)
- Involve supplier technical engineer early in the sourcing process

Minimum requirements for manufacturing are

- Participate in sourcing decisions, based on quality history and total cost
- Provide shipping requirements, shipment windows, and packaging requirements
- Monitor supplier quality deficiencies and communicate them to the suppliers and supplier technical engineer as well as purchasing
- Maintain historical data on supplier's quality and shipping performance
- Participate in quality reviews at the supplier sites
- Conduct quality reviews of selected suppliers at the customer's manufacturing plants

- Provide technical/manufacturing knowledge and assistance to suppliers that have major quality/productivity issues

Minimum requirements for engineering are

- Provide customer requirements, program scope, and quality and reliability objectives
- Provide clarification on proposed direction
- Establish a product assurance management process
- Monitor and manage supplier execution
- Participate actively in supplier technical and process capability reviews
- Lead regular and periodic risk assessments
- Lead problem resolution and continuous improvement efforts

Minimum shared responsibilities with purchasing are

- Achieve common understanding of supply base
- Recognize areas of concern
- Understand which suppliers are doing it right
- Work together to source new business
- Deal with ongoing performance issues

Specific Categories of a Supplier Technical Engineer

Whereas the basic requirements are very critical to a successful program between customer and supplier, the reader has noticed that all these requirements are very demanding in content and volume. Therefore, the responsibility of a supplier technical engineer (depending on the size of organization) may be divided into four categories with specific responsibilities. These categories are

1. Site engineers. This individual is primarily the process lead for any issues between supplier and customer. This means that the engineer
   a. Interfaces with supplier site
   b. Utilizes manufacturing site assessment to improve supplier quality
   c. Is responsible for ongoing part quality and product launch

   Specifically, a site engineer

   a. Serves as a single point of contact for assigned supplier sites, and is accountable for site performance for new product launch and ongoing part quality.

b. Deploys the basic quality system of the customer requirements and monitors its attainment status.
c. Leads the APQP and timing supplier engagement process.
d. Ensures PPAP compliance to customer-specific requirements.
e. Provides technical support to aid in improving quality systems and leads all supplier concern resolution that is, warranty, capacity constraints, capability verification, quality rejects, stop ships, stop builds, field service actions).
f. Drives supplier continuous improvement and supports Six Sigma and customer cross-functional matched pair initiatives.

These specific tasks may be separated into four groupings:

a. Sourcing/site development
   i. Sourcing for quality assessment
   ii. Basic quality system attainment planning and execution
   iii. Cross-functional matched pair activities
   iv. Manufacturing site assessment
b. Ongoing production
   i. Warranty issues resolution
   ii. Quality issues resolution
   iii. Supply/capacity risk mitigation
   iv. Campaign prevention
c. New product launch
   i. Site engineer lead for timing supplier engagement at priority sites
   ii. Advanced product quality planning to ensure flawless launch at each build milestone
   iii. Capability and capacity verification
d. Change management/resourcing
   i. Quality risk mitigation
   ii. Bank quality
   iii. Capacity planning and verification
   iv. New site assessment

2. Program engineers. A program engineer is responsible for the overall performance of the team efforts to make sure that launching is on time and without any issues for delay. This means that the program engineer
   a. Interfaces with customer program teams
   b. Supports program development and launch teams at customer sites
   c. Synchronizes site engineer activities with program/launch teams—single point of contact

Specifically, a program engineer

a. Serves as a site engineer single point of contact for assigned program team
b. Manages priority supplier selection, cross-functional buy-off, and supplier priority status communication
c. Manages consolidation of supplier APQP status reporting and interfaces with program team on elevation and resolution of supplier issues
d. Monitors and provides management reporting of supplier readiness for each milestone
e. Communicates emerging supplier quality issues during builds to site management team for timely resolution
f. Supports product/part builds to track, communicate, and report resolution status of supplier quality issues to program team

These specific tasks may be separated into four grouping:

a. Program coordination/priority supplier selection
    i. Site engineer point of contact for cross-functional coordination
    ii. Lead priority supplier selection process and communication
    iii. Communicate program milestone timing and updates
b. Timing (APQP) supplier engagement
    i. Track supplier engagement status
    ii. Collate supplier APQP status reports. Elevate issues for cross-functional intervention and resolution
    iii. Manage milestone review reporting of the APQP, part submission warrant (PSW) growth charts, and part change tracking
c. On-site product/part launch support
    i. Support product/part build events at pilot and final assembly plants
    ii. Communicate supplier quality issues to site engineer for timely resolution
    iii. Verify and drive containment of supplier quality issues at build locations
d. Postlaunch engagement
    i. Develop lessons learned document with launch team for continuous improvement
    ii. Track completion of capacity verification and status of final PPAP
    iii. Transition management of supplier quality issues to resident engineer

3. Resident/plant engineers. A resident/plant engineer is responsible for identifying problems, primarily quality issues, which involve suppliers, and communicating them to the appropriate site engineer and supplier management. This means that the resident/plant engineer
   a. Interfaces with customer manufacturing facilities
   b. Links site engineers to customer manufacturing plants
   c. Leads resolution of things gone wrong (TGW)/warranty concerns

   Specifically, a resident/plant engineer

   a. Serves as a single point of contact at the assigned customer plant, reporting directly to the resident engineer manager and indirectly to the plant validation team (PVT) manager
   b. Leads triage of supplier process warranty claims engaging the supplier and site engineer for concern resolution, ensuring that warranty projects are initiated and containment is in place to help reduce R/1000, cost per unit (CPU), and TGW warranty metrics
   c. Works with the plant to identify suppliers requested to attend incoming quality (IQ) reviews or product quality reviews, and ensures supplier presentations are available
   d. Assists suppliers and plants in resolution of plant quality reject (QR) disputes
   e. Engages site engineer during stop shipments and initiates stop ship management notification
   f. Assists in containment of supplier-related stop ship issues

   These specific tasks may be separated into four groupings:

   a. TGW/warranty concern resolution
      i. Lead triage of supplier process-related warranty projects
      ii. Engage suppliers and site engineer required for project completion
      iii. Represent site engineer on plant variability reduction teams (VRT)
   b. QR/chargeback
      i. Support QR/chargeback procedures as defined in customer's procedures
      ii. Review/approve threshold QRs
      iii. Place disputed QRs on hold
      iv. Manage dispute resolution
   c. Stop ship engagement
      i. Provide initial notification of stop shipments to all of purchasing organization

 ii. Provide support in containing supplier issues within the plant
 iii. Work with stakeholders to ensure proper root-cause owner identified
 d. Focus supplier improvement
 i. Work with PVT and IQ to properly identify focus suppliers
 ii. Drive supplier responsiveness by requesting adequate supplier support within plant
 iii. Engage site engineer or engineering matched pairs to drive focused site improvement
4. Technical specialists. Generally, a technical specialist is a recognized internal expert for a specific skill, problem-solving ability, commodity, process, or Six Sigma and Lean expertise. Because of their uniqueness in skill or expertise, they are considered subject matter experts (SME) and they may travel wherever their expertise is needed. Their time is dedicated 100% to technical services. Typical services are to
 a. Provide in-depth technical expertise to both the organization and its suppliers.
 b. Develop/deploy best practices.
 c. In addition to being recognized as corporate technical expert, they provide process-specific strategies to deliver results that are consistent with corporate goals and expectations. Quite often, they also create, maintain, and deploy standards as appropriate and necessary. Finally, they serve as teachers, mentors, and coaches for others to become experts themselves.

 Technical specialists may come from unique disciplines or difficult processes such as robotics, casting, electrical, heat-treat, plastic injection molding, welding, tire specialists, and others.

 Specifically, a technical specialist

 a. Prioritizes supplier on-site manufacturing assessments based on supplier performance, new model programs, and product/part criticality
 b. Ensures failure mode avoidance of systemic quality issues by deploying closed-loop manufacturing process standards
 c. Assesses supplier manufacturing risk and capabilities to the organization's standards and drives improvement plans
 d. Drives continuous improvement in suppliers to meet the organization's standards within area of process expertise
 e. Is an available resource for consultation on critical quality or capacity concerns or supplier manufacturing changes
 f. Communicates supplier capabilities to the technical engineer, purchasing, and product development (PD)

g. Develops and cascades ongoing technical training material for technical engineer, purchasing, PD, and suppliers within the area of expertise

Miscellaneous Functions

Quite often, the knowledge of supplier technical engineers gets them involved with responsibilities beyond their position that deal with warranty and overall strategy and business decisions—especially in the early stages of sourcing. In the case of warranty, it is not unusual to have a supplier technical engineer supporting supplier warranty recovery programs, including

- Total organizational lead for warranty reduction program (WRP)
- Support supplier-related warranty spike, field service action, and warranty chargeback processes
- Assist site engineers and suppliers in utilizing warranty improvement tools (e.g., supplier warranty reporting and supplier improvement metrics that improve overall quality and customer satisfaction)

Strategy and Business

Perhaps one of the most unusual yet important special functions of supplier technical engineers is their participation in strategy and business decisions. The earlier this is done, the better the decision will be. Typical activities of this participation are

- Lead for common organizational goals, development, and continuous improvement metrics, such as capability, timing, organizational quality standards, PPAP, QOS, APQP, and so on
- Develop reduction and process lead for stop/field
- Co-lead recognition programs and awards
- Develop business process development and standardization
- Provide current supplier information for strategy research and development needs
- Provide fundamental supplier information for sourcing for quality through historical data or MSA
- Provide/recommend standardization in reporting
- Provide an opportunity for vertical communication

TABLE 7.1

Summary of Roles and Responsibilities of the Lead

Role	Commitment	Contribution of Time	Responsibility
Process lead (PL)	Site supplier engineer identified to support tech Services (primary responsibility is site supplier engineer)	25%	Supports execution of process-specific business plan in the region Conducts assessment and reviews SREAs to improve customer satisfaction and prevent issues Supports technical training in the region Provides technical leadership to drive resolution of urgent supplier production issues (quality/capacity) Participates in monthly global process team meetings and maintains regional QOS
Process specialist (PS)	Some site responsibility—dedicated to technical service activities	50%–100%	Regional subject matter expert Executes process-specific business plan Conducts assessment and reviews SREAs to improve customer satisfaction and prevent issues Deploys technical training for the region Provides technical leadership to drive resolution of urgent supplier production issues (quality/capacity) Participates in monthly global process team meetings and maintains regional QOS
Technical specialist (TS)	Recognized technical expert by the organization Fully dedicated to global technical services	100%	Corporate technical expert Global leader for process-specific strategy to deliver results Creates, maintains, and deploys global standards Teach/mentor/coach all others in expertise area Conducts assessments and reviews SREAs to improve customer satisfaction and prevent issues Global consultant on emerging issues

The reader will notice that these activities require a lead and support. In Table 7.1, we provide a guideline as to who is supposed to do what, as well as the amount of contribution that it is expected to be consumed by each lead or support. In addition, we have generalized some of the responsibilities that each is accountable for. Obviously, the table provides a generalization, and the role, timing, and responsibilities addressed here are not exhaustive or absolute. Each organization must develop its own for the best results.

Section II

Supplier Responsibilities

It has been said many times in many situations, "garbage in, garbage out." In fact, it is known as the GIGO principle. In the case of launching a program on time, on budget, and with minimum interruptions (for whatever reason), that process is a function of the supplier doing the "best" it can all the time. Of course, in order for this to happen, the original equipment manufacturer (OEM) (customer) must have identified the requirements.

There is no argument that the behavior of the supplier is of paramount importance in this process. So let us identify the issues, concerns, and requirements that any supplier has and the antidotes for poor performance.

To be sure, nothing is perfect! However, perfection can be defined based on the set requirements that a customer requires of a supplier. Small or indifferent deviations may indeed be overlooked, but in the case of the supplier satisfying the customer, it is the supplier's responsibility to either eliminate or minimize the issues, concerns, or problems so that flawless launching takes place. To do this, an organization must have core business strategies, planning, and a vision for continual improvement. This must be a way of life and a relentless perseverance to perfection. A pictorial view of at least one approach to this journey are shown in the core business strategy chart on the next page. There are many variations of this, and every organization should develop its own.

In Section II, we are going to examine some of the requirements that every supplier must understand and practice so that the customer will receive the product and/or service flawlessly. Above all, by following the concepts on which we have focused, we are certain that organizations will improve dramatically. A survey in the *Modern Machine Shop Journal* (September 2014, 81–87) supports my contention. The results are summarized in the table on the next page. The article does not say what the overall sample was, but it does indicate that the sample was taken from aerospace, automotive, military, and equipment suppliers.

One approach to core business strategy

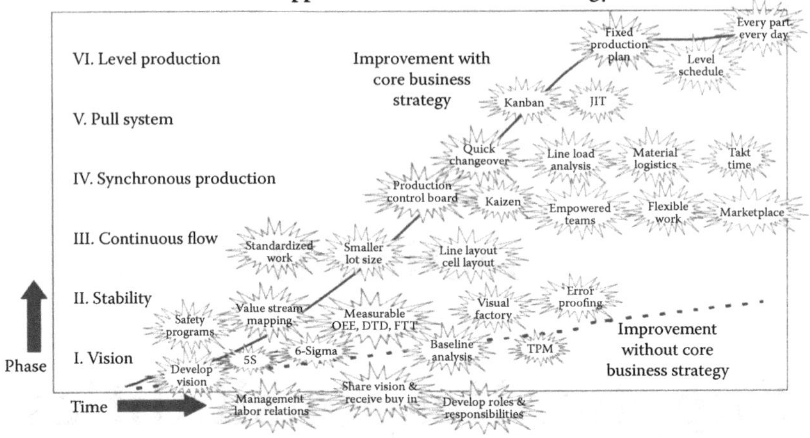

Top Five Improvement Methods			Top Five Supply Chain Practices		
Items	**Top Shops (%)**	**Others (%)**	**Items**	**Top Shops (%)**	**Others (%)**
Continuous improvement program	85	55	Customer satisfaction surveys	70	30
5S workplace organization	63	42	Access to customer forecasts	63	31
Cellular manufacturing	44	25	Just-in-time customer delivery	59	35
Just-in-time manufacturing	44	30	Sharing forecasts with suppliers	59	18
Quality certifications	59	45	Design for manufacturability	56	28

Furthermore, scientific evidence presented by Lakhal (2014) suggests that ISO 9000, total quality management (TQM) practices, and organizational performance have a direct significant relationship with excellence ($p < .01$). On the other hand, Jones (2014) has identified the critical factors of success as being leadership ($p < .0001$), measurement analysis, and knowledge management ($p < .0001$).

References

Jones, M. (2014). Identifying critical factors that predict quality management program success: Data mining analysis of Baldrige award data. *Quality Management Journal* 49–61.

Korn, D. (2014) What it takes to be a top shop. *Modern Machine Shop* September, 81–87.

Lakhal, L. (2014). The relationship between ISO 9000 certification, TQM practices and organizational performance. *Quality Management Journal* 38–48.

8

Commitment to Continual Improvement

Overview

The continual improvement (CI) process is an ongoing effort to improve products, services, or processes. These efforts will seek *incremental* improvements on a daily, weekly, monthly, or yearly cycle. Customer valued processes are constantly evaluated and improved in the light of their efficiency, effectiveness, and flexibility.

It must be emphasized here that CI process is not a managerial approach. It is a mindset that facilitates a culture of innovation and constant improvement at all levels of the organization. Specifically, it encompasses at least the following three elements:

1. Balanced scorecard: Goals and objectives; measurable metrics, that is, key behavioral or performance indicators, comparisons, or tracking of goals versus actuals on a month/day/week basis. A generic balanced scorecard may be considered a way to measure the most basic metrics in the organization. For example, financial by measuring the financial impact with a weight of 25%; customer by measuring the improvement for the customer with a weight of 30%; internal process by measuring improvement project completion as identified in the quality operating system (QOS) with a weight of 35%; learning and growth by measuring competency of appropriate tools and their implementation as well as overall results of improvement with a weight of 10%.
2. Fast response board (control visually): This is also known as the SQDC board. It is a board to document and visually control actions on Safety, Quality, Delivery, and Cost (SQDC). The focus here is on identify/plan, in process/do, complete/check, lessons learned/act.

   In essence, this board serves as a visual review of the various roles within a CI culture to provide the workforce with clear goals and expectations. The process improvement team helps in identifying and developing metrics and improvement targets that are *meaningful* and *doable* for the work force, as well as train the workforce in visual management tools and resources required to effectively and efficiently perform their task (job). The board serves as a management tool to expose problems, barriers, and concerns so that management can better understand the major causes and work to remove them. A typical SQDC board tracks the performance of safety, quality, delivery and cost, trend improvement against target improvement level, daily performance, and improvement activities, and provides a visual analysis of main problems.

In other words, the purpose of the board is to present a floor-level visual representation of defined metrics at all work centers for shift-to-shift tracking daily, weekly, and or monthly. The tracking of the metrics is used to identify gaps and opportunities for improvement. In fact, the scope of the SQDC board is to be utilized and tracked by work center employees as an informational stream showing the work center key performance indicators (KPI). To make sure that simplicity prevails, the SQDC chart is used visually with colors, charts, and graphs to show measurable actual against targeted performance within each critical metric of SQDC. Improvement opportunities are identified to close the gap between actual and targeted performance metrics. When the work center is not in operation for a particular day, the designation of DNR (did not run) should be recorded.

Typical definitions used on the SQDC boards are

a. Safety, quality, delivery, and cost. These are the key performances or behavioral indicators that must be monitored. (Some organizations have more.)
b. First time quality (FTQ). A metric that indicates to what extend parts are manufactured correctly the first time without need for inspection, rework, or replacement. This is generally calculated as: (Good Parts)/(Good Parts + Rejects).
c. Overall equipment effectiveness (OEE). An established method of measuring and then optimizing capacity utilization of a process performance or that of a whole manufacturing plant. This is generally calculated as (Good Parts)/(Total Target).
d. Unplanned downtime. Any event or downtime instance that was unplanned, resulting in lost planned production time at any work center. Downtime that does not count as unplanned is lunch, scheduled breaks, scheduled maintenance, or scheduled new product modification.
e. Pareto. It is a bar chart in which data is plotted to analyze what problems need attention first because of the taller bars on the chart. The more an issue is present, the taller that bar will be on the graph. This is a very effective way to see which issue happens most frequently.

3. Employees: Utilization of cross-functional process improvement teams; usage of data for action with the intent of results that reduce the gap identified in the balanced scorecard; develop, involve, and empower one another with both authority and responsibility.

Of these three elements, the employee element is very fundamental and without it, there would be no systematic improvement in the organization. The employees make up teams and through synergy, the teams come up with consensus decisions that are successful. Therefore, let us review some common issues dealing with teams.

Process Teams

Generally, they are between 5 and 9 individuals performing together for a common goal. Team members can be volunteers and or be appointed, given a specific expertise that may offer to the team discussion. A leader as opposed to a chairperson guides the team in the direction of the problem at hand. The effort here is to make sure that the team is able to collaborate and produce actionable actions that will reduce the gap. The reason for this collaboration and multidiscipline is to make sure that the cross-functionality will result in creating

- Creativity (out of box thinking)
- Collaboration
- Development
- Involvement
- Empowerment
- Reducing the gap of the problem

At this stage, one may wonder how a cross-functional and multidisciplinary team may accomplish all these results. The answer may be in the following activities, which the team and individuals may undertake because of their participation:

- They are self managed.
- They are focused on the safety, quality, delivery, and cost (SQDC) metrics, specifically on actions concerning the SQDC matrix. The metrics are monitored, tracked, and measured from week to week (or any other time reference as appropriate and applicable) so that improvements or lack of them will be discussed and shown on the scorecard.
- They are meeting together and at regularly scheduled times. These meetings are to review and update the course of action on the given

problem. Typically, a summary of the meeting will be posted on the fast response board (FRB) to reflect the improvement or opportunity and actions needed for closure. A typical FRB is shown in Figure 8.1.

- They provide opportunities for individuals to work on common goals.
- They provide opportunities for individuals to work on new and challenging skills.
- They provide encouragement to team members to help others, even those out of their own team.
- They improve communication for the entire organization through their openness.
- They recognize and diffuse opportunities and wins for the entire organization.
- They generate ownership and responsibility for opportunities, actions, and solutions.
- They create an atmosphere of rewarding work environment by reducing the gap together.
- They identify *best practices* as well as *lessons learned* and diffuse them in the organization.

The effectiveness of the team is based on the traditional Plan, Do, Check, Act (PDCA) model. The essence of the model is indeed continual improvement

The Visual Control by Team				
The Desired Behavior Is the Generation of Opportunities and Successes				
	Plan	**Do**	**Check**	**Act**
	(Identify)	**(In Process)**	**(Complete)**	**(Lessons Learned)**
S				
Safety				
Q				
Quality				
D				
Delivery				
C				
Cost				

FIGURE 8.1
Typical fast response board.

because it focuses on a perpetual cycle of improvement. Specifically, in the *Plan* stage, the team is engaged in

- Referencing the goal and objective
- Establishing the metric of the *success*
- Establishing the baseline for comparison
- Establishing the goal for *success*
- Identifying the opportunity or action that will assist in reducing the gap between the desired goal and the current state
- Establishing the objectives and processes necessary to deliver results in accordance with the expected output (target or goals)

In the *Do* stage, the team is engaged in

- Initiating action based on the opportunities identified
- Planning the start and finish dates for the actions
- Placing applicable actions in process
- Defining the leader for the action
- Defining the priority for the action
- Reviewing the status of action regularly
- Testing the result of the action

In the *Check* stage, the team is engaged in

- Verifying the results, in such ways as to
  - Study the actual results that were collected in the Do stage
  - Compare against the expected results (targets or goals from the Plan stage)
  - Monitor action versus result
  - Identify the monitored result of the action
- Identify the result as a win or opportunity

In the *Act* stage, the team is engaged in

- Continuing to evaluate the process
- Applying lessons learned to systems or other processes
- Recognizing team members for their *extra contribution* if it was a win
- Recognizing another opportunity to begin a new cycle of PDCA

In each stage of the model, one may use Figure 8.2 to generate or summarize the events of the individual stage.

Process Improvement Team:		
Opportunity:		
Item 1	**Plan**	**Actual**
Item 2		
Action:		
Leader:		
Priority:		
Results:		
Win or Opportunity		

FIGURE 8.2
Typical form to generate or summarize events of the PDCA model.

Key Items of Concern in the CI Process

One-Piece Flow

One-piece flow is a production technique in which the work moves from one operation to the next one piece at a time, continually. Ideally, there is no batch or build-up of parts at any given point in the process. The idea is to make one then move one. Why is this important? Because it will help to achieve

- Shorter lead time
- Better communication
- Improved quality
- Visual management
- Less space
- Reduced costs

The process of implementing the one-piece flow is based on a five-step approach.

Step 1

Identify product families and collect data. Typical items should be

- Process work sequence
- Required daily customer demand
- Operating time

- Manual times (load, unload, etc.)
- Machine cycle times
- Walk distance and times
- Tooling locations
- Tool change times

Step 2

Calculate Takt time. The reader should be familiar with two types of Takt time. The first one is spelled *Tact* and the second is *Takt*. Both are calculated the same way. However, the difference is that Tact time is based on either theoretical or surrogate data, and is used for preliminary analysis or planning. As the real data are generated, the theoretical or the surrogate data should be replaced with the actual data. Takt time is the time needed to produce one unit of a product to meet the customer demand. It is calculated as

$$\text{Takt Time} = \frac{\text{Operating Time}}{\text{Customer Demand}}$$

where

Operating time = (Shift Time per Day) − (Planned Downtime per Day)
Customer demand = Average Daily Volume (as determined from the master delivery schedule)

Takt time is very important and is used to verify both volume and capacity. As a result, this facilitates how the operator will work on the line and at what rate. It is customary that the Takt time and the operator's work and rate do appear on the Production Board and they should be the same as the calculated numbers. The reason for this display is to

- Highlight overproduction and waiting wastes
- Identify process difficulties quickly
- Reduce finish goods and work-in-process inventory
- Highlight imbalances

An example of Takt time is as follows:

Shift time per day = 2 × 8 h (3600 s/h) = 57,600 s

Planned downtime per day = 2 × 30 min (60 s/min) = 3600 s

Operating time = 57,600 − 3,600 = 54,000 s

Customer demand = 800 + 700 + 600 + 200 = 2300 pieces

Takt time = 54,000 s/2,300 pieces = 23.5 s

Always round down Takt time. In this case, it is 23 s. This means that the customer requires one part every 23 s. This is a very important number because it is used to set the production rate equal to the customer demand rate. It also helps in

- Synchronizing operations of all departments to customer demand
- Preventing overproduction
- Determining optimal staffing
- Driving capacity analysis and equipment utilization
- Creating the appropriate flow

Takt time is *not* affected by

- Associate speed, skill, or attendance
- Machine breakdowns
- Part/material shortages
- Quality problems

On the other hand, Takt time is affected by

- Number of shifts
- Shift lengths
- Relief/no relief, breaks, and so on
- What parts are assigned to which line
- Order volume

Because of these items, the Takt time must be periodically double-checked as demand changes or other factors may influence it.

Step 3

Determine the work elements (tasks) and time each element. If an element can be done repeatedly, it can be standardized. We should strive to standardize the best. Standardized work is generally accepted as using the most effective combination of employees, materials, and machines to perform an operation using the best method to meet the customer demand with as little waste as possible. Standardization is important because it

- Minimizes process variation introduced by the operator and eliminates unnecessary motion
- Helps to balance work load and labor to best meet customer demand
- Provides the baseline required for continual improvement

The standardized work elements are generally considered as

- Calculating Takt time. It serves as a baseline for customer demand.
- Standardized work sequence. An effective set sequence of work performed repeatedly by team members who are processing or assembling a product. (Special note: Standardized work sequence must be fixed and observed.)
- Standard work-in-process. This means using a minimum amount of work-in-process inventory so employees can complete their standardized work smoothly, repeatedly, in the same sequence, with the same movement, and so on To accomplish this, the following sequence of events must take place:
  - Develop data collection sheets. This must be done with actual data from the shop floor.
  - Develop a work combination table. A table with cross-reference of the operator, tasks, time, and any other pertinent information.
  - Develop a work chart. A summary chart of all pertinent information. For maximum effectiveness, it should be displayed in the work cell.
  - Develop a work balance sheet. The information in this chart will guide the cell into continual improvement.

Step 4

Create an operator balance chart. Make sure the appropriate cycle time is appropriately associated with the right work element and operator. This may be accomplished by understanding the process. There are two approaches for this: (1) traditional process flow diagram and (2) value stream map.

Traditional Process Flow Diagram

The purpose of the process flow chart is to identify, understand, and define the activities or steps in a process. Once the process is documented, it is important to determine the health of the process. A good process is effective, efficient, adaptable (flexible, robust), and in control. There are generally three phases to a process flow diagram: *as is, should be,* and *could be.*

As Is

As is indicates the current state of process. To construct this phase, the following are required.

Select a Process for Investigation or Study

When a process is selected for improvement, effort is needed to map, evaluate, and make changes to the process. It is important to select a process that, if improved, will increase the performance of the service. To do this, brainstorm a list of processes that you might like to work on. Then, select the *first* one to work on by asking the following questions:

- How much does the process affect the customer?
- Can the process be fixed?
- How is the process performing?
- How important is the process to the company's goals (critical success factors)?
- What resources are currently available to work on the process?

To be a candidate for study, the process should display one of the following:

- Internal and external customer complaints
- Missed deadlines
- High costs
- Inconsistencies
- Long cycle times
- Information has shown that there is a better way
- New technologies are available
- High incidences of rework
- Low productivity

Select or Identify the Process Owner and Team Members

The process owner and team members are chosen to ensure that the process will be effective, efficient, under control, and flexible. In essence, both the owner and members must be empowered to take action. It is imperative here to emphasize that especially the process owner must see the cross-functional perspective and realize that limited resources do exist in the organization. Having said this, the process owner must

- Define the scope of the process
- Select members for the process improvement team
- Ensure members of the team are trained
- Provide feedback to the team about any business changes
- Keep management abreast of progress the improvement team is making

Define the Scope of the Process

Here, the external limits of the process are identified as well as (1) the interface between the supplier and the process and (2) the interface between the process and its primary customer. The interface shows how work flows between different departments. On the other hand, the customer is the person, people, or process that uses the output—whether internal or external. An internal customer is someone inside the organization, whereas an external customer is outside the organization.

Define Process Boundaries

Here we establish the process boundaries by outlining where the process starts and ends, by defining what is included in the process, defining what is not included in the process, listing all the outputs, listing all the appropriate and applicable inputs to the process, and identifying all the departments involved. Once boundaries are established, it is important to be able to identify the outputs and the customers of the process under study. The output of the process should be stated in terms of deliverables of the process such as plans, procedures, orders, answers, and proposals, but never as an activity. If the process can have several outputs or different types of customers, you may need to separate them and summarize the information in a table.

Develop a Mission Statement

The mission statement outlines the intent of the project or what the team is trying to accomplish. It also aids in communicating between team members and with individuals outside of the team. A good mission statement should be short, define the scope of activities, state what will be accomplished, specify performance improvement targets, and include a completion date.

Define the Process

The process may be defined in two steps: the first is based on a macroanalysis (big picture), and the second is based on a microanalysis (more detailed). The type selected depends on the overall objective of process mapping.

Should Be

This is how the process should be running. A good *should be* process map must take into consideration the evaluation of the *as is* process. This means a good walk-through and a thorough analysis of the current process. Once this is complete, the team is ready to develop the *should be* process map. This implies a streamlining of the process. In essence, this means that any inefficient step (delays, redundancies, rework, changes, motion, or extra reviews, inspections, and so on) should be identified and eliminated. To construct this phase, the following are required:

1. Evaluate control points. Control points are areas in the process where rework is produced or where actions verify the output. Without control points, the process is reactive, relying on customer feedback to determine quality. Finally, a control point determines process characteristics to be measured based on either internal (organizational) or external (customer) requirements or both.
2. Conduct a process walk-through. The purpose of the walk-through is to evaluate the accuracy of the process map and to interview the individuals working on the process. During this interview, the people in the process are able to identify barriers to the process and provide suggestions for improvements. Typical questions on the walk-through are what do you need to do the job (time, materials, tools, equipment, and so on)? Were you trained? How? By whom? Was training appropriate? Do you feel adequately trained for this job? Who are your customers? Is there a disconnect between actual and perceived customers? What do you do if your operation produces defective work? What is your biggest problem? Do you receive appropriate and applicable feedback from the next process or from other sources? What suggestions do you have for process improvement?

   As you proceed with the walk-through, the next step is to make sure that each person performs tasks appropriate and applicable to his or her job (this takes the process from the activity level to the task level to obtain more detail). Typical questions at this stage are (a) How is this activity performed? (b) Why is it done that way? (c) Can it be combined with another activity?
3. Identify areas of improvement. During the *process walk-through*, information is collected on the areas of the process that may need improvement. In addition, opportunities for waste reduction can be identified in either or both categories of effectiveness and efficiency. In the case of effectiveness, the focus is on how well the output of the process meets the requirements of the customer whether internal or external. In the case of efficiency, the focus is on how well the resources are used to produce the output. Typical metrics are cycle time (the time from input to output) and cost (the expenses of the process), respectively.
4. Draw the *should be* process map. Once the process analysis is complete, the team needs to look for opportunities to streamline the process and develop the *should be* process map. Drawing the *should be* process map requires streamlining the process and the removal of waste.

5. Streamline the process. Streamlining implies that the process is carefully analyzed to remove waste and improve efficiency and effectiveness. The process is examined to provide the smoothest flow and the least resistance, ensuring that the process can meet its overall objectives with minimum effort. Specifically, streamlining includes the following steps:
   a. Standardization: Procedures are critical as they outline how work should be done. They provide a means to standardize the process to ensure the consistency of the output. Procedures should
      i. Clarify responsibilities
      ii. Be realistic and be developed by the owners of the process
      iii. Be easy to understand and not be open to interpretation
      iv. Define minimum performance standards
      v. Outline training requirements

      Therefore, standardization is a way to
      i. Find the best way to perform an activity
      ii. Ensure that all individuals performing an activity use the same methods
      iii. Help identify primary deficiencies
      iv. Create and follow procedures
      v. Review and update outdated procedures for effectiveness
   b. Value-added assessment: A value-added assessment involves analysis of the process to determine the time for value added and the total cycle time for the process. In other words, it accounts for all tasks to be evaluated to determine impact on customer requirements. Questions that need to be answered are
      i. Can the value-added activities be improved?
      ii. Can the value-added activities be performed at a lower cost and a shorter cycle time?
      iii. How can non-value-added activities be eliminated or minimized without affecting the customer or the integrity of the product or service?
      iv. How can business value-added activities be minimized?

      Based on these questions, actions should be taken to
      i. Eliminate bureaucracy: Remove any unnecessary approval steps or paperwork.
      ii. Eliminate duplication: Identify reasons for redundancy and correct underlying causes.

iii. Simplify steps: Combine tasks where possible and ensure that work instructions can be easily understood.

c. Upgrading: Upgrading requires at least the evaluation of
   i. Equipment (Is it defective or old?)
   ii. Work environment (Is it safe, clean, and comfortable?)
   iii. People in the process (Are they properly trained and do they have the opportunity to contribute to process improvement?)

   One may start by asking simple questions such as
   i. Do employees have the right equipment to perform their tasks?
   ii. Is the office properly laid out for the type of work that needs to be done?
   iii. Does the office have the right lighting for the jobs that need to be done?

d. Optimizing the process: This means that the supplier must find the "best" way for the process to run. This implies that appropriate setup procedures exist and there are error-proofing approaches that prevent defects from happening. Error proofing sometimes is called mistake proofing, which is the equivalent of the Japanese concept of poka-yoke (ポカヨケ). It is defined as building into the process ways to not make the mistake in the first place.

e. Reducing cycle time: Cycle time is critical to business processes. The longer the cycle time, the higher the overall cost. This means that cycle time must address issues that involve
   i. Identifying tasks that could reduce overall cycle time
   ii. Identifying serial versus parallel activities
   iii. Determining sequence of activities
   iv. Reducing interruptions
   v. Improving timing
   vi. Reducing output movement
   vii. Identifying location, location, location

   Typical questions are
   i. Where should the review of the process be located?
   ii. Can the sequence of activities in the process be changed to reduce the overall cycle time?
   iii. Is the process performed in the right building, city, state, or country?

f. Simplification: Simplification means to reduce complexity where possible. When simplification is applied to business processes, every element is evaluated to ensure that it is less complex, easy to understand, information is readily available, and the process can be easily managed. In essence, simplification looks for ways to reduce the complexity of the process and address the two fundamental questions of any business. Can we do whatever we are supposed to do faster and easier? To answer these questions, we look for

i. Duplication or fragmentation of the process
ii. Complex work flows or bottlenecks
iii. Memos and other correspondence that may indicate possible or future issues
iv. Meeting summaries that may identify discussions of pending or current problems
v. Opportunities for reducing handling, refining standard reports, eliminating unused data, and unnecessary copies

To facilitate this investigation, specific questions may include the following:

i. Is the process properly documented? If not, it may indicate that the process is not understood.
ii. Are errors made in filling out the forms? If yes, it may indicate that the forms are complicated or they do not allow for appropriate data.
iii. Could activities be combined? If yes, what would it take? Would the combination have a positive effect on the customer?
iv. Are all the documents in the right location? If not, why not?
v. Can a particular activity be eliminated? If so, how and at what cost to the customer?
vi. Are instructions self-explanatory? If not, why not?

6. Measure process improvement. It has been said that if you cannot measure it, you cannot control it; and if you cannot control it, you cannot manage it. Therefore, before you make actual process changes and improvements, you should measure process performance. When measures are tracked after improvements have been made, then it will be possible to evaluate the effectiveness of the changes. However, one must always remember that measurement by itself is worthless if it is not part of a feedback system to improvement. Therefore, the objective of process measures should be to evaluate the health of the process by

a. Tracking process performance. To establish where measurements should be taken, examine the process map and identify the activities that have the strongest influence on effectiveness or efficiency. Measurements should be taken by the people in the process. In fact, there are three major process measurements. They are

i. Effectiveness: how well the output meets the requirements of the customer. This primarily deals with accuracy, performance, costs, and timeliness. A synonym for effectiveness is quality having the right output at the right time and at the right place. Some effective measures may include percentage of reports completed on time, number of reports per report, amount of rework in the process, number of times the report is late, and so on.

ii. Efficiency: how well resources are used to produce the output. This primarily deals with processing time, resources (people, time, money, space, material, machine) per unit of output, wait time per unit, and value-added cost per unit. Whereas effectiveness measures are easy to see and measure, poor efficiency may be hard to recognize. However, one of the most meaningful efficiency measures is cycle time.

iii. Adaptability: the ability of the process to handle changing customer expectations and special requests. This primarily deals with the number of special requests handled per month and the percentage of special requests granted. Adaptable processes are those designed to meet the changing needs of the customer. Adaptability provides a sense of how flexible the process can be. Another way of saying this is to ask the question of how robust the process is.

b. Monitoring effective use of resources as well as the progress of change. One of the best and easiest ways to monitor a process is with pictures or charts. They help to see data in a much easier format rather than raw data. The charts used most often are

i. Bar charts or histograms: They show performance in a bar format. The bars are rectangular with lengths proportional to the values that they represent. A histogram is a special bar chart where the intervals of the bars are equal.

ii. Trend charts: A trend chart is used to display performance over time. This enables quick evaluation and reaction to the measurement. It is very important that any trend chart should include only one measurable, use a line graph, not a bar chart, have a line indicating the goal, and use an arrow to show favorable direction.

iii. Pareto charts: The Pareto chart helps to determine which problem to work on first. It is developed using year-to-date information. This provides data on the major issues that may be driven by a system in need of change. The Pareto chart is a simple bar chart that ranks related issues in decreasing order of occurrence. One should be very careful of the x-axis because it can be manipulated to show the important item of concern depending on *a* particular point of view.

iv. Painter charts: The painter chart is used to demonstrate the effectiveness of any process improvements. The categories are the same as on the Pareto chart and the number of complaints for each category is tracked for each month. The process improvements are also noted on the chart and should alert the team if the corrective action has had an effect (positive or negative) on any other issue.

Could Be

What is the optimal level of the process? There are many ways to approach this question. However, the most efficient is to look at benchmarking. The reason is very simple. Benchmarking allows the implementation of an existing "best" to be transferred into the organization's practice. To address this phase, the following must have occurred:

1. Current performance of processes is not adequate.
2. Process team unable to identify alternate methods for improvement.
3. The process team needs some additional ideas to improve efficiency and effectiveness.

Once these conditions have been met, then the team is ready to tackle the *could be* improvement process. The steps are

1. Review the performance of the process. This will identify where the process is. It will be used to compare the new data from the benchmarking study. It is the baseline of the current process.
2. Select benchmark subjects. Here you select the *best*. It may be another department, division of your own corporation, competitor, or even someone outside of your industry.
3. Collect data.
4. Analyze the performance gaps.
5. Develop the *could be* process map. Remember, the best process is the one with only value-added activities. Obviously, no process can be with only value-added activities. Therefore, the goal here is to minimize non-value-added activities.
6. Implement changes

Value Stream Map

Value stream is all the process steps required to bring a product to the market, starting from raw material and ending with the customer. In other words, the value stream is all the process steps to transform (add value) an item into a product. On the other hand, a value stream map is a tool that displays the flow of material and information as a product makes its way through the value stream as well as the calculated lead time. In other words, it is a display tool to show value stream and all information.

The value stream map may be applied to any manufacturing process or service and it is encouraged that all employees with ownership in the process being evaluated participate no matter what level they are or what job functions they perform. The reason for such open encouragement for participation is that this tool helps us to visualize all wastes in the current situation and transform current state to a leaner future process with an effective implementation plan. Why is it important to create a value stream map? There are at least six reasons. They are as follows:

1. Helps to visualize more than the single process level
2. Links the material and the information flows
3. Provides a common language
4. Provides a blueprint for implementation
5. Provides a simple and useful quantitative tool
6. Ties together Lean concepts and specific techniques

To develop a value stream, the following are necessary:

1. Identify the product family. Here is where you categorize the products, organize them in a family, and then identify the individual steps.
2. Create the current state map. It is imperative here to identify the process as is, not as you would like it to be. Items of concern here are all the relevant information about the process such as process steps, information, supplier, customer, materials (raw, work in process, and finished goods), lead time (total time to bring one piece of product to customer through the whole process). A classic approach is to follow the supplier-input-process-output-customer (SIPOC) model.
3. Create a future state map. Based on the current map, evaluate the findings and identify opportunities. To create the future state map, consider the level of production, pull system and supermarket, one piece flow, performance improvement, and reduction of non-value-added time.
4. Develop an action plan to address the opportunities from the previous step.

A typical value process flow chart is shown in Figure 8.3.

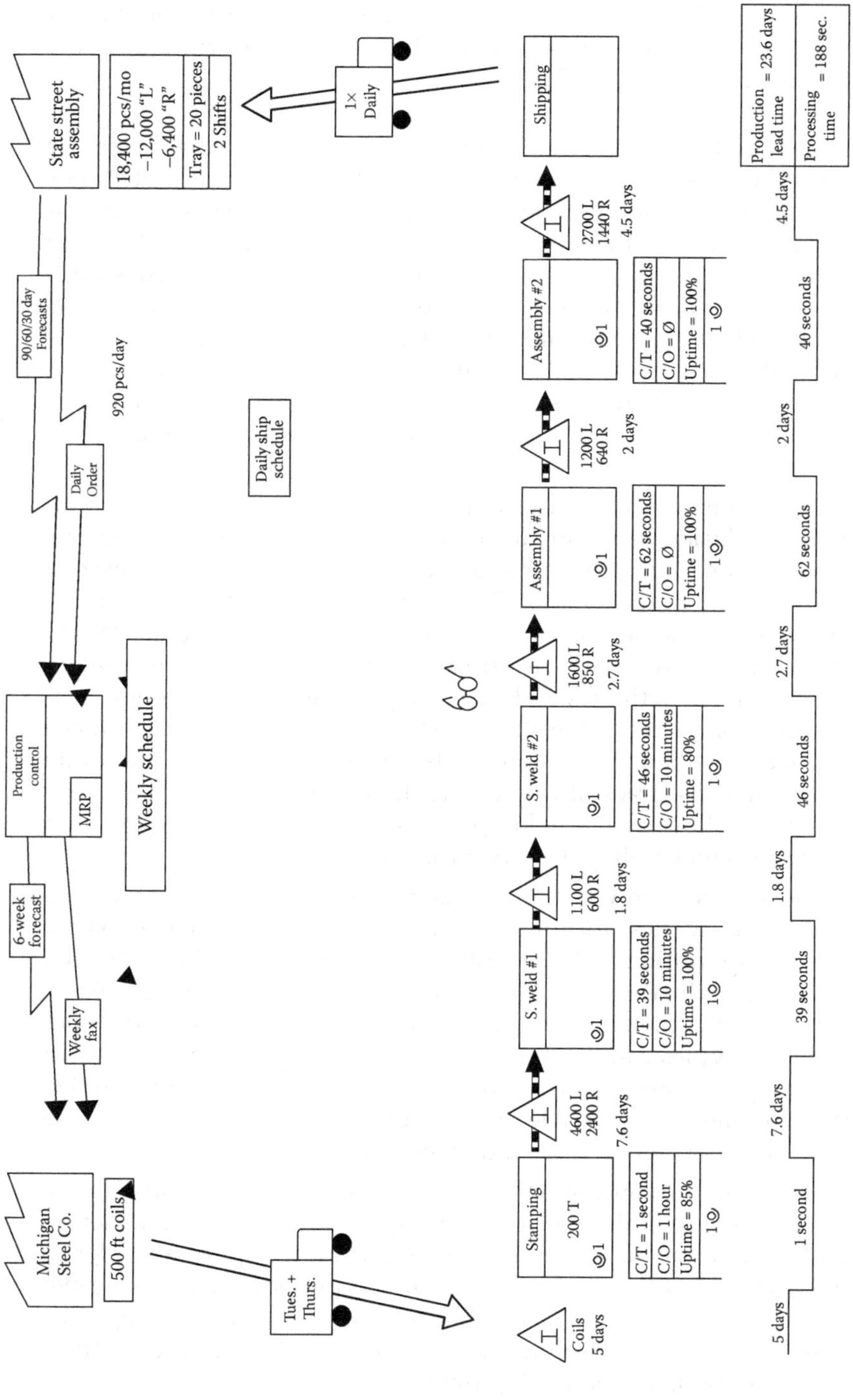

FIGURE 8.3
Typical value stream process flow chart.

Step 5: Create a Lean layout: Arrange the machines, workstations, and material presentation devices as if only one operator makes the product from start to end—even if you will never run the cell this way. Here there are at least four principles that should be considered.

1. Understand the concept of value and non-value-added activities: A value-added activity is the process step that transforms material or information into products and services that the customer requires. It is the activity for which the customer is willing to pay. On the other hand, non-value-added activity is the process step that consumes time, space, or other resources, while not directly contributing to the final product or service. It is the activity for which the customer is not willing to pay. Since this is an important and distinct difference, how should we distinguish it? The answer lies in the effectiveness of each element (task) in the process that defines whether the activity is of value.

   This is important because value-added and non-value-added is an indicator inside value stream mapping telling us how to improve processes and how we can reduce wastes to better achieve or exceed the customer's expectations. Keep in mind that on the value stream map, all activities are transformed to time in days or seconds. We do this to make sure we focus on reducing or eliminating non-value-added time or activities and increasing value-added time or activities to improve the value stream. This is of paramount importance and we must be diligent about it because it is very easy and quite common to lose track of why we are doing this. In fact, a classic mistake is to follow traditional improvement schemes and not focus on the wastes that really affect the process.

   The classic non-value-added activities have been categorized into eight wastes forming the acronym TIM WOODS, where T = Transportation, I = Inventory, M = Motion, W = Waiting, O = Overproduction, O = Overprocessing, D = Defects, and S = Skills. It must be mentioned and, in fact, emphasized that not all non-value-added activities can be eliminated immediately and some of them will not be eliminated at all. If that is the case, we try to minimize them as much as possible. Typical examples in this category are transportation, movement, and inspection.
2. Understand the concept of lead time: Lead time is the elapsed time for one item to make it through the system from the initial step to customer shipment (in manufacturing, from receipt of an order to the shipment of the product). Lead time is also known as *throughput* or *turnaround time*. It is calculated by adding all process step time used for transforming a product or service, including waiting time, transportation, and delivering time.

Lead time is made up of three distinct categories and all of them are reported in units of seconds, minutes, hours, or days. The categories are

a. Preprocessing lead time: This is also known as *planning time* or *paperwork*. It represents the time required to release a purchase order (if you buy an item) or create a job (if you manufacture an item) from the time you learn of the requirement.

b. Processing lead time: This is the time required to procure or manufacture an item.

c. Postprocessing lead time: It represents the time to make a purchased item available in inventory from the time you receive it (including quarantine, inspection, etc.).

Lead time may be captured when creating a value stream map. A value stream map will help in understanding the process and lead time and help identify potential improvements quickly by distinguishing value and non-value-added activities. In fact, the analysis will help to separate value and non-value activities, reduce the time to deliver products to the customer, increase the future business capacity, increase production quality, reduce the inventory, improve customer satisfaction, and reduce cost. The focus of all of these is to minimize or eliminate all non-value activities from the process that consume time, space, or other resources, while at the same time have no direct contribution to the final product or service. Remember, we want to maximize the activity for which the customer is willing to pay. Typical value-added expectations are work in parallel, eliminate loops, eliminate handoffs, eliminate steps altogether (whenever possible), compress the work, and increase capacity.

The responsibility for lead-time analysis generally falls in the cross-functional team made up of manufacturing/production, quality, material, maintenance, and industrial engineering personnel.

3. Understand the concept of single minute exchange of dies (SMED): SMED is a theory and set of techniques that make it possible to perform changeover operations in fewer than 10 min. In essence, SMED is all the activities done in changing over from the last good part of one product to the first good part of another. It is a very powerful technique and may be applied in situations that deal with tool and die (i.e., dies, molds, drill bits, blades), programming (computer numerical control [CNC], software), assembly line (supplies, parts, equipment, dies), and general setups (arranging equipment, 6S tasks, projects).

In dealing with SMED, one must recognize that there are two elements of concern:

a. Internal: Changeover activities that can only be done when the machine is shut down from making a good product.

b. External: Changeover activities that can be done while the machine is still operating.

   Obviously, the SMED is very important for any organization because by reducing changeover time, the organization is able to

   i. Make smaller lots
   ii. Make every part every day
   iii. Eliminate waste
   iv. Shorten lead time
   v. Provide faster deliveries, better quality, greater flexibility, lower costs, and higher productivity

To be successful in implementing SMED methodology, there are four essential steps.

a. Preparation and clean up. From our experience, 30% of the total time should be spent in this step.
b. Change mold. From our experience, 5% of the total time should be spent in this step.
c. Measurement or settings. From our experience, 15% of the total time should be spent in this step.
d. Trial runs or adjustments. From our experience, 50% of the total time should be spent in this step.

The percentages given here may be tempered depending on the process and product. However, given these four steps, how do we reduce the changeover time? There are many options here. The most common and critical in any approach of reduction is to follow these guidelines:

- Form a changeover improvement team. The team must be cross-functional and multidiscipline and, more importantly, must have ownership of the process. Furthermore, the team must be empowered with both authority and responsibility for decision making.
- Analyze the changeover. A good way to start this is to record the entire changeover operation. The focus should be on where the actual changeover activities are taking place. Once the recording is complete, show it to the team for detail study. Record any comments appropriately. Typical recordings should be changeover procedure steps and highlights of opportunities for improvement. If possible, data should be displayed in graphical format.
- Separate the internal and external changeover activities. This is very important because each activity will require different analysis.
- Convert internal changeover to external changeover. This approach will optimize your improvement because all changes

will be completed while the machine is in operation. To do this, one needs to evaluate the true functions and purpose of each operation in the current internal changeover. Then, find ways to convert these internal changeover steps into external changeover. Finally, record and indicate which steps can be converted on the changeover chart or log.

- Streamline all aspects of the changeover operation. There are two options here:
  - Improve internal setups (include adjustments). This is accomplished by using specially designed carts to organize tools, using quick-release fasteners instead of bolts and nuts, using stoppers to quickly position the jigs, using rolling bolsters instead of cranes, using overhang mechanisms to handle heavy jigs, using locating pin and holes (socket) to eliminate the adjustment, and using standardized die height.
  - Improve external setups. This is accomplished by using a checklist to avoid omission, using specially designed carts to help organize tools, applying visual control principles, and organizing workplace (6S) approaches to reduce search.

4. Understand the concept of overall equipment effectiveness (OEE): There is no doubt that OEE is a shop floor tool. Lately, it has come to be used as a benchmark for performance and capacity (for more information, see Chapter 19). OEE has become a metric to measure and evaluate capacity based on the product of quality, equipment availability, and performance efficiency.

We already know it is impossible to have 100% effectiveness in the long run. There are always going to be downtimes, maintenance factors, setups, and disruptions. When evaluating traditional machines, the guideline is 85% effectiveness as the norm. This is generated by a machine producing 99% of the product right the first time, operating at a speed of 95% of the theoretical maximum speed, and running 90% of the operating time (99% quality × 95% speed × 90% running time = 85% effectiveness).

This 85% is misunderstood and misused by many. There is nothing secret about this 85%. It is a guideline. That is it. What is important is that the initial OEE must be the baseline for the specific machine's performance and from that point on it should be getting better. The question then becomes is there a consistent standard for OEE? The answer is a definite *no*! However, the practical answer is that the OEE must be better as time goes on. The designation 100% OEE means the theoretical maximum capacity of the equipment. This 100% implies perfection every time. However, we know the standard parameters for OEE can easily be broken. If you have a performance over 100%, it becomes evident that the standard of choice—usually cycle time—is too low. The focus then is removed from identifying and removing losses. For a very detailed discussion on OEE, see Stamatis (2010).

Whereas these five steps are very simple and easy to follow, there are still three fundamental questions with which we should be concerned. They are

1. Did we account for fake flow? Fake flow is (a) stations that operate independently as isolated islands, (b) inventory that accumulates between operations, (c) operators that process batches, and (d) unaccounted extra material, floor space, and people.
2. Is the appropriate equipment available and needed? Attention should be given to (a) the appropriate machine cycle time (it should be less than the Takt time) and (b) the process or task must be small, simple, mobile, reliable, autonomated, and dedicated.
3. Is autonomation at the right level? Autonomation describes a feature of machine design to effect the principle of jidoka. In a sense, it is a technological innovation that enables machines to work harmoniously with their operators by giving them a human touch. This leads to improvements in the processes that build in quality by eliminating the root causes of defects.

Therefore, in order to optimize improvement, it is necessary to make improvement decisions based on facts and figures describing the whole productivity process. Always remember that world-class manufacturing does not accept any losses. Management must focus on this and must have the will to go further than mere window dressing and scratching the surface. There is more to it than just spending money. Companies that do consider the whole process discover repeatedly that there still lies a nearly unlimited potential for improvement.

Visual Factory (6S)

It is important to understand the difference between visual displays and controls so that one may choose the most effective tool to meet the visual factory need in a particular floor situation. In some cases, simply sharing information using visual displays can solve a problem. In other cases, a visual control will be needed to maintain on-site control of work activities. The visual factory is another way to align the behaviors of management to the core principles of continual improvement. Visual factory is the use of controls that will enable an individual to immediately recognize the standard and any deviation from it, creating process control. In Table 8.1 one can see an overview of the enterprise alignment and continual process improvement.

The items in Table 8.1 may be separated into two categories:

1. Visual control: Communicate information in such a way that activities are performed according to standards. For example, painted walkways, machine guards, and color-coded lubrication inlets all direct and control specific human behaviors. Visual controls often fall into the top four levels of the visual control pyramid. These controls are considered to be (a) warning systems about abnormalities such as alarms, (b) items to stop abnormalities such as items that prevent a defect from moving on and, (c) items to prevent abnormalities such as mistake proofing.
2. Visual display: Communicate important information or build controls into the workplace and process but do not necessarily control what people or machines do. For example, information may be posted on a safety chart that influences behavior but does not control it. Visual displays often fall into the first two levels of the visual control pyramid. These displays are generally to (a) share information, (b) share standards at the site, and (c) build standards into the workplace.

Both of these have as a basic requirement visual management, which supports Lean systems implementation, allowing for

TABLE 8.1

Overview of Enterprise Alignment and Continual Process Improvement

Enterprise Alignment	Continual Process Improvement
• Decide on long-term success	• Integrate improvement into work
• Systems thinking	• Focus on the value stream
• Team utilization	• Level load
• Thorough validation	• Create continuous flow and pull
• Ensure capability	• Control visually
• Reduce gap of problem	• Generate the right result the first time
• Challenge the status quo and help employees and suppliers to excel	• Go to the source
• Be innovative	• Standardize work and processes
• Expect excellence	• Adapt best practices
The alignment must be viewed as the application of some of the most inherent tools of Lean manufacturing, such as measurable metrics, standardized work, 6S/visual factory, Kanban, quick changeover total productive maintenance, error proofing, and Kaizen.	To optimize the continual process improvement, the organization must be willing to evaluate and remove the following wastes: correction (inspection or rework), conveyance, overproduction, overprocessing, inventory, waiting, motion, and talent.

- Quick detection of normal and abnormal conditions
- Line of sight presentation of information
- Information to be visually interpreted within 3 min
- Managers to see what is happening in one or two locations at once
- Process management to become highly effective

and the fundamental initiative of 6S. The six fundamentals are

1. Remove the clutter with *Sort.*
2. Organize with *Set to order.*
3. Clean with *Shine.*
4. Set your routine with *Standardize.*
5. Motivate with *Sustain.*
6. Protect your employees with *Safety.*

This is shown in Figure 8.4. Information sharing through visual display is shown in Figure 8.5. Therefore, the entire concept and application of 6S is to detect and prevent abnormalities from happening. If they do, the 6S approach will facilitate the particular action for correcting that abnormality.

Specifically with *Sort,* we focus on how to

- Organize items according to what is needed and what is not needed.
- Examine the area of concern and determine what is needed, get rid of what is not.
- Reduce the amount of items until you have what is actually needed.
- Find appropriate locations for all items according to their usage, urgency, frequency, size, and weight.

To start the process of sorting, we must keep in mind two basic rules: (1) everything has only one place and (2) every place holds only one thing. The significance of these simple rules is that it will help in a variety of ways

	Type					
	1	2	3	4	5	6
Location	**Sort**	**Set in Order**	**Shine**	**Standardize**	**Sustain**	**Safety**

FIGURE 8.4
Typical 6s/visual factory matrix.

Item	Activity Description	Responsibility	Department	Start/ Finish	Plan Date	Actual Date	Comments
1				S			
				F			
2				S			
				F			
.							
.							
.							
n				S			
				F			

FIGURE 8.5
Typical 6s action item list.

including the disappearance of chaos from the process in a very short time. The result will be to force the process owners—with the active support of both Management and Engineering—into the consideration of asking at least the following questions:

- Is the placement of every object according to the logic of the process?
- Does the placement of the object eliminate waste?
- Is the object placed for optimal quality, safety, ergonomics, and environmental outcomes?

These questions generally will be taken care of with a four-step approach.

Step 1: Select an impound area for red tag (something obsolete and not used) items. This selection is based on (1) the identification of the "keeper" of the impound area, (2) the definition of the process for placing and removing items in the impound area, and (3) the determination of a process for unneeded items such as scrap, rework, return items, and so on.

Step 2: Examine every item in the work area. This examination is based on (1) taking a focused look at the target work area, (2) determining what is needed, and (3) removing what is not needed.

Step 3: In this stage, there are two issues to be concerned with. Identify unneeded items and discard them. This is done with the identification of the *red tag*. By definition, a red tag is something that is not being used. By extension of this, a red tag may be interpreted as an item that may be moved within a week to an impound area for scrap. In evaluating the red tag information that may be collected ask the following questions: Where did the item come from? How many items are available? Who needs to be involved in the rationalization of disposal? When was the item tagged? What was the date moved? What was the location item moved from? What was the place identified for storage? Identify the good (needed) items and tag them with a *green tag*. By definition, the green tag is a good item and it needs to have a location as well as the date. Sometimes a yellow tag is issued, and that signifies that the item is acceptable, however, only if there are certain additional requirements to be met.

Step 4: Select a central storage area. This selection is based on the best available place to store each infrequently used item. Infrequently here means that is used less than every cycle. It is also important in this stage to make sure that a permanent identification tag is attached to every movable item designating where it is located when not in use and where it is located when in use, and display a color-coded map for quick location of items.

With *Set in order* we focus on a place for everything and everything in its place. This means that we are interested in (1) determining the best location for all necessary items, and (2) determining how many of each item will be stored in a given location (min/max) and set limits on the space allocated. Standardize reorder points for consumable supplies based on usage and lead time to replenish (3) increasing job efficiency by making it easier to find, use, and return items, and (4d) using tool boards and foot-printing techniques. These four items may be completed by following a four-step approach.

Step 1: A place for everything and everything in its place. This means good organization. In effect, one can quickly and easily find the items so their function and location become intuitive. By doing so, non-value will be reduced and the ability to see what items and how many are displaced will be increased. Furthermore, the advantage of having all items set in a designated place provides more time for procession and customer satisfaction rather than search and retrieval. After all, when items are set in a designated place it is easy to (1) control the correct amount of items, (2) determine if anything is missing, and (3) rapidly retrieve and return objects. To facilitate these three basic concerns, one must design a permanent identification tag and attach one to each moveable item that remains in the area, indicating (1) where the object is located when not in use and (2) where the object is located when in use.

Step 2: Color-code items and locations for intuitive retrieval and usage. This means that the application of selected colors must identify specific functions or meanings in a workplace. The assignment of colors to functions must also be standardized across an organization.

Step 3: Determine how to mark location of (1) equipment, (2) material, and (3) aisles.

Step 4: Practice self-discipline to follow the rules. You may have the *best* rules but if they are not followed, you have no rules. Many problems in both manufacturing and service organizations are the result of not following rules and especially procedures and instructions.

With *Shine* we focus on five key characteristics. They are

1. Eliminate all forms of contamination.
2. Find ways to keep workplace clean.
3. Adopt cleaning as a form of inspection.
4. Make cleaning part of everyday work.
5. Develop processes that do not generate extra work of any kind.

These five items may be completed by following a four-step approach.

Step 1: Clean your workplace. This means that you are the responsible person who controls your workstation. Therefore, keep it clean. Of course, the initial cleaning is important. However, do not be surprised if other problems surface during this activity. Always remember that problems actually are opportunities to improve. They must never be viewed as something to hide.

Step 2: Remove dust, dirt, and debris. Make it a habit to clean on a daily basis. In fact, if the organization is truly committed to the 6S philosophy, it must be part of the daily routine and scheduled.

Step 3: Paint, if necessary. The advantage of a 6S approach is that it can accomplish major results toward a Lean system implementation without costing a lot of money.

Step 4: Expose health, safety, and maintenance issues for both people and equipment.

With *Standardize* we focus on maintaining and monitoring the first three S's. In essence, we develop a 6S schedule board to control regular intervals of 6S duties (daily, weekly, monthly, and yearly) with the intent to check, standardize, maintain, and improve the process. We do this by following a four-step approach.

Step 1: Define normal conditions. This is perhaps the most difficult step in the entire philosophy of the 6S. This is so because it requires real effort, patience, and process knowledge from all process owners to define and agree upon the real *operational definitions*. It can exist in verbal, written, or a pictorial form, but be aware of personal perceptions of understanding. Once the *normal* of the process is agreed on, it will be easier to define what is right and what is wrong. Always remember that *there is no process management without standards*.

Step 2: Communicate the standard. To communicate means that everyone must be trained appropriately and be committed to creating a process that allows for documentation and communicating continual improvement.

Step 3: Build the standard into regular work processes. This means to create and communicate a plan to keep an area in defined *normal* condition by establishing an employee developed cleanliness standard that defines what the area looks like when it is clean. In addition, beyond the definition, the standard must be implemented so that cleaning schedules that use frequent, short amounts of time for everyone on every shift will be incorporated in the process task without any inconvenience or objections from either management or employees.

Step 4: Develop and post a reaction plan for any discovered abnormality. Step three may indeed end up creating confusion as to the specific responsibility of a task. This final step clarifies that responsibility by focusing on specificity of the following: (1) Who is to be contacted first? (2) Who needs to be communicated with on the abnormality? (3) Can the abnormality be corrected immediately?

With *Sustain* the focus primarily is on five items. They are

1. Procedures exist that support 6S and are part of the actual process (be it manufacturing or service [office]).
2. Workers are properly trained in 6S duties and schedules.
3. Everyone has adopted visual factory techniques and concepts and understands their value.
4. The workplace is well ordered according to agreed-on procedures.
5. Managers, operators, and engineers are deeply committed to the 6S philosophy.

We do this by following a four-step approach.

Step 1: Be committed to being persistent, regular, and organized (PRO) in housekeeping by

- Maintaining PRO housekeeping, which can only happen after implementing the first 4S's.
- Cleaning up for important visitors, which is a symptom that the process is *not* working.
- Sustaining and improving processes, which is a common task of both management and workers.
- Developing and following standards for all employees in the factory and making sure that everyone follows them. To make sure that everyone follows the appropriate standards, audits and feedback must be scheduled on a regular basis to ensure compliance.

Step 2: Respect standards and their maintenance. This means that if groups show respect for standards and the pursuit of continual improvement, the relationship between workers and management should change. Furthermore, applying the standards allows everyone to see the logic behind all activities. Change should be understood as a normal part of work and *not* judged as unfair or added pressure.

Step 3: Demonstrate trust and respect for each other's abilities and knowledge. All of us must recognize that every achievement is recognized as valuable. Management is able to objectively rate performance of workers, and workers can objectively rate performance of the system because every task is now clearly defined.

Step 4: Respect standards and their maintenance. That means focusing on the PRO by

- Making sure that cleaning is intuitive to maintain the elimination of waste
- Making sure that monitoring and enforcement of the 6S principles, processes, and tools exists
- Practicing self-discipline in all areas of the 6S methodology, following two practices:
  - continually maintain standard
  - never do a special clean up for a visitor

With *Safety* the focus primarily is on the organization to create a safe environment for everyone. In fact, safety is indeed a priority and everyone's responsibility. So much so, that it has been added to the five pillars of the 5S's. Typical things that may be done to follow the 6S methodology by everyone in any organization are

- Make sure that the employees have the correct personal protection equipment (PPE) for their job (safety glasses, gloves, and steel safety shoes are some typical items required in the warehouse or manufacturing complex).
- Train the staff to use the equipment correctly (heavy equipment, electronics, forklifts, and power tools are all dangerous to those who use them incorrectly).
- Use easy to understand labels, signs, and space markings, which will contribute a great deal to creating a safe, visual workplace.
- Train the staff on exactly what to do in an emergency. From the seemingly small incidents like cuts and bruises, to the larger ones like injuries and fire, each employee should know what he or she needs to do or whom he or she needs to notify for the appropriate source for action.

At this point, one may wonder how this differentiation helps the organization to improve. The answer is in being able to separate two of the fundamental areas of Lean manufacturing.

1. Waste: Anything excess. Activities that consume time, motion, resources, or space, but do not contribute to satisfying customer need.
2. Value add: Changes the fit, form, or function of the product to meet customer requirements.

Of course, for this differentiation to take place, management must develop systems and controls to enable an individual to recognize the *standard* and any *deviation from the standard*. It is also a way of marking or displaying information so that people are immediately aware of the what, who, why, and how of a given work area. Typical tools for this are

- Hourly production board.
- Andon (アンドン, あんどん, 行灯), a manufacturing term referring to a system to notify management, maintenance, and other workers of a quality or process problem. The centerpiece is a signboard incorporating signal lights to indicate which workstation has the problem.
- 6S.
- Error proofing.
- Foot-printing, layout.

The benefits of all this is in four major areas:

1. Validates process condition
2. Highlights problems as they occur
3. Initiates support and urgency for problem solving
4. Supports continual improvement

Therefore, one may summarize the philosophy of a visual factory in four fundamental principles. They are

1. Provide effective and efficient process monitoring and control.
2. Validate standardized work activities.
3. Sustain a clean, safe, and self-maintaining environment.
4. Recognize that everything has its place and is in its place.

The results of this philosophy are to

- Create process control and stability.
- Eliminate waste by exposing waste that consumes effort and resources.
- Signal abnormality by establishing standards that are visible in such a way that deviation or abnormality can be instantly detected and corrected.
- Encourage safety by promoting a clean work environment, which of course equals a safe work environment.

- Promote prevention by encouraging proactive approaches to problems as opposed to reactive
- Share information by making visible all information needed for effective control of storage, operations, equipment, quality, and safety. This open sharing of information drives the continual improvement into the organization.

Strategies of Control

To be successful in implementing any program within an organization, planning is very important. The 6S philosophy is no exception. It also needs a strategy, a strategy that is easy to implement as well as monitor. For example, the flow of implementation should be in training all employees in the 6S methodology followed by appropriate displays and finally making sure that there is some type of visual control.

On the other hand, the monitoring should (1) follow a daily walk-through by management to audit process stability by line, (2) integrate quality data to the performance board by evaluating process trends to drive problem-solving activity and establishing 80% tolerance triggers to initiate immediate action in correcting these process trends, and (3) develop process control inspection. This means utilization of performance boards to report and track process measurables.

Visual Management and Relationships

Thus far, we have addressed visual management several times but we have not officially defined it. When we talk about visual management, we mean the use of displays and controls to enable anyone to immediately recognize (1) prescribed standards and (2) any deviation from standard. In other words, visual management marks or displays information using visual controls to (1) make everyone aware of the what, why, where, who, and how of a given working area, (2) create a common "visual language," and (3) be applicable beyond the factory floor.

Why is visual management important in the journey of continual improvement? The reason is that visual management is the link between all tasks and the organization. It is the mortar holding the bricks together. In the final analysis, 6S/visual factory is the first step toward implementing an effective production system that everyone can use. Figure 8.6 shows this relationship.

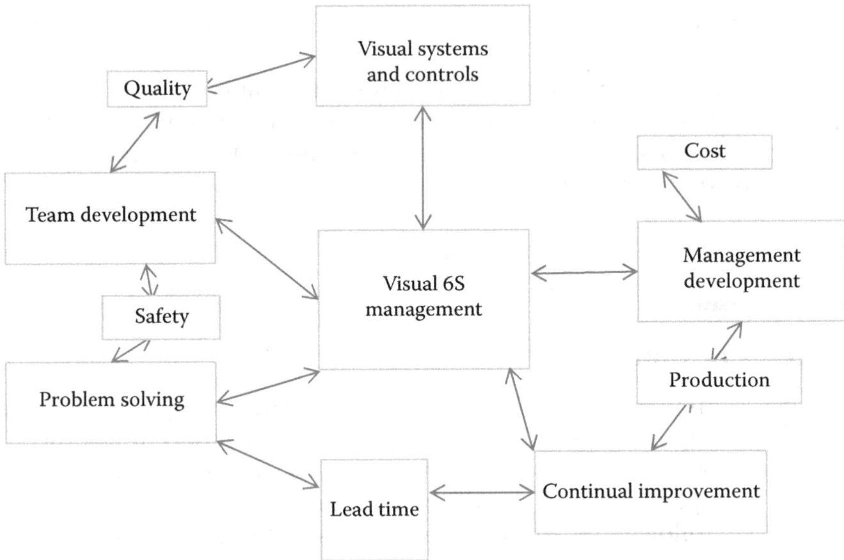

FIGURE 8.6
Visual management and relationships.

Reference

Stamatis, D. (2010). *OEE Primer.* Boca Raton, FL: St. Lucie Press.

Selected Bibliography

Baban, V., D. Fertuck, and K. Vermilye. (2009). *The 5S Desktop (PC) Pocket Handbook.* Chelsea, MI: MCS Media, Inc.

Byron, J.S. and P.V. Bierley. (2003). *Working with Others.* Hope, ME: Lowrey Press.

Ford, H. (1922). *My Life and Work.* Retrieved January 30, 2014, from http://www.gutenberg.org/etext/7213.

Hirano, H. (1990). *5 Pillars of the Visual Workplace.* Portland, OR: Productivity Press.

Smith, J. (2014). Checks and balances: a technique to sync organizational vision and strategy. *Quality Progress,* August, p. 64.

Smith, J. (2014). Quality professional need to remember Crosby's message. *Quality Progress,* August, p. 18.

Spenser, K. and S. Helfer. (2014). Need for speed: Developing a fast and flexible supply chain that leaves competitors in the dust. *Quality Progress,* June, pp. 22–26.

Taylor, F.W. (1911). *The Principles of Scientific Management.* Retrieved January 30, 2014, from http://www.gutenberg.org/etext/6435.

9

Lean Manufacturing

Overview

In the last chapter, we introduced the concept of a visual factory as a fundamental approach to improvement. In this chapter, we will discuss the notion of removing waste (nonvalue activities) not only from individual processes but also from the entire organization. The terminology for this methodology is *Lean manufacturing*.

Even though the consensus is that Lean manufacturing as applied today is derived from Toyota's principles of Lean production methods, the roots of the process are in Henry Ford's integrated philosophy in the Rouge Complex in Dearborn, Michigan, in the early 1900s. The difference between Ford's approach and Toyota's is that Ford's method was static, whereas Toyota's became a dynamic process.

The term Lean manufacturing was coined by Krafcik (1988) and made popular by Womack, Jones, and Roos (1991). If the reader is interested in more details about the history of Lean, Holweg (2007) gives a complete historical account.

The reader will notice that we identified Lean as a methodology and not a tool. That is because Lean incorporates many tools to accomplish its mission, which is elimination or reduction of waste—however defined. Some of these tools are Single Minute Exchange of Dies (SMED), Value Stream Mapping (VSM or VM), 6S, Kanban (pull system), Poka yoke (error/mistake proofing), Total Productive Maintenance (TPM), rank order clustering, Statistical Process Control (SPC/control charts), redesigning working cells, multiprocess handling, single point scheduling, and many others.

In essence, these and many other tools within Lean are implemented with the intention of increasing quality while at the same time reducing production time and costs. This is accomplished by separating the waste (*muda*) into three categories, which are based on Toyota's system of preventive maintenance. They are

1. *Muda* (non-value-adding work). Focusing on the system, not reduction per se in individual pockets of a process. Generally, it is a *reactive* activity of identifying waste and is seen through variation in output.
2. *Muri* (overburden). Focusing on the preparation and planning of the process, or what work can be avoided *proactively* by design.

3. *Mura* (unevenness). Focusing on production *leveling* by utilizing a *pull* system such as the Kanban or the Heijunka box (SMED) approaches. This focus is predominantly on how the work design is implemented and the elimination of fluctuation at the scheduling or operations level, such as quality and volume.

To recognize and do something about these wastes, management must take an active role. Its basic role is to examine the *muda* in the processes and eliminate the deeper causes by considering the connections to the *muri* and *mura* of the system. The *muda* and *mura* inconsistencies must be fed back to the *muri*, or planning, stage for the next project. To optimize this link, there are several assumptions that must be considered so that the three categories of waste will work. The assumptions are

1. Pull processing
2. Perfect first-time quality
3. Waste minimization
4. Continuous improvement
5. Flexibility
6. Building and maintaining a long-term relationship with suppliers
7. Autonomation (the basic idea of *jidoka* 自働化) may be described as intelligent automation or automation with a human touch
8. Load leveling
9. Production flow (just-in-time, JIT)
10. Visual control

In essence, the results of understanding these three waste categories are to get the right things to the right place at the right time in the right quantity to achieve perfect workflow, while minimizing waste and being flexible and able to change. In order for that flexibility and change to take place within an organization, they have to be understood, appreciated, and embraced by the actual employees who build the products (or carry out the service) and therefore own the processes that deliver the value. It is of paramount importance that in addition to the employees being on board with the implementation process, the culture and management of the organization must be ready to embrace the change enthusiastically and be committed to removing waste. It is worth mentioning here to practice Toyota's mentoring process of *Sempai* (senior) and *Kohai* (junior), which has proved to be one of the best ways to foster Lean thinking up and down the organizational structure—including suppliers. Without a strong commitment, nothing will happen in the end.

Womack and Jones (1996) suggest an additional strategy for encouraging companies, organizations, and teams to seek outside, third-party experts, who can provide unbiased advice and coaching via the concept of a *Lean Sensei*.

Lean Goals and Strategy

It is very interesting to note up front that the goals of Lean manufacturing systems differ between various authors and in some cases organizations. The main differences are on the perspective of how Lean is viewed. Some see it as an internal methodology to the organization to increase profit. Others see it as a methodology focusing on improvements that should be undertaken for satisfying the customer.

While some maintain an internal focus, for example, to increase profit for the organization (Liker 2004, Feld 2001, Ohno 1988, Monden 1998, Schonberger 1982, Shingo 1984), others claim that improvements should be done for the sake of the customer (Womack et al. 1990, Womack and Jones 2003, Bicheno 2004, Dennis 2002, Schonberger 1982).

However, whatever the perspective, everyone agrees on the fundamental goals and strategy of Lean, which are to

1. Improve quality: An inherent characteristic to stay competitive in today's marketplace. This is very important in today's globalization endeavors because unless an organization understands its customers' wants, needs, and design processes to meet their expectations and requirements, they will fail in the market.
2. Eliminate waste: In order to improve quality, one must understand and eliminate waste. Waste is any activity that consumes time, resources, or space but does not add any value to the product or service. In statistical terms, waste is variation. Therefore, as we reduce variation, we shrink the waste.

Therefore, in order to improve quality by any metric, one must address waste. Waste is diminished by at least focusing on the two following areas:

1. Reduce time: This means reducing the time it takes to finish an activity from start to finish is one of the most effective ways to eliminate waste and lower costs. It is indeed the cycle time.
2. Reduce total costs: This means to minimize cost. A company must produce only to customer demand. Overproduction increases a company's inventory costs because of storage needs.

When one discusses waste, one is reminded of the original seven deadly wastes defined by Ohno (1988) as part of his Toyota Production System (TPS) and reported by Womack and Jones (2003, 352). The original wastes are

- Transport (moving products that are not actually required to perform the processing)

- Inventory (all components, including work-in-process and finished product not being processed)
- Motion (people or equipment moving or walking more than is required to perform the processing)
- Waiting (waiting for the next production step, interruptions of production during shift change)
- Overproduction (production ahead of demand)
- Overprocessing (resulting from poor tool or product design creating activity)
- Defects (the effort involved in inspecting for and fixing defects)

As time passed, more wastes were added, which are described as manufacturing goods and services that are not necessarily meeting customer demand or specifications. They include waste of unused human talent. For example, in the Six Sigma methodology, waste of skills is identified as a waste and is referred to as underutilizing capabilities and delegating tasks with inadequate training (Stamatis 2003, 2004). Space is also recognized as a waste (Stamatis 2011).

Mika (1999, 2006) added three more forms of waste that are now universally accepted: (1) the waste associated with working to the wrong metrics or no metrics, (2) the waste associated with not utilizing a complete worker by not allowing him or her to contribute ideas and suggestions and be part of participative management, and (3) the waste attributable to improper use of computers, that is, not having the proper software, training on use, and time spent surfing, playing games, or just wasting time. For a complete listing of the "old" and "new" wastes, see Bicheno and Holweg (2008).

The strategic elements of Lean can be quite complex and comprise multiple elements. Out of this complexity, Pettersen (2009) has been able to identify four distinct notions of Lean. They are

1. Lean as a philosophy (Lean thinking)
2. Lean as a continuous change process (becoming Lean)
3. Lean as a fixed state or goal (being Lean)
4. Lean as a set of tools or methods (doing Lean/toolbox Lean)

Steps to Achieve Lean Systems

As it was mentioned earlier, Lean is a methodology and it takes time to be implemented in any organization. There have been many suggestions and recommendations in the literature as to how to go about introducing, implementing, and sustaining Lean in a given organization. Here we are going to summarize the approach that Womack and Jones (1996, 247–271) have suggested and a much later approach recommended by Akinlawon (n.d.). In both cases, we recognize there are limitations, but we believe they are somewhat the extreme approaches. Individual organizations may adjust these to fit their own cultures and timing.

In both cases, the organization must be willing to change the culture and that may take longer than some as the change agents are either experienced or committed and involved in the process of change. A second prerequisite is for management not to focus on a policy of *quick fixes* but rather to understand the real change at hand and build a solid foundation for improvement using Lean.

The recommended approach by Womack and Jones is very lengthy and time consuming as they assume that the organization starts from ground zero. Their approach is the following:

1. First 6 months: This is where the organization begins. It means to find an appropriate change agent, educate the organization with Lean knowledge, map value stream, expand the knowledge as appropriate, begin focusing on both fundamental as well as radical changes to a production system (your operating system or a service system), and define the scope.
2. Six months to Year 2: This is the beginning of the new transformational change of the organization. It means organize or reorganize categories of products, create a Lean function based on the value mapping, define a policy for excess workers, define a growth strategy, and instill a zero defect (perfection) mind set.
3. Years 3 and 4: This is the time period for installation of the business systems to make a difference. It means a new process for accounting, change base payment to performance metrics, implement transparency, make sure the appropriate and applicable tools are available to the workers, demonstrate Lean successes, continue Lean training, and initiate policy deployment to encourage Lean approaches throughout the organization.
4. By the end of Year 5: This is where the transformation is complete and functioning. It means the beginning of cascading the principles of Lean to your customers and suppliers, developing global strategy, if applicable, and making sure that the transformation of change is driven by bottom-up rather than top-down improvement.

On the other hand, Akinlawon (n.d.) approaches the implementation from a holistic perspective. His focus is not so much on the timing but rather the concepts and activities associated with Lean. Fundamentally, he has suggested three steps for a successful implementation. They are

1. Design a simple manufacturing system. A fundamental principle of Lean manufacturing is demand-based flow manufacturing. In this type of production setting, inventory is only pulled through each production center when it is needed to meet a customer's order. The benefits of this goal include
   a. Decreased cycle time
   b. Less inventory
   c. Increased productivity
   d. Increased capital equipment utilization
2. Recognize that there is always room for improvement. The core of Lean is founded on the concept of continuous product and process improvement and the elimination of non-value-added activities. The value-adding activities are simply only those things the customer is willing to pay for; everything else is waste and should be eliminated, simplified, reduced, or integrated (Rizzardo and Brooks 2003, 2008). Improving the flow of material through new ideal system layouts at the customer's required rate would reduce waste in material movement and inventory.
3. Continuously improve the Lean manufacturing system design. A continual improvement mindset is essential to reach the company's goals. The term *continual improvement* means incremental improvement of products, processes, or services over time, with the goal of reducing waste to improve workplace functionality, customer service, or product performance (Suzaki, 1987). In other words, continual improvement may be characterized as a never-ending pursuit of excellence.

Easier Application for Lean

To be sure, both approaches just mentioned are available for any organization to use in their implementation of Lean. After all, Lean transformation cuts across organizational boundaries of department and job description:

- The aim is to create smooth-flowing consumption streams (customers with needs) matched up with smooth-flowing provision streams (production or services) so that customers get what they need, where they need it, and when they need it, without waiting.

- Consumption and provision streams run across organizational departments.
- Most improvement efforts have been aimed at particularly deficient single points within discrete departments along the stream rather than viewing the whole stream.

This means that the implementation must be easy to understand and be implemented. Whereas the two approaches assume that the organization has no or little quality initiatives on board, the reality is that most organizations start with some good quality practices in their organization. As such, a more practical approach is recommended. We believe that a more realistic method is available and is based on the following principles.

In any endeavor, no one tool will help; it may be necessary to invent one's own. However, just start doing something. Even if it is wrong, learn and start again. Because Lean naturally looks at the whole stream or process, it requires a chief process officer (CPO). This person does not have to be managerially locked into the pathway. He or she must report to the chief executive officer (CEO) because the CPO's plans will require major organizational change. Above all, the strategy developed *should not* be formulated around the processes and structure currently in place. One is looking for improvement with a change in the process. Therefore, the old has to be sidetracked and new ways of approaching the process must be thought of. There is another way of viewing this:

- The strategy of a Lean transformation is to create smooth-flowing consumption streams matched up with smooth-flowing provision streams.
- The processes must be reorganized to achieve this.

Organizational structure must change to support the new processes supporting the new strategy. It is a misconception that Lean is a tool or a toolbox. Rather, as we have already mentioned, it is a methodology that provides a way of thinking about work via six principles (Stamatis 2011):

1. Philosophy as the foundation. It is imperative that management recognize that successful Lean implementation depends on making management decisions for the long-term goal of adding customer value. This is easier said than done. However, practical things to do include
    a. Change the focus of all board meetings to look at key value streams. All other considerations are secondary, including performance targets.
    b. Board members must ask for new-style board reports that inform them of the performance of the whole process from the

customer's perspective. A good start is to utilize quality operating system (QOS).

c. Board members must devise and promulgate a simple, clear message or mission statement that encapsulates the aims of the Lean transformation that all staff locally understands.

d. This is usually intrinsic (i.e., not "we must meet targets," but rather, "customers will experience no unnecessary waits or errors for their products or services").

2. Level out workloads. Have stability in all work processes so that problems are easily seen and continuous improvement is possible. Understand deeply the nature of the demand placed on the process and the capacity it has to meet that demand. Seek to match the two as a basis for all other improvement efforts. Practical things to do include

   a. Perform capacity and demand analysis by day, hour, or whatever is applicable and appropriate, using run charts. Feedback should be given to all relevant managers and reasons for variation discussed.

   b. Analyze actual versus planned capacity and root-cause reasons for any difference.

   c. Analyze demand: source, type, urgency, and appropriateness.

3. Work on flowing work faster and without batching or delays. Processes should work at the rate of demand placed on them. Queues or waiting lists are signals from a process that this is not happening. Process steps should quickly and automatically signal to each other along the value stream; they should act and think in a coordinated, customer-centric way. Practical things to do include

   a. Map the process, adding timings to each step and waiting list counts.

   b. Involve everyone touching that process and ask why queues develop.

   c. Look to where demand is coming from and seek to smooth that; eliminate inappropriate demand.

   d. Foster ownership of a value stream, rather than just process steps, with a value stream group, customer group, chief process owner, and value stream-oriented reports rather than specialty- or department-oriented reports (i.e., total journey time measurement, demand versus overall capacity reports).

4. Get quality right the first time. Stop and fix problems right now. Do not ignore them or invent work-arounds. Management must invite alerts to problems and be prepared to solve them as quickly as possible. Foster a culture of designing-out problems from occurring in

the workplace with visual clues, for example. Practical things to do include

a. Never blame the person; blame the process.
b. Encourage staff to fix the problem, learn, and disseminate the information.
c. Have a method agreed to by all staff to report problems easily and quickly.
d. Have an agreed on escalation protocol so that everyone knows who is responsible for diagnosing and fixing.
e. Enable quick offline testing and remedying.
f. Keep a local problems log and have regular reports on errors and fixes.
g. Celebrate and advertise successes.

5. Standardize tasks. This is not to be confused with rigidity. Use standardized procedures simply as the commonly understood base point to improve the process from now on and to ensure sustainability as people move on. Practical things to do include
    a. Standard work does not start with writing, laminating, and hanging up standard work charts. That comes last.
    b. Start discussion on what the repeatable, standard elements of work are.
    c. Establish a lead person in charge of identifying these, measuring the variation, and investigating why. Do all this from the standpoint of wishing to improve and help the staff have dependable processes.
    d. Investigate the relationship between the variation in demand coming in and work processes. Is variation due to the complexity of customer demands (specifications) or poor processes (i.e., missing materials, lack of space, etc.)?
6. Grow Lean leaders and managers. Leaders instill the intrinsic motivation for this new way of thinking. Managers are doing it every day. Managers must be highly visible and near the value-adding work. Managers are there simply to improve the value stream. They should make improvement decisions through consensus and implement swiftly. Practical things to do include
    a. Make other site visits; provide short-term Lean consultant support to management.
    b. Encourage study, training, and education in this area. Create a succession plan; aim to keep promising leaders.
    c. Improve the management selection process and learn from others doing it well.

d. Encourage experimentation, allow mistakes (from which you can learn something), and see who takes up that challenge.

A key tenet in Lean thinking is that no matter how many times a process is improved, it can be further enhanced. The idea of perfection rests on the notion of continual improvement through incremental change based on outcomes (Tsasis and Bruce-Barrett 2008). Use of the plan-do-check/study-act (PDC/SA) cycle helps in pursuing the idea of perfection. In any process improvement initiative using Lean thinking, a small incremental change is recommended based on study of the process. The recommendation is put into place and then studied to determine its impact. If the impact is positive, the change is incorporated in the process, and the cycle begins anew. (We are not suggesting that big changes do not occur. They do, but they are infrequent. Big changes come with re-engineering. Small changes are cumulative and come with the Kaizen approach.)

Thus far, we have been talking about the general concepts associated with Lean thinking. We have not talked about specific tools. Part of the reason is that we cover some of the most important ones (6S, value stream mapping, SMED, and Kanban) in Chapter 8, and part of it is that there are so many specific tools that one can use depending on the organization and product that it is beyond the scope of this book to cover all of them. However, the reader is encouraged to see Stamatis (1996), Brassard and Ritter (1994), and Tague (2005) for a plethora of tools to be used in both implementation and sustainment of Lean practices. In addition, Appendix V provides many forms and tools that one may use in pursuing continual improvement.

Warning

One criticism of Lean perennially heard among rank-and-file workers is that Lean practitioners may easily focus too much on the tools and methodologies of Lean, and fail to focus on the philosophy and culture of Lean. The implication of this for Lean implementers is that adequate command of the subject is needed in order to avoid failed implementations. Another pitfall is that management decides what solution to use without understanding the true problem and without consulting shop floor personnel. As a result, Lean implementations often look good to the manager but fail to improve the situation (Hopp and Spearman 2011).

Let us all remember that only in the dictionary is success ahead of work. In the real world, however, we all must strive to drive excellence all the time in all we do (work) and the result will be an unequivocal success. After all, any way you look at it, Lean is a way of life to do the right thing with what you have and then improve on it. It does sound very easy, but it is indeed very hard to do!

References

Akinlawon, A. (n.d.). Thinking of lean manufacturing systems. http://www.sae.org/manufacturing/lean/column/leandec01.htm. Retrieved on March 17, 2014.

Bicheno, J., (2004). *The New Lean Toolbox: Towards Fast, Flexible Flow.* Buckingham: PICSIE Books.

Bicheno, J., and M. Holweg. (2008). *The Lean Toolbox,* 4th ed. Sandton, South Africa: PICSIE.

Brassard, M., and D. Ritter. (1994). *The Memory Jogger II.* Salem, NH: GOAL/QPC.

Dennis, P. (2002). *Lean Production Simplified: A Plain Language Guide to the World's Most Powerful Production System.* New York: Productivity Press.

Feld, W.M. (2001). *Lean Manufacturing: Tools, Techniques, and How to Use Them.* Boca Raton, FL: St. Lucie Press.

Holweg, M. (2007). The genealogy of Lean production. *Journal of Operations Management.* 25(2): 420–437.

Hopp, W., and M. Spearman. (2011). *Factory Physics: Foundations of Manufacturing Management,* 3rd ed. Long Grove, IL: Waveland Press.

Krafcik, J.F. (1988). Triumph of the Lean production system. *Sloan Management Review.* 30(1): 41–52.

Liker, J.K. (2004). *The Toyota Way: 14 Management Principles from the World's Greatest Manufacturer.* New York: McGraw-Hill.

Mika, G. (1999). *Kaizen Implementation Manual.* Dearborn, MI: The Society of Manufacturing Engineers (SME).

Mika, G. (2006). *Kaizen Implementation Manual,* 5th ed. Dearborn, MI: The Society of Manufacturing Engineers (SME).

Monden, Y. (1998). *Toyota Production System: An Integrated Approach to Just-In-Time.* London: Chapman & Hall.

Ohno, T. (1988). *Toyota Production System: Beyond Large-Scale Production.* Portland, OR: Productivity Press.

Pettersen, J. (2009). Defining Lean production: some conceptual and practical issues. *The TQM Journal.* 21(2): 127–142.

Rizzardo, D., and R. Brooks (Eds.). (2003, 2008). *Understanding Lean Manufacturing.* College Park, MD: Maryland Technology Extension Service (MTES) Tech Tip. www.mtech.umd.edu/MTES/understand_lean.html. Accessed on April 9, 2014.

Schonberger, R. (1982). *Japanese Manufacturing Techniques: Nine Hidden Lessons in Simplicity.* New York: Free Press.

Shingo, S. (1984). *A Study of the Toyota Production System from an Industrial Engineering Viewpoint.* Tokyo: Japan Management Association.

Stamatis, D.H.(1996). *Total Quality Service.* Boca Raton, FL: St. Lucie Press

Stamatis, D.H. (2003). *Six Sigma for Financial Professionals.* New York: John Wiley & Sons.

Stamatis, D.H. (2004). *Six Sigma Fundamentals: A Complete Guide to the System Methods and Tools*. New York: Productivity Press.

Stamatis, D.H. (2011). *Essentials for the Improvement of Healthcare using Lean and Six Sigma*. Boca Raton, FL: CRC Press.

Suzaki, K. (1987). *Manufacturing Challenge: Techniques for Continuous Improvement*. New York: Free Press.

Tague, N. (2005). *The Quality Toolbox*. Milwaukee, WI: Quality Press.

Tsasis, P., and C. Bruce-Barrett. (2008). Organizational change through Lean thinking. *Health Services Management Research*. 21(2): 192.

Womack, J., and D. Jones. (1996). *Lean Thinking: Banish Waste and Create Wealth in Your Corporation*. New York: Simon and Schuster.

Womack, J., and D. Jones. (2003). *Lean Thinking*. Revised and updated. New York: Free Press.

Womack, J., D. Jones, and D. Roos. (1990). *The Machine That Changed the World: The Story of Lean Production*. New York: Rawson Associates.

Womack, J., D. Jones, and D. Roos. (1991). *The Machine That Changed the World*. New York: Harper Perennial.

Selected Bibliography

Bailey, B. (June 2014). No weak links: Use Lean and quality tools to strengthen global supply chain performance. *Quality Progress*. 14–21.

Conner, G. (2001). *Lean Manufacturing for the Small Shop*. Dearborn, MI: Society of Manufacturing Engineers.

Flibchbaugh, J., and A. Carlino. (2005). *The Hitchhiker's Guide to Lean: Lessons from the Road*. Dearborn, MI: Society of Manufacturing Engineers.

Hounshell, D.A. (1984). *From the American System to Mass Production, 1800–1932: The Development of Manufacturing Technology in the United States*. Baltimore, MD: Johns Hopkins University Press.

MacInnes, R.L. (2002). *The Lean Enterprise Memory Jogger*. Salem, NH: GOAL/QPC.

Maskell, B., and B. Baggaley. (2003). *Practical Lean Accounting*. New York: Productivity Press.

Mika, G.L. (1999). *Kaizen Event Implementation Manual*, 5th ed. Dearborn, MI: Society of Manufacturing Engineers.

Nichola, J., and A. Soni. (2006). *The Portal to Lean Production*. New York: Auerbach Publications.

Page, J. (2003) *Implementing Lean Manufacturing Techniques*. New York: Productivity Press.

Ruffa, S.A. (2008). *Going Lean: How the Best Companies Apply Lean Manufacturing Principles to Shatter Uncertainty, Drive Innovation, and Maximize Profits*. New York: AMACOM.

Spear, S., and K. Bowen. (September 1999). Decoding the DNA of the Toyota Production System. *Harvard Business Review*. http://hbr.org/1999/09/decoding-the-dna-of-the-toyota-production-system/ar/1. Retrieved on March 15, 2014.

10

Quality Operating System (QOS)

Overview

An organization with a solid quality operating system (QOS) has an excellent chance of being a superb supplier. The reason for this is that a good and solid QOS determines the direction and culture of any organization. It means commitment of the management to improve continually and satisfy the customer. It means that management knows what the key processes of the organization are and what the key indicators for those processes are. Mandatory elements that management must understand in order to generate a good QOS are

- Customer integration: All the customer's needs are integrated into the products and relevant processes. This is achieved through personal interrelationships with customers as well as by handling all customer information in a professional way.
- Embedded quality in processes/projects: Quality standards in processes and projects lead to improved quality because of standardized processes, quality gates, and preventive measures. Process discipline and stability are key prerequisites for quality.
- Consequent supplier management: There is absolutely no room for compromise in supplier management. Inadequate supplier quality can only be corrected by systematic management, that is, selection, assessment, classification, development, and phasing out.
- Business-driven quality planning: Planning quality means analyzing future developments and potential problems and taking appropriate preventive steps. Precise objectives, activities, and quantitative indicators are specified with the aid of benchmarked targets.
- Focused quality reporting: The current quality status is the subject of regular reports, identical in structure to the planning. Reporting is based on a focused summary of key performance (process) indicators (KPIs), concrete quality incidents, and cause-effect analysis of possible quality problems.

- Broad qualifications on quality issues: All employees must have skills and abilities appropriate to the activities they perform. Gaps in experience are analyzed and are subject to competence management. Experience is reinforced to ensure know-how over the long term.
- Continuous improvement: Objective of continual improvement is to continuously optimize the market and cost position. Pragmatic proven tools include CIP, Kaizen, Lean, Six Sigma, and similar approaches. Employees from all organizational units are involved.
- Spirit by management involvement and commitment: Managers bear full responsibility for and are committed to quality, monitoring quality and taking part personally in solving critical quality problems. Examples of both superb and unsatisfactory quality are openly published internally.
- Control and support role of quality manager: Quality managers support, control, and deal with conflicts. The independence of the quality department and individual quality managers in the processes and projects is guaranteed and heavily supported by executive management.
- Commitment to measurement: Management must be committed to data-driven decisions. In order for this to happen, the data must be appropriate and applicable to the situation. Furthermore, the data must be analyzed with the appropriate and applicable method. If you do not measure, you do not know where you are, and if you do not know where you are, you cannot improve.

Of course, for these elements to be part of the culture, there must be a *passion* for excellence throughout the organization by everyone, beginning with the management leadership. Typical issues recognizing excellence should be at least the following:

- Zero tolerance for defects: All activities must be directed toward rigorous avoidance of any failure in the products and processes. The focus must be on zero defects and 100% customer satisfaction as a general goal on everything that the organization does.
- Customer satisfaction: All activities must be customer focused and committed to developing a successful partnership from the very beginning, applying effective project and process management throughout the whole lifecycle. Long-term customer satisfaction is *the* standard of quality. This translates to loyalty and, by extension, increasing business.
- Continual improvement process: The principles of continual improvement must always be present and relevant for the competitiveness of the organization. Deep root-cause analysis, fast and

systematic improvement, and best practice sharing as well as innovations are the foundation to increase quality and productivity. Without the attitude for continual improvement, the organization will not improve consistently over the long run.

- Top issues of waste in process (WIP): Waste, rework, and rejects in process must be identified and either eliminated or minimized—especially the top 10. Initiatives and benchmarking as well as all kinds of improvement activities must be supported and implemented. The implementation must be with clear measurables and be systematically tracked to monitor the improvement.
- Entrepreneurial spirit, empowerment, and involvement with a strong commitment: The organization that aims for excellence must encourage the entrepreneurial spirit, empowerment (delegation with both authority and responsibility), involvement, and commitment by management to all employees by continuously and systematically developing and utilizing their knowledge, experience, and skills.
- Conscious use of resources: Organizations aiming at excellence must be cognizant of their own resources. Since all resources are finite, appropriate and applicable allocation of these resources must be developed for *optimum efficiency*. Resources are not limited to material. In fact, they may include employees, shareholders, suppliers, communities, administrations, and other relevant parties to build relationships as well as to generate an attractive return on investment (ROI).

Quality Operating System

Thus far, we have discussed the need for quality and some key items of concern for the quality discipline as we move into the twenty-first century. Whereas these are indeed important, we must also focus on how we bring all these together. The focus that brings everything together is indeed the QOS.

A QOS is a systematic, disciplined approach that uses standardized tools and practices to manage business and achieve ever-increasing levels of customer satisfaction. A generic model of a QOS is shown in Figure 10.1.

As complicated as this definition sounds, in fact it is not. All organizations have some type of a QOS, but it may be known as

- Total quality control (TQC)
- Total quality management (TQC)
- Total control system (TCS)
- Quality management system (QMS)

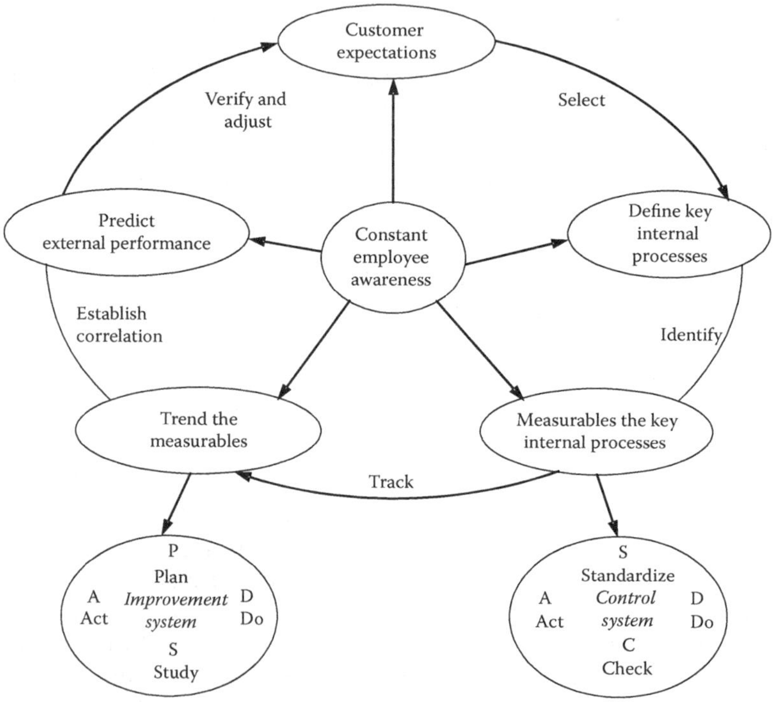

FIGURE 10.1
Generic QOS model.

- Quality reporting system (QRS)
- Quality improvement process (QIP)

In essence, QOS is a system that helps organizations to improve both internally and externally—internally by identifying appropriate metrics for improvement in their key processes and externally by focusing on their customer's needs. In terms that are more generic, QOS is a way of doing business consistently, a systematic approach that focuses on both common and assignable causes to problems, used in all areas of the organization with a focus on improvement, a way to correlate process measurables to customer measurables, a way to implement preventive actions, a system that encourages data utilization for decisions, and, above all, it forces the organization to work as a team rather than individual entities. QOS is an *optimization* approach to management rather than a *suboptimization* approach for certain departments at the expense of others.

As important as the QOS system is in any organization, it must be recognized that unless management believes and exercises empowerment at the source, QOS cannot be implemented. Empowerment, of course, is personal ownership and commitment to take initiatives aligned with the objectives of the organization.

Having given an overview of the definition of QOS, let us look at the components of the model itself for a better understanding. We start with customer expectations. It is very easy to conceptualize what the customer wants, but in order to know for sure, the organization must know who the customer is. Any organization has many customers and everyone has a unique situation or requirement for their product or service. Sure, we can say that the customer wants *defect-free* product, but that usually is a minimum requirement. Some other requirements may be value received, timing (was the product delivered in a timely fashion?), and many more. All these depend on whether the customer is perceived as an immediate, intermediate, or ultimate customer. Each one has its own needs and requirements and, depending on the definition of customer, the expectations and feedback will differ.

The second component of the model is to select key processes. The operative word here is *key*. This means that only selected processes must be identified and monitored. This is the Pareto principle (80/20 rule). The process may have many subprocesses, but it is imperative that we identify the critical processes that will make the product or service to the satisfaction of the customer. For example, in manufacturing we may identify the assembly, machining, and plating. For staff processes, we may identify training, attendance, and accounts receivable.

The third component of the model is to identify the measurables for these key processes. A measurable is a way to gauge how well a specific aspect of the business is doing and the degree to which the processes of the business meet customer needs and requirements. The focus of this component in the model is to be able to standardize and then control the measurable. We standardize the measurable so that we have consistency, stability, and predictability. Once those measurables are standardized, we move into the control mode to make sure that they stay consistent. The two most common types of measurables are

1. Process (production or operation focus)
2. Result (customer focus)

A variety of examples of potential QOS measurable are shown in Table 10.1.

The next key item in the model is tracking the trend of the measurable that was identified in the previous step. This is a very important step in the sense that the control process began in the previous item, continuous; however, the monitoring is more formal and systematic. One may use many specific tools. Here we mention the most common:

1. Flow chart: To make sure that we understand the process and we follow it.
2. Check sheet: To make sure that data are appropriately collected and recorded. One must remember that without data, you are just another person with an opinion.

TABLE 10.1

Potential QOS Measurables

Category	Measurable	Quantifier
Generic	Customer rejects or returns Process capability Reliability testing Absenteeism Cleanliness Claims Training Cost of quality	Return per 100,000 Parts per million Rejects over time Percentage significant characteristics in control Percentage significant characteristics capable Percentage control plans completed Mean time between failure Time to failure Field failures Percentage workforce missing Percentage of excused hours Bathroom Work area Hallway Claims for part 1 Claims for part 2 Number of hours per person Number of classes Percentage attended class Productivity improvement Prevention cost Appraisal cost Percentage of sales Internal failure costs External failure costs
Service	Turn days (new business order or repeat order) Quote turn time Order acknowledgment Order entry Customer surveys Reliability to acknowledge ship date Reliability to request ship date	Average number of days from point of order entry until shipped Number of quotes completed in 48 hours as a percentage of total quotes processed Number of orders that we acknowledge ship dates in less than 48 hours as a percentage of total orders acknowledged Number of orders entered the same day as received as a percentage of total orders received Response received from customers Number of orders shipped complete and on time to acknowledge date as a percentage of total orders shipped Number of orders shipped complete and on time to the customer's requested ship date as a percentage of total orders shipped

(Continued)

TABLE 10.1 (Continued)

Potential QOS Measurables

Category	Measurable	Quantifier
Manufacturing	Scrap Rework Run hours per day Throughput per hour Setup hours Downtime Overall equipment effectiveness	Percentage of scrap as related to the total order Percentage rework as related to the throughput or the total order Number of hours per day spent running good product Average good production per hour produced by process Hours spent on setup by process Hours down by reason code by process Overall equipment effectiveness by machine
Cost/ productivity improvement	Project teams Specifications	Quantifiable improvements in areas of team objectives Annual target $100M measure quarterly
Capital investment	Annual capital plan systems	Rate of return on investment Improved performance in meeting customer requirements
Quality	Process capability Audits Customer quality reports	C_{pk} or P_{pk} values of all key processes C_{pk} or P_{pk} values reported to customer Quantified score from customer #1 Quantified score from customer #2 Summary score from customer #1 Summary score from customer #2
Financial	Inventory turns Sales/employee Customer complaint incidents	Annualized cost of sales/total inventory for current month (sometimes this is measured based on more recent 3 months) Net sales/total employees Number of complains written in a monthly period—administrative and production
General	Turnover Attendance Safety	Total number of terminations, including voluntary terminations Total number of days missed Absenteeism documented on a daily basis Injury rate is documented and based on incidents per 100 employees

3. Brainstorming: To make sure that everyone is participating and contributing to ideas of improvement.
4. Nominal group technique: To make sure that legitimate issues are recognized by everyone.
5. Pareto chart: To make sure that the significant few are identified as opposed to the trivial many.

6. Cause and effect: To make sure that the items of concern have been identified and are really related to the problem at hand.
7. Run chart: To make sure consistency is followed.
8. Stratification: To make sure appropriate groups are identified correctly.
9. Histogram: To make sure that the data are normally distributed.
10. Scatter diagram: To make sure that there is a relationship between variables.
11. Control charts: To make sure that the process is in control and stable.
12. Capability: To make sure that the product produced meets the requirements of the customer and, more importantly, to make sure that the machine is capable of producing what the customer wants.
13. Force field analysis: To make sure that the problem is clearly understood.
14. 5 Whys or 3 × 5 Whys: To make sure you have taken a deep dive for the root cause of the problem.
15. Paynter chart: A chart that identifies the defect, identifies the corrective action, and then monitors the performance. One may say it is a combination of an attribute chart and a time line (trend) chart.

The last key item in the model is to identify and establish correlation and begin to predict external performance. When you have reached this level of analysis in the model, you have indeed identified the key processes as well as the key metrics of those processes. Indeed, you are capable of separating common from assignable cause and you are on your way to achieving your quality improvement targets. Furthermore, because you are able to predict performance, now you can actually adjust the process or processes to new customer expectations. At this point, the model repeats itself.

QOS Implementation

Now that we know what QOS is all about, let us examine the implementation process in a given organization. Generally, there are seven steps as shown in Table 10.2. Table 10.3 shows the responsibilities for each step.

QOS Meeting Summary

It has been said, "What gets measured gets done." QOS is no different. To have a successful QOS, regular meetings must take place to review the key

TABLE 10.2

QOS Implementation Steps

Step	Implementation	Feedback	Model item
1	Steering committee formation		
2	Steering committee orientation		
3	Customer expectations	Steering committee into continual improvement loop	Customer expectations
4	Strategic goals and objectives	↑	Internal key processes
5	Measurable and processes	↑	Measurable for key internal processes
6	Quantifier/ measurement selection	↑	Trend measurables
7	Action plan formation	Start cascading throughout the organization to various teams	Predict external performance

processes and measurables and provide effective and applicable feedback. The path to success begins with the current state of your organization—no matter what it is. Generally, there are four steps:

1. Create a work group to assemble a listing of all performance measurements used to operate the organization.
2. Prioritize the most important and create "trend" reports and goals.
3. Establish the internal measures, which correlate directly to customer data.
4. Establish periodic meetings to review the data and the action plans. At the start of the QOS program, it is not unusual to have two meetings per week and eventually one per month.

The flow of the QOS meeting is generally very simple and straightforward. Generally, it consists of six items in the following order:

1. Meeting dates
2. List of a few vital measurables
3. Prior meeting action items
4. Overview (trend charts for both process and results)
5. Supporting data for each measurable with applicable charts or statistics
6. Lessons learned (both things gone wrong and things gone right)

TABLE 10.3

Requirements of Each Step

Step	Implementation	Requirements
1	Steering committee formation	• Secure top management commitment and involvement • Explain the QOS process to top management • Explain the implementation steps • Select a steering committee • Explain steering committee responsibilities • Explain champion responsibilities • Schedule meetings
2	Steering committee orientation	• Review company mission statement and business plan • Develop steering committee mission • Identify all customers
3	Customer expectations	• List customer expectations • Group customer expectations • Rank customer expectations • Benchmark
4	Strategic goals and objectives	• Establish strategic goals and objectives • Select internal key processes • Assess team members' knowledge of statistics and variation • Assess team members' additional training needs
5	Measurable and processes	• List potential key processes • Brainstorm measurable • Identify measurable for key internal processes • Assign measurable champions • Gather existing data on measurable assignment • Create data management plans
6	Quantifier/measurement selection	• Review continual improvement tools for appropriateness • Consider consensus • Correlate quantifiers and measurements to customer expectations • Use problem-solving disciplines • Track trends of measurables
7	Action plan formation	• Present improvement opportunities • Study opportunities • Predict external performance • Identify strengths and weaknesses • Plan for the future • Verify and adjust

During the meeting, the following are discussed:

- Review prior meeting action items
- Review trend charts and establish meeting agenda
- Review adverse trends using problem-quantifying tools, make assignments to achieve improvement or correction
- Rotate meeting chairperson to promote team ownership

- Revalidate measurable, quantifiers, and quantifying tools on an ongoing basis
- Issue meeting minute action items
- Keep an ongoing log of lessons learned and breakthrough experiences

Example

To demonstrate the process and flow of QOS, let us show an example of generating one. Our example is a baseball QOS.

Step 1. Customer expectations: What do the fans (customers) want of their sport team?
Answer: Some possibilities are win the World Series, win American or National league pennant, finish in top of division,* or have a winning season. These and many more may be identified through (a) market research, (b) focus groups, (c) satisfaction surveys, (d) field experience, (e) benchmarking, and so on.

Step 2. Internal key process: Draft, trading of players, pitching/hitting,* defense.
Answer: Key question here is what is important to the (a) pitcher, (b) batter, (c) coach, and (d) owner. Therefore, if we choose the key process as pitching and batting, the measurables have to reflect those characteristics. A measurable for a pitcher is pitching and for the batter is hitting. Now we are ready to proceed with the identification of the quantifiers.

Step 3. Measurables for key internal processes: In our example here, we decided on pitching and hitting.
Answer: Depending on how we answered Step 2, our quantifiers will reflect that selection. For this example, we have chosen pitching and hitting as key processes. Therefore, the quantifiers are for pitching (ERA,* walks, and SO) and for hitting (team average,* HR, power index, and RBI).

Step 4. Trend measurable: Trend with charts the key quantifiers.
Answer: You may do a team ERA run chart or a composite comparison chart with last year's performance; you may also do a run chart on the team batting average or a composite comparison with last year's performance. It is very interesting to note here that baseball has more statistics than just about anything that you can imagine not only for the team, but also for individual players. The reason you do a trend here is to be able to identify any special (sporadic) as opposed to common (chronic) variation.

Step 5. Predict performance: Given the understanding of the key measurables, we can now predict performance.
Answer: We may want to estimate correlation of ERA and attendance. We may want to use scatter diagrams or correlation analysis to find the relationships that interest us.

Note: All items designated with an * have been identified as key measurables of the process at hand. Of course, these may be changed depending on what the objective is and who the customer is.

Selected Bibliography

Ahmed, A. (2014). Solid base: Improve supplier assessments by incorporating process factors into checklists. *Quality Progress*. June, 28–33.

Machado, H. (2014). Plan of attack: Managing the anatomy of your key projects is fundamental to organizational success. *Quality Progress*. May, 32–37.

Ramu, G. (2014). Knowledge at work: Unleashing the power of knowledge workers. *Quality Progress*. May, 52.

Ramu, G. (2014). Pay it forward: Anyone can invest in a culture of quality. *Quality Progress*. May, 50.

11

Certification to the International Standard and Customer-Specific Requirements

Overview

Every organization worldwide is expected to follow some kind of basic quality standards. In fact, it has become so critical that the International Organization for Standardization has developed such a system called ISO 9001. This standard has been revised several times since the late 1980s and currently we are at the 2008 revision. It is expected that a new version will be published by late 2015.The standard does not guarantee product quality, but rather establishes the foundation for a good quality system. Because of this weakness, several industries have added requirements to make the ISO 9001 system into a product-oriented quality system. Examples of this are the ISO/TS 16949 (2009) (automotive), the AS 9100 (2009) (aerospace), ISO/IEC 17025 (2005) (testing and calibration), ISO/IEC 9126 (2001, 2003) (software), and many others.

In this chapter, we are going to give an overview of the requirements without very much detail. The reason for the cursory review is that there many sources (books, articles, blogs, and pamphlets) that the reader may find helpful and as such it will be redundant effort on our part to reiterate what is already available.

The significance of the standard, however, for any organization is profoundly important, and we believe that it should be covered here as an overview. With the overview, we hope the reader will understand the importance and rationale of what the standard is all about. The general structure of the standard follows the Plan, Do, Check, Act (PDCA) model.

It begins with noncertified items such as the *scope* that defines the boundary of the standard; it is followed by the *normative references*, which were used to generate the standards as well as how they relate to the actual text; and the explanation of key *terms and definitions* and how they are used throughout the text of the standard.

Once these informational items are presented, the standard moves into the certifiable elements. This means that the organization must be certified as adopting and following the requirements as specified in the standard through a third-party auditor recognized by the ISO organization. In general terms, the structure follows the PDCA cycle.

- Plan: In this stage, we see the coverage of the overall quality system and how it relates to the individual organization as well as how management responds to it. Specifically, it covers the
  - Context of the organization: Here the standard focuses on the big picture of the organization and its context as well as the needs and expectations that are necessary for improvement. These are expected to be appropriately defined, identified, and communicated throughout the organization. Furthermore, in this first item of the PDCA model, the inclusion of both scope and quality management system/processes are defined.
  - Leadership: Here the standard addresses the leadership responsibility and commitment to quality. It focuses on having a policy; roles, responsibility, and authority are clearly defined and cascaded throughout the organization.
  - Planning: Here the standard focuses on how the organization defines the risks and opportunity actions that bring about improvement or need attention. This is done by making sure that specific plans and planning changes are identified and incorporated in the organization. In other words, they become part of the organizational culture.
  - Support: Here, the standard addresses the need for defining, developing resources, competence for needed skills, awareness, and communication programs to make sure appropriate personnel are notified of the needs of the organization to improve, and needed documented information exists to support management's commitment to continual improvement.
- Do: It is the stage where the standard is demanding some action to verify that the planning activities are being carried out as planned. Therefore, the focus here is on operations.
  - Operation: By far the most important element of the PDCA model is the verification and validation of what is being done and how this action compares with what was planned. Its significance is based on the verification of what has been defined and implemented in the planning stage. Without verification and validation, the planning stage is not worth very much. Items recognized as important are
    - Operational planning and control, that is, operational definitions of requirements
    - Determination of requirements that satisfy the customer
    - Design and development of appropriate and applicable requirements

- Control of externally provided products and services (i.e., from subsuppliers)
- Production and service provisions, that is, alternative or contingent plans
- Release of products and services, that is, meeting appropriate milestones on time
- Control nonconforming process outputs, products and services, that is, third-party inspection, defining quarantine space, and disposition practices of the nonconforming item

- Check: In this stage of the PDCA model, the need for measurement and how effective this measurement is are discussed in measuring the process, product, or service. Specifically, the item of concern is performance evaluation. This is evaluated by
  - Monitoring, measurement analysis, and evaluation, that is, gauge R&R, ANOVA, etc.
  - Conducting internal audit at frequent and predetermined timeline, that is, layer audits
  - Conducting regular management review of quality indicators and quality improvement, that is, QOS regular reviews, customer feedback, warranty issues, etc.
- Act: The final stage of the model addresses the closure of the continual improvement process by directly addressing
  - Nonconformities and appropriate corrective action
  - Continual improvement documentation
  - Comparative analysis of before and after corrective actions implementation

ISO's Key Elements

So now that we recognize the structure, how can we understand the elements (clauses) of the standard? In a summary form, one can review the standard and see the following eight elements interwoven throughout its context. They are

- Customer focus. Every organization exists because of their customers. Therefore, management needs to understand who their customers are, what they need, what excites (delights) them, and then aim to meet their needs. The strategy for a successful organization must be to always try to understand the customer needs, and provide satisfaction, loyalty, and trust. If the customer is not appreciated and does not get the satisfaction that is expected, then the organization will either go out of business or remain in a status quo. A fundamental principle here is to understand the KANO model and its significance in customer satisfaction over time. The parameters of the KANO model must be evaluated on a continual basis. Those parameters are the basic characteristics (items taken for granted), performance characteristics (items that meet minimum market requirements), and excitement or delighted characteristics (items that are a surprise or unexpected to the customer). Here we must emphasize that over time, the surprise or unexpected items become basic. Also over time, new characteristics become the new surprise and unexpected characteristics because of customer behavior changes, competition innovations, or new technology.

- Leadership. Leadership is the engine of any organization. Management's leadership guides the organization with the right strategy, direction, and ultimate success for growth. The management's leadership style molds the culture of the organization via a clear communication vision and purpose. Without a good leadership, the organization will fail in due time.
- Involvement of people. Every organization depends on people. In fact, the best asset that any organization has is its people. They make the difference between excellence and just so-so performance whether in sales, operations, management, engineering, customer service, or behind the scenes. It is imperative that in order for the people to be engaged in continual quality performance, they must be empowered to do what it takes to complete the task at hand. This means that they have to be appropriately *authorized* to carry out the assigned task but they also have to be *responsible* (accountable) for decision making within the boundary of their position.
- A process approach. It is a fact that most business (or organization) activities are part of one or more processes. Therefore, these processes should be viewed as a system rather than individual silos. The results will be much better because the holistic approach will examine individual processes but also their interaction.
- A systems approach to management. All businesses are systems, made up of smaller parts. Therefore, the management should view their organization as an integrated system and make decisions based on optimization of the entire organization rather than suboptimization (individual department maximization at the expense of some other department).
- Continual improvement. Improving is essential—a permanent element of any organization that wants to do well. The ISO standard is full of both direct and indirect references to continual improvement throughout.

- A factual approach to decision making. Decisions should be based on facts and evidence. In other words, organizations must be *data driven*. This implies that the data must be defined (operational definitions) in clear terms, must be appropriate, and must be applicable to the task. Sampling, frequency, and analysis are of paramount importance for a factual decision.
- Mutually beneficial supplier relationships. This implies a win-win relationship. That is, seeing the supplier/customer relationship as an interdependent relationship that can provide benefits to both parties by focusing on an effective and efficient managing strategy.

This is a simple summary of what the ISO 9001:2008 standard for quality management system says. To be certified to ISO 9001, you must meet all the applicable requirements (things you must do or have). Rather than explaining the requirements in detail, I will summarize them. The literature offers an abundant resource for the explanation and the requirements of the standard in books and articles as well as the Internet. For a good search, start with www.amazon.com or www.ASQ.org.

Individual Certifiable Clauses

In the previous section, we summarized some of the key threads of the standard. Here we will identify each category and give the specific requirements of each. We have simplified the language for an easier understanding. However, it should be pointed out that the standard is very specific in some words that are used. For example,

- The word *shall* is a legal term and it means that it must be done.
- The word *should* implies recommendation.
- The expression *such as* is used to provide examples.
- The word *note* is used to provide explanations.

The actual certifiable standard by the third party is Sections 4–8 and covers the following items.

Section 4: General Requirements for the Management System

4.1 Establish a quality management system: Develop it, use it, and continue to improve it.

4.2 Document your system.

4.2.1 Create whatever documents you need to describe, manage, and operate your system (must include a quality policy, objectives, procedures, and records).

4.2.2 Have a quality manual.

4.2.3 Have a systematic way to control your documents to make sure they are accurate and current.

4.2.4 Create and maintain records that show that your system does work.

Section 5: Requirements for Management

5.1 Demonstrate commitment to your quality management system; communicate how important quality is; and have a policy on quality plus objectives you can measure. Plan and do reviews of your system. Make sure you provide adequate resources for the system: people, plant, equipment, tools, IT, etc.

5.2 Focus on your customers. Identify their requirements. Aim to enhance their satisfaction.

5.3 Have and use a quality policy. Define a policy suitable for you. Include a commitment to meet all relevant requirements—ISO 9001, legal and statutory related to your services/products, etc.—and to continual improvement. Communicate your quality policy.

5.4 Plan for quality.

5.4.1 Set measurable objectives (aims/goals) for your quality system and its services/products.

5.4.2 Plan for how to achieve them.

5.5 Define and communicate who does what.

5.5.1 Identify who has what responsibility and authority in your system.

5.5.2 Have someone at senior management level be responsible for the quality system overall.

5.5.3 Have effective methods for communicating internally, including about the quality system.

5.6 Review the quality management system (management review).

5.6.1 Schedule and do regular, periodic reviews.

5.6.2 Include in your reviews customer feedback, how your processes and services/products are performing, failures and weaknesses (nonconformity, etc.), results from quality audits, actions from previous reviews, changes (whether internal or external), and planned or suggested improvements.

5.6.3 Keep records from your reviews, which include actions and decisions.

Section 6: Resource Requirements

6.1 Decide what resources you need and provide them. Include the resources needed to operate your system, meet customer requirements, and improve and enhance customer satisfaction.

6.2 Have competent people. [Author's note: This can be through education, experience, training, on the job training, or any combination of these.]

6.2.1 Make sure whoever does work that affects the quality or conformity of your services or products is competent for that work. [Author's note: Have an employee matrix of their qualifications.]

6.2.2 Decide what competencies you need; make sure you get and maintain them.

6.3 Provide whatever infrastructure you need. Identify what you need, and then provide and maintain it.

6.4 Provide the work environment you need for your products/services to meet requirements.

Section 7: Requirements for Services or Products (Requirements for Realization)

7.1 Plan and develop the processes needed for your services/products.

7.2 Have effective processes related to customers.

7.2.1 Identify the requirements that apply to your products/services.

7.2.2 Before promising to supply, make sure you can.

7.2.3 Have effective arrangements to communicate with customers.

7.3 Have effective processes for designing and developing your products/services.

7.3.1 Plan for design.

7.3.2 Define the inputs needed (includes all relevant requirements).

7.3.3 Have your design (plan, model, prototype, etc.) in a form that can be verified against its requirements (see 7.3.5).

7.3.4 Review your plan/design at suitable stages.

7.3.5 Make sure the design meets the input requirements (verification).

7.3.6 Validate the design (where practicable).

7.3.7 If changes to design occur, manage them.

7.4 Control any purchasing of goods, materials, or services that affect product/service conformity; this includes outsourcing and subcontracting.

7.4.1 Specify your requirements and make sure you get what you specified. Assess and monitor your suppliers.

7.4.2 Describe appropriately what you intend to purchase.

7.4.3 Verify that what you purchased meets the requirements.

7.5 Control your operations.

7.5.1 Have effective ways to control what you do, that is, your services/products.

7.5.2 Validate production/service provision where required.

7.5.3 Identify and track your services or products where required or needed.

7.5.4 Take care of any property supplied by customers, including confidential information.

7.5.5 Look after your products/components during storage/ delivery, etc.

7.6 Control measuring and monitoring equipment. Identify if you need monitoring/measuring equipment. If equipment is required, choose and use suitable equipment; ensure the results you get are valid and accurate.

Section 8: Requirements for Analysis, Measurement, and Improvement

8.1 Have processes that are suitable for ensuring your services or products meet the requirements of your customer, your quality management system including improvement, and any others that apply (e.g., legal, statutory, contractual, etc.).

8.2 Monitor/measure to determine if you got the results you intended and planned for.

8.2.1 Monitor the satisfaction of your customers' satisfaction.

8.2.2 Plan and implement a program of internal audit.

8.2.3 Monitor and measure your processes appropriately.

8.2.4 Monitor and measure your services/products at suitable stages, and only release your final service/product once requirements are met.

8.3 Control nonconformity. Have suitable methods to identify and control product/service nonconformity. This includes written procedures.

8.4 Analyze information from your system. Define the information needed to show your system is working and improving. Collect, analyze, and use that information.

8.5 Continually improve.

8.5.1 Improve your system.

8.5.2 Have a systematic approach to fix nonconformity and stop it from recurring.

8.5.3 Have a systematic approach to prevent potential nonconformity from happening.

Pending Changes for the 2015 Revision

As we already have mentioned, the ISO structure allows for revision of the standard every five years. However, sometimes the 5-year cycle is not met and extensions are given by the ISO. This is the case with the 2008 version. The pending change was supposed to be during the 2013 year, but has been changed twice and the anticipated new date is for late 2015.

As of this writing, the 2015 revision—at least the published draft—does not present any major changes from the last revision. A cursory overview of the high-level changes follows:

- The process approach is now embedded in Clause 4.4, which specifies requirements *considered essential to the adoption of the process approach.*
- Risk management is in, preventive action is out. Risk management is a requirement. However, preventive action has been removed. It is considered to be replaced by planning, risk management and be a part of a management system.
- The structure has changed. Whereas in the 2008 version we had 1–3 as not certifiable and 4–5 as the actual certifiable standards, the 2015 version will have 10 main clauses. The 1–3 remain the same, but the requirements now will be 4–10. The rationale for the change seems to be that the new standard will be in line with a future harmonized and consistent structure that may be used by all management systems. It is important to note here that even though the requirements are changed to a new clause, they are still present but in a different location. For example, requirements for both management review

and internal audit are now under Performance Evaluation (clause 9), and most of what used to be called Management Responsibilities is now under Clause 5 Leadership.

- There is a new addition to Clause 4. Context of the Organization is added to it. It requires the organization to consider itself and its context, and to determine the scope of its quality management system.
- Five 'Documented Information' replace both procedures and records. As of this writing, this pending change is somewhat controversial and may be changed by the publication date of the standard because it does not recognize *any* mandatory procedures.
- Terminology changes. The term *product* has been replaced with *goods and services*. This is to make it more generic and applicable to service fields, and remove the inherent manufacturing bias. 'Continual' has been dropped from the phrase 'continual improvement' in favor of just 'improvement'.
- Now there are only seven quality management principles. The existing version of 9001 is based on eight principles. Upon reviewing them by the experts of the ISO/TC 176, they decided on two main changes. The first one deals with the principle of *A systems approach to management*, which has been dropped because it was considered covered by the act of having a management system. The second one deals with the principle of *Relationship Management*, which replaces the old phrase *Mutually beneficial supplier relationships*.

All of these pending changes are based on 10 clauses. They are divided into two categories of noncertifiable and certifiable. They are

1. Noncertifiable clauses: Clause 1: Scope; Clause 2: Normative references; Clause 3: Terms and definitions
2. Certifiable clauses: Clause 4: Context of the organization; Clause 5: Leadership; Clause 6: Planning; Clause 7: Support; Clause 8: Operation; Clause 9: Performance evaluation; and Clause 10: Improvement

References

ISO 9100. (2009). *Quality Management Systems—Requirements for Aviation, Space, and Defense Organizations*. Rev. c. Troy, MI: Society of Automotive Engineers (SAE).

ISO/IEC 17025. (2005) (E). *General Requirements for the Competence of Testing and Calibration Laboratories*. 2nd ed. Geneva, Switzerland: International Organization for Standardization.

ISO/IEC 9126. (2003). *Software Engineering—Product Quality, Parts 2 and 3*. Geneva, Switzerland: International Organization for Standardization.

ISO/IEC 9126-1. (2001). *Software Engineering—Product Quality—Part 1: Quality Model.* Geneva, Switzerland: International Organization for Standardization.

ISO/TS 16949. (2009) (E). *Quality Management Systems: Particular Requirements for the Application of ISO 9001:2008 for Automotive Production and relevant service part organizations*. 3rd ed. Southfield, MI: Automotive Industry Action Group (AIAG).

Selected Bibliography

Cox, H. (July 2014). Calibration reports: A document without data is not much of a report. *Quality*. 13.

Diesing, G. (July 2014). Changes ahead for AS 9100. *Quality*. 16–18.

Dimeria, J. (August 2014). Improving your business with management system standards. *Quality*. 49–52.

Grachanen, C. (May 2014). Pass or fail: Understanding calibration content and requirements. *Quality Progress*. 44–45.

Hampton, D. (March 2014). A step forward: ISO 9001 revision focuses on reducing risk and applying the process approach. *Quality Progress*. 38–43.

Nichols, A. (October 2012). ISO 9001: Internal quality system audits. *Quality*. 39–41.

West, J., and C. Cianfrani. (August 2014). Managing the system: Revision introduces focus on organizational operating conditions. *Quality Progress*. 53–56.

12

Statistical Process Control (SPC)

Overview

Statistical process control (SPC) is a methodology for understanding, analyzing, and improving the process. By definition, it requires an understanding of its components, which are

- Inputs: Material or data transformed by process to create output. The inputs are manpower, machine, method, material, measurement, and environment (5M&E).
- Process: Any activity that transforms and adds value to the inputs to result in the desired outputs.
- Outputs: The result of transformation of inputs to the desired result.
- Feedback: The information from the process or customer that is taken into consideration for adjustments of the inputs or process.

Model of SPC

Many books in the market have defined, explained, and shown different implementation plans for an effective SPC program in many industries. Classical references are Feigenbaum (1961), Grant and Leavenworth (1980), Ishikawa (1982), Juran and Godfrey (1999), and Shewhart (1980). In this chapter, we are not focusing on the mechanics of SPC, but rather on the rationale and an overview of the SPC philosophy. Readers who are interested in the actual mechanics and calculations of control charts, capability, and all the details of implementing an SPC program in their organization are encouraged to see Gitlow et al. (1989), Gulezian (1991), Wheeler (2010), Montgomery (1985), Stamatis (2003), and others in the literature.

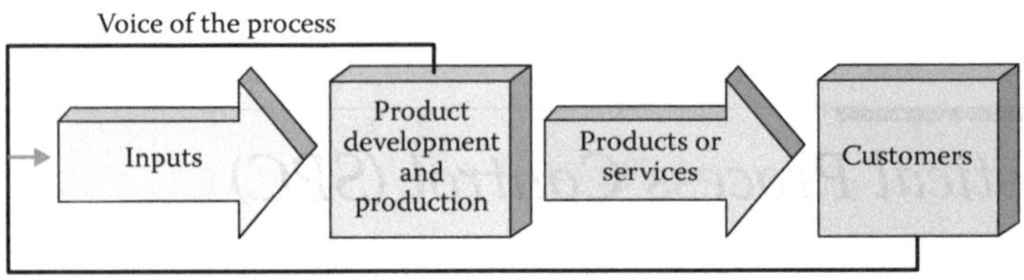

FIGURE 12.1
Classic SPC model.

The classic SPC model for a generic process is shown in Figure 12.1. The reader will notice that the product or service goes directly to the customer without any kind of handling. This is what is called the *first time throughput capability*. Obviously, the product development and production is the PROCESS. We include product development because that is also a process and it may have issues with variability as well. The products or services are the outputs of the process that the customer is willing to pay. That presupposes that all the specifications have been met. The feedback is given by the customer to both inputs and the process itself, so that improvements may be made as soon as possible. A much more detailed discussion on this is given by Stamatis (2003).

These four items (inputs, process, outputs, and feedback) are essential in any endeavor of SPC as they focus on improvement of the process rather than the product. The emphasis is on the following:

- Describe components of a process
- Document a process
- Describe features/applications of process improvement tools
- Describe the concept of variation
- Identify opportunities for process improvement
- Describe statistical measures of location, dispersion, and skewness
- Assess process performance in terms of capability indices

These items identify a systematic approach that must be managed by management. The systematic approach is to

- Understand the process from start to finish
- Define the process goals
- Understand how to achieve the defined goals
- Focus on key elements (indicators) of the process
- Monitor, evaluate, and improve the process

On the other hand, the management approach is a way to define, develop, and cascade to the organization the overall strategy of continual improvement.

Fundamentally, this approach begins with the quality operating system (QOS) (see Chapter 10) and is followed by the SPC methodology in the following steps.

Step 1. Identify and Define the Process

Perhaps this is the very most important item of starting an SPC program. This step addresses three concerns:

- Focus on the process. What and why is it important? Are consistency, stability, and normality present?
- Understand the process flow. How is this process contributing to the flow of improvement? This may be evaluated in parallel with the existing flow chart, a graphic depiction showing sequential steps of the process.
- Develop a process model. If there is no current flow chart, model what really happens, not what should happen. A process flow diagram is a good starting point for this early development. This will help in (a) subsequent process modeling based on adjusted, modified, or changed parameters, (b) simplification of the process, and (c) selected method of studying the process or work measurement.

The fundamental concepts and principles that guide the three concerns may vary from organization to organization, but some of the most common are the recognition that

- Every business is made up of processes.
- Every business exists at several levels.
- Every business has core processes.
- Every business has major processes.
- Every business has subprocesses.
- Every business has unique activities (tasks).

All these are functions of people. As a consequence, it must be understood that

- Every person manages a process.
- Every person is a customer and a supplier.
- Every person has customers and suppliers.

Step 2. Establish Process Ownership and Responsibilities

Here we must identify the owners of the process and the responsibility of the owners of the particular selected process in the continual improvement process of the organization. The reason for the significance of the ownership to be identified early on in the implementation process is that there are specific requirements for proper assignment of responsibilities. Some of the key ones are

- Develop the process model
- Establish and maintain appropriate process controls
- Monitor performance, initiate reviews and audits
- Ensure customer satisfaction
- Locate resources
- Stimulate suppliers to provide first time throughput quality
- Initiate and manage process improvement
- Establish roles and responsibilities and organize the team

Once the ownership has been identified, then the criteria must be defined. Items of concern that should be addressed are

- Criteria for process ownership
- Process owner's responsibilities
- Process team's roles and responsibilities
- Process owner's role in improving quality and productivity
- Understanding process from beginning to end
- Authority to effect changes
- Responsibility for process results
- Commitment to continuous improvement
- Ability to inspire creative teamwork

In addition to the direct ownership, a complete and efficient SPC program depends on the inclusion of a team. The team must be cross-functional and multidiscipline as well as having some relation to the process. Typical team roles and responsibilities are

- Ensure everyone understands the process model
- Ensure everyone is fully aware of his or her expected contributions
- Create a supportive working environment
- Identify and resolve issues
- Identify, instigate, and manage quality improvement

The reader should note that all these concerns imply a strong intervention of management commitment and recognition that the work involved within the SPC jurisdiction is more often than not a process approach. For the management commitment, it is imperative that all management levels be familiar and exercise the spirit of Deming's 12 management points (Deming 1986) and simplified by Scherkenback (1988). Deming's original management points are

1. Create constancy of purpose
2. Adopt new philosophy
3. Cease dependence on inspection
4. End lowest tender contracts
5. Constantly improve systems
6. Institute training
7. Institute leadership
8. Drive out fear
9. Break down barriers
10. Eliminate slogans, exhortations
11. Eliminate work standards
12. Permit pride of workmanship
13. Encourage education
14. Transformation is everybody's job

As for the team process, key questions that should be asked or discussed are

- Is the proper definition of a team addressed?
- Do we have a group or a team?
- Have the five steps of team development been recognized (forming, storming, norming, performing, and adjourning?)
- Have review milestones and feedback process been defined and agreed upon? Typical issues here should be
    - Makes explicit what is happening during team process
    - Describes characteristics of team process
    - Determines what needs to be changed in order to facilitate continuous improvement
        - What roles need to be filled?
        - What tools will be needed?
        - What methodologies can be used?
        - What are possible hurdles?
        - Is the Plan, Do, Check, Act (PDCA) model applied to the team process?

Given the answers to these questions, the team will develop an overall plan and define the requirements of the review process to make sure the goals are being met. Typical questions may be

- What precisely did you notice?
- How did you feel?
- What (if anything) would you like to do differently?

Step 3. Define and Establish Process Controls

Once the first two steps have been completed, then the owners of the process should begin the determination of sampling, frequency, and calculating the voice of the process (control limits). The notion of determining the appropriate and applicable sampling and frequency is based on determining the correct control on critical and significant process parameters as well as establish dominant process factors that influence process output.

The process of defining and establishing controls is very important and depends on sampling and frequency, as well as the data involved in the process. For sampling and frequency, the selection is based on the assumption that both reflect the current process. This means that the sample and frequency are representative of the process pattern and a reflection of the population. As for the data, consideration must be given to what type of data is used. There are two types:

1. Variable or measured: Examples are length, viscosity, weight, time, temperature, pressure, dimension, and so on.
2. Attribute or countable: Examples are pass/fail, go/no go, present/absent, yes/no, and so on.

The preferred choice of data is variable because there is a greater sensitivity of variability and there is greater flexibility for analysis. With attribute, both the sensitivity and flexibility are missing. Depending on what data the process generates or the process owner decides to choose, the selection of control charts will be different. For example, with variable data, the X bar, R chart, medium chart, individual moving range, standard deviation range chart, CUSum, and other charts may be utilized. With attribute charts, the p, u, np, c, and other charts may be used. The goal here is to remove all assignable causes, and develop a consistent and stable process so that capability may be calculated. The calculated control limits will identify whether the process is consistent and stable enough to pursue capability study. Only random variation (common cause variation) is allowed for studying capability of the process. Unless the process is normally distributed (all assignable causes have been removed), capability cannot be calculated. If data is not normally distributed, sometimes by transforming the data (via Wilk's lambda [λ] or

Box-Cox functions or some other transformation function) we can proceed to capability as the transformed data become normal. If, after transforming the data, the distribution does not follow normality, you should start all over again and make sure you review the steps to find what is really going on with the process or maybe the selection of the data.

Yet another benefit of establishing the voice of the process is to assure that the defined standards of performance are related to the customer's demands as well as the organization's demands. Therefore, *performance* is monitored for all critical/significant process parameters and product characteristics via (a) *variable data/charts* (the preferred approach) to establish target value and minimize variation, and (b) *attribute data/charts* to define/demonstrate the minimum acceptable and unacceptable levels of nonconformance. The preferred goal is zero defects and zero nonconformities. These approaches depend on two very important perspectives: (a) *technical*—ideas, tools, and methodologies used to achieve customer needs and expectations, and (b) *historical*—experiences based on things gone wrong (TGW) and things gone right (TGR) of past practices on the same or similar processes. In essence, these are comparative studies that relate controls to past process events. Typical relationships of control to process events may be divided into four activities.

- *Before the event* activities, such as
  - Quality Function Deployment (QFD). Translating the customer wants into engineering requirements
  - Design of Experiments (DOE). Identifying and quantifying within and between variation as well as interaction between factors
  - Failure Mode and Effects Analysis (FMEA). Identifying potential failures and suggesting appropriate controls
- During the event activities, such as
  - SPC on process parameters. Monitor a process for vital indicators.
- After the event activities, such as
  - Inspection
  - SPC on product
  - Characteristics reports
- Activities that may be performed *before, during, after the event*, such as error proofing, which is 100% prevention or detection at source, contact/non-contact devices.

No matter where they are performed, the purpose of these activities is to establish an accurate, appropriate, and applicable system for measurement and reporting of performance. This system should be as simple as possible, such as a technician taking simple measurements (e.g., temperatures, length, etc.). It must be noted here that quite often these activities may be complex, such as, plant-wide SPC system to customer-specific standard or requirement, a pattern of running the process as is and taking measurements, or interpreting results at least with signals that indicate the possibility of out of control conditions. The signals used most often are

1. Out of control points
2. Run of seven points (in a run, the points do not cross the center line)
3. Trends of seven points (in a trend, the points cross the center line)
4. Points too close to the limits (process repeats every so often but the points are creating gaps; it is an indication of a mixing issue)
5. Cycles (process repeats every so often; points are continuous)
6. Hugging (points suddenly are hugged around the center line or have shifted above or below the center line without adjustment of the control limits)

Perhaps one of the most important and critical issues in this step is to recognize and make sure that the measurement system used is verified as valid. Chapter 16 discusses measurement system analysis (MSA) in more detail. Here, we emphasize that the measurement verification and validity is an issue of

- Verifying the validity of a signal or message that the measurement indicates
- Evaluating significance of a signal or message
- Evaluating options for improvement and taking actions as needed on
    - Action on process to bring out-of-control conditions into control
    - Action on output to prevent nonconforming output from reaching the customer

Step 4. Continuously Improve Process Performance

The final step to SPC is to make sure that the variation is controlled and minimized. As such, the process of continual monitoring is a never-ending one, unless the variation is zero, which is impossible. Because of this never-ending journey, the effort to reduce variation with the systematic approach that we just discussed is continuous. That approach is

- Understand process from start to finish
- Define process goals
- Understand how to achieve goals

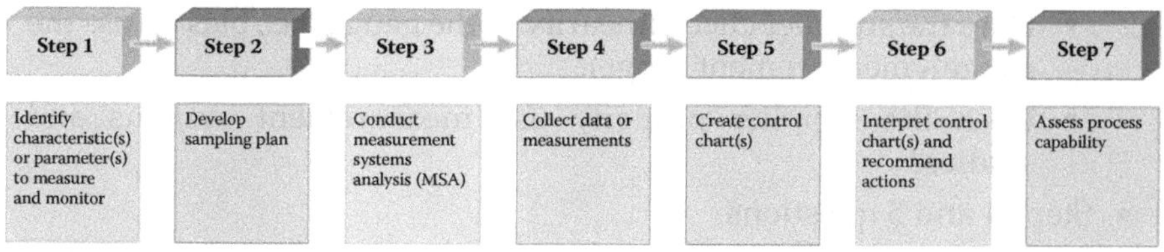

FIGURE 12.2
Seven-step approach to SPC implementation.

- Focus on key elements
- Monitor, evaluate, and improve

Now that we know the rationale of the SPC, let us summarize the steps for its implementation. Everyone dealing with SPC must be aware of seven distinct steps. They are shown in Figure 12.2. Of special importance is Step 7, which assesses process capability. Indeed, it is the final step of any SPC endeavor because before capability is established, the process must be consistent, stable, in control (no special causes), and normally distributed. If any one of these requirements is not met, capability cannot be calculated. For a detailed discussion on this issue, see Stamatis (2003), Wheeler (2000a,b, 2010), and Bothe (1997).

To make sure the supplier follows this approach, we have some key questions that must be answered for every step. They are by no means exhaustive, but hopefully they will be helpful. For individual organizations and specific processes, the list may be modified to reflect their specific requirements.

- Step 1 questions
    - How does the supplier select SPC characteristics (or process parameters)?
    - Are the supplier's SPC procedures and work instructions adequate?
- Step 2 questions
    - How does the supplier determine sampling frequency?
    - How does the supplier determine subgroup size?
    - Are control charts appropriate for the processes and the data collected?
    - Under what circumstances does the supplier recalculate control limits?
- Step 3 questions
    - Does the supplier employ, at a minimum, gauge calibration (bias), gauge repeatability and reproducibility (R&R) analysis?

    - Does the supplier check linearity of the gauge (i.e., bias across the gauge's measurement range)?
    - Does the supplier properly use measurement systems and studies?
- Steps 4 and 5 questions
    - Does the supplier apply an appropriate process for collecting data and creating control charts?
    - Is the supplier using statistical control limits? Statistical control limits are the voice of the process. They should never be confused with customer specifications. They are not the same.
- Step 6 questions
    - How many of the applicable out-of-control tests is the operator using?
    - What is the recommended reaction to out-of-control signals?
    - Does the chart indicate that corrective actions are effective?
    - When the supplier finds a signal, how does the supplier quarantine production? How far back (in time or in number of pieces) does the supplier quarantine product? Is it effective?
- Step 7 questions
    - Are the supplier's process capability and/or performance indices valid? Make sure they are using P_{pk} and not C_{pk}. The difference is that P_{pk} is using actual data for calculating the standard deviation, whereas C_{pk} is using estimated data for the standard deviation.
    - Do the supplier's process capability and/or performance indices meet customer expectations? Minimum expectation for the PPAP approval is P_{pk} of 1.67 or greater and for long-term full production, P_{pk} is expected to be no less than 1.33. In both cases, the bigger the number, the better it is.

SPC and Six Sigma

As we have seen, SPC is a very powerful yet simple methodology to identify special causes, remove them, and generate stability in a process to eventually calculate capability. There is also a much more powerful yet difficult methodology that focuses on more demanding problems of the process and that is the Six Sigma approach. As powerful as the methodology is, however, it uses SPC in every stage of its evaluation for both the traditional Six Sigma model (DMAIC) and the more advanced version of the design for the Six Sigma model (DCOV).

TABLE 12.1

Summary of SPC Utilization in the DMAIC Model

Model	SPC Utilization
Define and measure	SPC may be used to determine the magnitude of the problem and the baseline capability.
Analyze	SPC may be used to evaluate consistency and stability in the process.
Improve	SPC may be used to confirm the improvement.
Control	SPC may be used to monitor process improvement and stability and to confirm process capability. It is also important here to use SPC as a tool to confirm infusion of the improvement into other similar processes in the organization and therefore to substantiate replication.

TABLE 12.2

Summary of SPC Utilization in the DCOV Model

Model	SPC Utilization
Define	SPC may be used to determine the problem's magnitude in an existing process and the baseline capability for an external or internal product benchmark.
Characterize	SPC may be used to assess capability of an existing process. This process may be a surrogate for the future process.
Optimize	SPC may be used to confirm the success of the improvement.
Verify	SPC may be used to verify long-term process capability and to continuously look for changes in the process.

It is beyond the scope of this book to discuss both Six Sigma models, but Tables 12.1 and 12.2 summarize where exactly the SPC is used.

References

Bothe, D. (1997). *Measuring Process Capability*. New York: McGraw-Hill.

Deming, W. (1986). *Out of the Crisis*. Cambridge, MA: Massachusetts Institute of Technology.

Feigenbaum, A. (1961). *Total Quality Control*. New York: McGraw-Hill Book Company.

Gitlow, H., S. Gitlow, A. Oppenheim, and R. Oppenheim. (1989). *Tools and Method for the Improvement of Quality*. Homewood, IL: Irwin.

Grant, E. and R. Leavenworth. (1980). *Statistical Quality Control*, 5th ed. New York: McGraw-Hill.

Gulezian, R. (1991). *Process Control*. New York: Quality Alert Institute.

Ishikawa, K. (1982). *Guide to Quality Control*. White Plains, NY: Kraus International Publications.

Juran, J. and A. Godfrey. (1999). *Juran's Quality Handbook*, 5th ed. New York: McGraw-Hill.

Montgomery, D. (1985). *Statistical Quality Control*. New York: John Wiley.
Scherkenback, W. (1988). *The Deming Route to Quality and Productivity*. Milwaukee, WI: Quality Press.
Shewhart, W. (1980). *Economic Control of Quality of Manufactured Product*. Reprint of the 1929 original work. Milwaukee, WI: Quality Press.
Stamatis, D. (2003). *Six Sigma and Beyond: Statistical Process Control*. Boca Raton, FL: St Lucie Press.
Wheeler, D. (2000a). *Understanding Variation: The Key to Managing Chaos*. Knoxville, TN: SPC Press.
Wheeler, D. (2000b). *Beyond Capability Confusion*. Knoxville, TN: SPC Press.
Wheeler, D. (2010). *Understanding Statistical Process Control*, 3rd ed. Knoxville, TN: SPC Press.

Selected Bibliography

Blanchard, K. and E. Parisi-Carew. (2009). *The One Minute Manager Builds High Performing Teams*, New York: William Morrow.
Brassard, M. and D. Ritter. (1994). *The Memory Jogger II*. Salem, NH: GOAL/QPC.
Concourse. (1988). *Statistical Methods for Improving Performance: Participant Manual*. Newton Square, PA: Concourse Corporation.
Tuckman, B. (1965). Developmental sequence in small groups. *Psychological Bulletin*. 63(6): 384–399.
White, A. (2009). *From Comfort Zone to Performance Management*. Baisy-Thy, Belgium: White and MacLean.

13

Problem-Solving Methodology

Overview

Perhaps the most misunderstood area in any organization is the understanding and selection of a problem-solving methodology. It is misunderstood because there are so many options to choose from and depending on whom you ask, the answer of the *best* approach is quite different. However, *all* organizations in the modern world must have a method—preferably standardized—across their facilities whether local, national, or international.

What is a problem-solving methodology? Problem-solving methodology is a *strategy* with specific steps that one would use to find the problems that are in the way to getting to one's own goal. Another way of saying it is that problem solving consists of using generic or ad hoc methods, in an orderly manner, for finding solutions to problems. Others may have different definitions, for example, calling it a *problem-solving cycle* (Bransford and Stein 1993). In this cycle, one will recognize the problem, define the problem, develop a strategy to fix the problem, organize the knowledge of the problem cycle, figure out the resources at the user's disposal, monitor one's progress, and evaluate the solution for accuracy. Although called a cycle, one does not have to do each step to fix the problem. In fact, in some cases if the problem is simple, not all steps may be necessary. However, the reason it is called a cycle is that once one is completed with a problem, usually another will surface.

Blanchard-Fields (2007) looks at problem solving from one of two facets. The first is looking at those problems that only have one solution (like math/engineering problems, or fact-based questions), which are grounded in psychometric intelligence. The second is addressing socioemotional issues that are unpredictable with answers that are constantly changing (like your favorite color or what you should get someone for Christmas). To show the diversity of problem-solving strategies, we provide the reader with a sample of techniques that are available. By no means is this an exhaustive list.

- Abstraction: Solving the problem in a model of the system before applying it to the real system
- Analogy: Using a solution that solves an analogous problem

- Brainstorming (especially among groups of people): Suggesting a large number of solutions or ideas and combining and developing them until an optimum solution is found
- Divide and conquer: Breaking down a large, complex problem into smaller, solvable problems
- Hypothesis testing: Assuming a possible explanation to the problem and trying to prove (or, in some contexts, disprove) the assumption
- Lateral thinking: Approaching solutions indirectly and creatively
- Means-ends analysis: Choosing an action at each step to move closer to the goal
- Method of focal objects: Synthesizing seemingly nonmatching characteristics of different objects into something new
- Morphological analysis: Assessing the output and interactions of an entire system
- Proof: Trying to prove that the problem cannot be solved; the point where the proof fails is the starting point for solving it
- Reduction: Transforming the problem into another problem for which solutions exist
- Research: Employing existing ideas or adapting existing solutions to similar problems
- Root-cause analysis: Identifying the cause of a problem
- Trial-and-error: Testing possible solutions until the right one is found

Therefore, even though there are many strategies to approach a problem, one must recognize that no matter what the strategy is, all of them have advantages and disadvantages. Therefore, the organization must select a methodology that fits its culture and processes. Some common methodologies are

- Applied problem solving (APS): The model is based on the classic approach of solving problems, which is
  - Comprehension
  - Representation
  - Planning, analysis, and synthesis
  - Execution and communication
  - Evaluation
- Comprehension: A model where making sense of the problem is by using strategies such as retelling, identifying relevant information, and creating mental images. This can be helped by encouraging participants to reread the problem several times and record in some

way what they understand the problem to be about (e.g., drawing a picture or making notes).

- Representation: A model of "homing in" on what the problem is asking solvers to investigate. Central to this stage is identifying what is unknown and what needs finding. Typical questions may focus on
  - Can they represent the situation mathematically (i.e., Y = f(x, n) where Y = depended variable, x = independent variable, n = noise)?
  - What are they trying to find?
  - What do they think the answer might be (conjecturing and hypothesizing)?
  - What might they need to find out before they can get started?
- Planning, analysis and synthesis: A model about planning a pathway to the solution, assuming that the participants have understood what the problem is about and established what needs finding. It is within this process that you might encourage participants to think about whether they have seen something similar before and what strategies they adopted then. This will help them to identify appropriate methods and tools. Particular knowledge and skills gaps that need addressing may become evident at this stage.
- Execution and communication: A model where participants might identify further related problems they wish to investigate. They will need to consider how they will keep track of what they have done and how they will communicate their findings. This will lead to interpreting results and drawing conclusions.
- Evaluation: A model where participants can learn as much from reflecting on and evaluating what they have done as they can from the process of solving the problem itself. During this model, participants should be expected to reflect on the *effectiveness* of their approach as well as other people's approaches, justify their conclusions, and assess their own learning. Evaluation may also lead to thinking about other questions that could now be investigated.
- Goal, reality, obstacles/options, and way forward (GROW model): The particular value of GROW is that it provides an effective, structured methodology that helps set goals effectively *and* is a problem-solving process. G = Goal = This is the end point (where the organization wants to be). The goal has to be defined in a way that is clear to the organization when it has achieved it. R = Reality = The current reality (where the organization is now). What are the issues and challenges, and how far are they away from their goal? O = Obstacles = The items the organization wants to remove. If there were no obstacles, the organization would already have reached its goal. O = Options = Once

obstacles have been identified, the organization needs to find ways of dealing with them if it is to make progress. W = Way forward = The options then needed to be converted into action steps, which will take the organization to its goal.

- How to solve it: A mathematical approach to problem solving that focuses on four steps developed by Polya (1945, pp. 6–15).
    - First, you have to understand the problem.
    - After understanding, then make a plan.
    - Carry out the plan.
    - Look back on your work. Ask how it could be better.

  If this technique fails, Polya advises, "If you can't solve a problem, then there is an easier problem you can solve: find it."
- Kepner-Tregoe problem solving and decision making: A model (some call it a technique) that refers to a process of weighing alternatives in which a person lists and assigns a numerical weight to a series of values (noting that some may be absolute requirements), gives each alternative a numerical rating according to each value, and computes a numerical score for the alternative (as the dot product). The technique allows that an absurd result may indicate an error in the weight and the rating to be solved by adjusting and iterating. A variation of this is the IS and IS NOT approach.
- OODA loop: This is the observe, orient, decide, and act model designed by Boyd. The actual model according to http://www.danford.net/boyd/essence.htm is shown in Figure 13.1.

According to Boyd, decision making occurs in a recurring cycle of observe-orient-decide-act. An entity (whether an individual or an

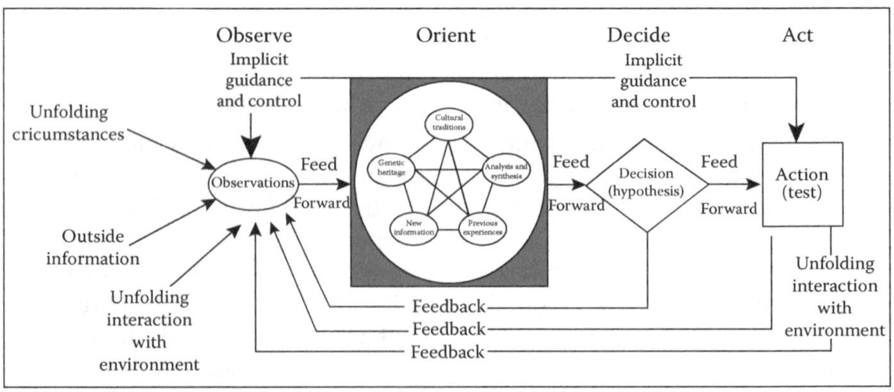

FIGURE 13.1
John Boyd's OODA loop.

organization) that can process this cycle quickly, observing and reacting to unfolding events more rapidly than an opponent, can thereby *get inside* the opponent's decision cycle and gain the advantage. Osinga (2006), however, argues that Boyd's own views on the OODA loop are much deeper, richer, and more comprehensive than the common interpretation of the rapid OODA loop idea.

Boyd developed the concept to explain how to direct one's energies to defeat an adversary and survive. Boyd emphasized that *the loop* is actually a set of interacting loops that are to be kept in continuous operation during combat. He also indicated that the phase of the battle has an important bearing on the ideal allocation of one's energies.

Boyd's diagram shows that all decisions are based on observations of the evolving situation tempered with implicit filtering of the problem being addressed. These observations are the raw information on which decisions and actions are based. The observed information must be processed to orient it for further decision making. This model has become an important concept in litigation, business, and military strategy.

Whereas the OODA loop focuses on *strategic* approaches and has been adopted by business and public sector entities, the difference with the PDCA or Shewhart cycle is that the PDCA model focuses on the *operational* or *tactical* level of projects.

- Plan, Do, Check, Act (PDCA) model: This is the classic model of improvement used by many organizations worldwide.
- Rapid problem resolution (RPR): A specific model for identifying diagnostically the root cause of IT problems (Offord 2011). Specifically, it deals with failures, incorrect output, and performance issues, and its particular strengths are in the diagnosis of ongoing and recurring *grey problems* (a term for an IT problem where the causing technology is unknown or unconfirmed). Common grey problems are (1) intermittent errors, (2) intermittent incorrect output, and (3) transient performance problems. The model is made up of two approaches. They are
  1. Core process: Defines a step-by-step approach to problem diagnosis and has three phases:
     a. Discover
        i. Gather and review existing information
        ii. Reach an agreed understanding
     b. Investigate
        i. Create and execute a diagnostic data capture plan
        ii. Analyze the results and iterate if necessary
        iii. Identify root cause

    c. Fix
        i. Translate diagnostic data
        ii. Determine and implement fix
        iii. Confirm root cause addressed
  2. Supporting techniques: Detail how the objectives of the core process steps are achieved and cite examples using tools and techniques that are available in every business

- TRIZ (in Russian, Teoriya Resheniya Izobretatelskikh Zadatch; in English, theory of solving inventor's problems): A problem-solving analysis and forecasting tool derived from the study of patterns of invention in the global patent literature (Hua et al. 2006, Barry et al. 2010, Shang and Kok-Soo 2010).
- 8D: A model that is used heavily in the automotive industry and is gaining favor in many other industries either directly or in a modified version. It is a model used to approach and to resolve problems, typically employed by quality engineers or other professionals. Its purpose is to identify, correct, and eliminate recurring problems, and it is useful in product and process improvement. It establishes a permanent corrective action based on statistical analysis of the problem (when appropriate) and focuses on the origin of the problem by determining its *root cause* as well as the *escape point* of the problem. Although originally it was comprised of eight stages or disciplines, it was later augmented by an initial planning stage. 8D was developed by Ford Motor Company and it is used widely all over the world in many manufacturing and nonmanufacturing organizations. The augmented version in Ford Motor Company is called Global 8D (G8D), but other users continue to call it 8D. Both the 8D and G8D follow the logic of the PDCA model. A detailed discussion will be given in Chapter 21.
- 5 Whys: The 5 Whys is an iterative question-asking model/technique used to explore the cause and effect relationships underlying a particular problem. The primary goal of the model/technique is to determine the root cause of a defect or a problem. It is equivalent to the "pealing the onion" approach. The assumption is that if you keep asking questions, eventfully you will get to a question to which you do not know the answer. At that point, you declare it as *the problem*. After the definition of the problem, you pursue the analysis and hopefully the right decision will be reached.
- 3 × 5 Whys: A variation of the 5 Whys. The difference between the two is that the 3 × 5 Whys addresses issues on both technical and system levels. A typical form is shown in Figure 13.2.
- Lean methodology: It is a method that focuses on waste reduction by using many tools, but the essence is on three approaches

1. Problem Statement								
What was wrong?								
Where was it detected?								
When was it detected?								
Who was impacted?								
2. Nonconformance Problem Statement	Why 1: Why the problem occurred	Why 2	Why 3	Why 4	Why 5	Why 6: Technical root cause	Why 7: System root cause	Corrective action
	Therefore	Therefore	Therefore	Therefore	Therefore	Therefore		
3. Nondetection Problem Statement	Why 1: Why was the problem undetected	Why 2	Why 3	Why 4	Why 5: Technical root cause	Why 6: System root cause	Corrective action	
	Therefore	Therefore	Therefore	Therefore	Therefore	Therefore		

FIGURE 13.2
Typical 3 × 5 Why form.

to reducing any type of waste in a given organization. The three approaches are

1. Stabilize: To establish stability, excess movement of work and variation of processes is removed.
2. Standardize: Once the stability has been reached, then formal standards via work rules, charts, and visual controls must be used to further reduce variation.
3. Simplify: Simplicity means leveling the workloads among staff to ensure effective and efficient service is provided to the customer.

Definition of a Problem

There are many ways to define a problem. However, here we use five of the traditional generic definitions.

1. A problem is a symptom, not a cause or a solution.
2. A problem is a gap between what is happening and what you want to happen.
3. A problem is something that keeps you from achieving your goal.
4. A problem could be a mistake, a defect, an error, a missed deadline, a lost opportunity, or an inefficient operation.
5. A problem may be thought of as an opportunity to do better.

The idea here is that by defining the problem correctly in your problem-solving methodology, you will be led to improvement. That improvement may be materialized through Kaizen or breakthroughs. Kaizen is gradual, relentless, incremental process improvement over time, while breakthrough is a sudden, significant improvement in process performance achieved in a very short interval of time (re-engineering). Obviously, continual improvement is best achieved by a combination of both. Ideally, an organization should employ and encourage both Kaizen and breakthrough in its improvement strategy; however, Kaizen is much easier to accomplish.

To optimize that improvement, one must recognize that there are two types of problems: the sporadic and the chronic. A sporadic problem is a sudden, elevated departure of the level of a problem from its historic, status quo level. On the other hand, a chronic problem is the difference between the historic level and an optimum or zero level; a chronic problem is a *chronic disease*, which can be eliminated economically. A chronic problem is something that keeps repeating over time. In order for both to be eliminated, management must have the appropriate attitude. This means that problems (1) must *not* viewed as burdens, (2) must be recognized and dealt with rather than ignored or pushed down the stream of processes, and (3) must be accepted as challenges or opportunities to improve.

These three attitudes are controlled by management, and as such it is management's responsibility to

- Create a *motivating environment* conducive to an effective, company-wide problem-solving methodology
- Create a real *opportunity* for problem solving
- Create a *systematic approach* (roadmap) for effective problem solving
- Create the applicable *knowledge* of the tools for problem solving via education and training

In addition to these requirements, six essential ingredients need to be in place for positive results. They are

1. *Awareness* of the importance of eliminating errors and the cost of errors to the business
2. *Desire* to eliminate errors
3. *Training* in proven, effective methods for solving and preventing problems
4. *Failure analysis* to identify and correct the real root causes of problems
5. *Follow-up* in tracking problems and action commitments
6. *Recognition*, giving liberal credit to all those who participate

Unfortunately, many organizations talk positively about their problem-solving methodology, but when they discuss implementation procedures, they find many roadblocks (excuses) for not implementing an effective methodology. Some of the key roadblocks are

1. Lack of *time*
2. Lack of problem *ownership* (if everybody is responsible, no one is responsible)
3. Lack of *recognition*
4. Errors have been accepted as *a way of life* (this is how we do it here)
5. Ignorance of the *importance* of a problem
6. Ignorance of the *opportunity cost* of a problem (potential savings)
7. Belief that *no one* can do anything about some problems
8. *Poor balance* by upper management among cost, schedule, and quality
9. People who try to *protect* themselves (CYA) (fear of intimidation and ridicule)

10. *Head-hunting* by management (selecting team members who are incompetent for the task at hand and are not willing to pursue real improvement)

Problem-Solving Cycle

Any problem-solving methodology is a cycle because it repeats itself, even though the problems may differ. Then why use a structured problem-solving approach? A structured, systematic, problem-solving approach helps us to

- Illustrate the relative *importance* of problems
- Show the real *root causes* of problems
- Guide problem solvers to become *unstuck*, know *where* they are at, keep on *track*, and *persevere* to results
- *Keep* problems solved
- Establish *accountability* and a *motivating* environment for problem solving
- Produce consistently *better* solutions to problems than unstructured approaches

As mentioned earlier, an organization may follow many options to optimize its improvement and remove defects from its processes. In fact, successful, continuously improving companies around the world use structured problem-solving systems. Some specific approaches in use in the United States are shown in Table 13.1.

Fundamentally, all of them use a variation of the basic steps for any problem solving. They are

1. Select the problem
2. Find the root cause
3. Correct the problem
4. Verify the correction
5. Prevent recurrence ... then, recycle.

These steps (see Figure 13.3) are based on three fundamental premises: (1) they are easy to understand and easy to remember, (2) they are designed to be completed in sequence, and (3) there are also designed to have several iterations or feedback loops that may have to be followed to properly complete the cycle (task). Because they are important steps, let us examine each of them separately.

TABLE 13.1

Comparison of Several Problem-Solving Methodologies

Xerox's 6 Steps	GM's 4 Steps*	Daimler Chrysler's 7 Steps	DoD's 6 Steps	Classical 7-Step Approach	Ford's 9 Steps
1. Identify and select the problem	1. Identify the problem	1. Problem: Poor logo quality	1. Identify the problem	1. Identify the problem	See Chapter 21
2. Analyze the problem	2. Analyze the problem	2. Containment: Verify process and final product Quality alert	2. Gather facts/make assumptions	2. Contain the problem	
3. Generate potential solutions	3. Identify decision criteria	3. Root cause: Machine setup, old equipment, needs adjustment	3. Define objectives, goals	3. Define root cause	
4. Select and plan a specific solution	4. Implement	4. Action: Give maintenance, modify preventive maintenance plan	4. Develop best solutions	4. Implement permanent action	
5. Implement that solution	5. Evaluate results	5. Verification	5. Analyze, compare solutions	5. Verify action success	
6. Evaluate the solution		6. Control	6. Select and implement solutions	6. Implement process controls	
		7. Prevention: Apply to similar processes/products, modify FMEA		7. Implement preventive measures	

*GM also uses the 7 diamond approach as a preliminary process for immediate reaction to internal quality issues. The first four steps are used to quickly determine if an out of standard condition (special cause) exists. The last three focus on whether engineering support is required.

1. Select the problem
   - ⇒ Prepare problem list
   - ⇒ Collect data
   - ⇒ Prioritize problems
   - ⇒ Select target problem
   - ⇒ Prepare problem statement
2. Root-cause analysis
   - ⇒ Identify possible root causes
   - ⇒ Collect data
   - ⇒ Select most likely cause
3. Correct the problem
   - ⇒ Make temporary fix
   - ⇒ Develop alternative solutions
   - ⇒ Select best solution
   - ⇒ Develop implementation plan
   - ⇒ Establish measuring method
   - ⇒ Gain management approval
   - ⇒ Implement solution
4. Verify correction
   - ⇒ Measure impact of solution
   - ⇒ Decision: solution effective?
   - ⇒ If necessary, loop back
   - ⇒ Remove temporary fix
5. Prevent recurrence
   - ⇒ Mistake proof system
   - ⇒ Alter systems & procedures
   - ⇒ Train involved personnel
   - ⇒ Publicize results
   - ⇒ Continue to monitor problem
   - ⇒ Recognize team's success
   - ⇒ Prepare final project report

Recycle

FIGURE 13.3
Summary of the 5-step model.

Step 1: Select the Problem (or Opportunity) That Represents a Waste or a Bottleneck in the Area

- Prepare a problem list.
- Collect data to determine the magnitude of each problem.
- Prioritize the problems to determine which problems should be worked on first. Usually, a vital few problems account for the greatest proportion of the losses.
- Select the target problem, generally the problem whose solution will have the greatest beneficial impact.
- Prepare a specific problem statement (e.g., reduce shipping errors from 3.1% to under 1.5% in the next 12 months). This represents the specific problem assignment or project committed to. A complete problem must have three components in order to be complete: (a) the item, (b) measurement, and (c) criteria. In the example of the reduction of scrap, we have the item of concern is scrap, the measurement is from 3.1% to 1.5%, and the criterion is the next twelve months

Step 2: Find the Root Cause of the Problem

This analysis step is usually *the* crucial work of the project. Problems *must* be solved at the *root-cause* level and must be doable within a reasonable time and cost if they are to remain fully and permanently corrected.

- Identify all possible causes of the problem.
- If necessary, collect data to ascertain the failure mechanism.
- Select the most likely cause. (Remember, there are three options here: probable cause, possible cause, and likely cause.)

Generally, a root cause cannot be fixed unless we find the escape point, that is, the place that we could have caught the problem but did not. The escape point indicates a system failure.

Step 3: Correct the Problem

Correct the problem by following a plan that will eliminate the problem or at least reduce it to a level compatible with the team's goal.

- Make a temporary fix, if appropriate, to protect the customer until a permanent fix can be implemented.
- Develop alternative solutions for correcting the problem.
- Select the best possible solution by narrowing down the alternatives using a priority setting approach and making a final, *consensus* decision.

- Develop an implementation action plan, which includes a time schedule for implementation. Everyone who is involved in carrying out the solution should be consulted and should approve the plan. The plan should answer the basic questions what, why, where, when, and how. Avoid the who.
- Establish the method for measuring the success of the proposed solution. Remember, the criterion for success was established as part of the problem statement in Step 1.
- Gain management approval, if necessary for implementation.
- Implement the action plan. (Assign a responsible person for the action plan and make sure there is a date of completion for the task.)

Step 4: Verify Correction

Verify the correction using the criterion for success established in Step 1 and the measuring the method established in Step 3.

- Measure the impact of the solution.
- Make a decision. Is the problem solved to a level compatible with the team's goal? Have any unforeseen new problems been created by the solution?
- If the correction is unsatisfactory, the team must go back to an earlier step, selecting an alternative solution or possibly even selecting an alternative root cause, and then proceeding.
- If the correction is satisfactory, any temporary protective fix should be removed.

Step 5: Prevent Recurrence of the Problem

This is an often-overlooked aspect of problem solving, the benefits of which cannot be overstated.

- Develop innovative methods for mistake proofing the system (i.e., Japanese poka yoke).
- Modify systems, FMEAs, instructions, and procedures to reflect the new, optimal practices.
- Train all involved personnel in the new methods.
- Publicize the results. Apply the knowledge gained to the rest of the product line and to other company activities with similar conditions.
- Continue to monitor the problem.

- Recognize the team's success.
- Prepare a final project report, which describes the problem, the methods used to correct it, the quality, and the productivity gains achieved. In addition, a final team report might be presented to management.

Recycle

Once the above five steps have been completed, the team should return to Step 1, select another significant problem for correction, and continue improvement efforts.

Tools for Problem Solving

In any problem-solving process, teams need to learn and apply a number of quality tools, that is, analytical techniques, to aid them in their working through the five steps of the problem-solving cycle. These tools are primarily used for collecting, analyzing, and understanding data or information. To be sure, numerous quality tools may be applied to problem solving. Some of these are very complex, advanced, mathematical techniques, while others are quite simple to learn and to apply. Table 13.2 shows an inventory list of 38 of the most commonly used tools. Appendix V also has many specific tools and their uses in any problem-solving endeavor.

If the problem-solving process demands an understanding of relationships, then the focus is on a slightly different approach to find the following: (1) What is the problem? (2) What can I do about it? (3) Put a star on the best plan to pursue. (4) Do the plan. (4) Did the plan work? To evaluate this situation, one may use the seven quality control (QC) management tools, which allow the team to study the relationships. The tools are

1. Affinity diagram
2. Relationship diagram
3. Tree diagram
4. Matrix chart
5. Matrix data analysis chart
6. Arrow diagram
7. Process decision program chart

TABLE 13.2

Inventory List of 38 of the Most Commonly Used Tools

Quality Tools Inventory	Basic 7	Selection	Cause	Correction	Verification	Prevention	Data
ANOVA		X	X			X	V
Audit		X			X		A
Brainstorming		X	X	X			A
Cause and effect	*	X	X				A
Check sheet	*	X	X		X		A
Control charts[a]	*	X	X		X	X	VA
Customer complaint log and tracking		X			X	X	A
Customer feedback survey		X			X	X	A
Design of experiments			X				VA
Evolutionary operation (EVOP)			X				V
Failure mode and effects analysis (FMEA)		X				X	A
Failure analysis		X	X				A
flow chart	*	X	X				A
Force field analysis		X					A
Goal setting		X					A
Group vote		X	X	X			VA
Histogram	*		X		X		V
Key characteristic selection			X				A
Measurement system evaluation (MSE)		X	X				V
Mind map		X					A
Mistake proofing						X	—
Nominal group technique		X	X	X			A
Pareto chart	*	X	X		X		A
Pilot run					X		—
Process capability study		X	X		X		VA
Process log/ events log		X	X		X	X	A
Process potential study		X	X		X		VA

TABLE 13.2 (Continued)

Inventory List of 38 of the Most Commonly Used Tools

Quality Tools Inventory	Basic 7	Selection	Cause	Correction	Verification	Prevention	Data
Process standardization or qualification						X	—
Cost of quality analysis		X			X		V
Quality function deployment (QFD)		X				X	—
Regression analysis			X				V
Response surface methodology (RSM)			X				V
Scatter plot	*		X				V
SPC		X	X	X	X	X	VA
Stratification			X				VA
Taguchi methods			X				V
Team meeting guidelines		X	X	X	X	X	—
Test of hypothesis			X				VA

Note: V, variable data; A, attribute data. For each tool listed, each step of the problem-solving cycle in which that tool can be utilized is indicated. For example, the Check Sheet may be used in Step 1: Problem Selection, Step 2: Root-Cause Analysis, and Step 4: Verification of Correction.

[a] Control charts = Variable charts, attribute charts, (run charts, multivariable charts, box plots, precontrol and so on).

*These tools (1) can be easily learned and used by everyone in the organization, (2) are very effective in achieving basic problem-solving success, and (3) are essential to any properly designed improvement strategy. Basic instructions describing these tools and their use may be found in any SPC book or basic statistics book as well as in the *Memory Jogger*.

Process Improvement Cycle

There is a huge difference between problem-solving methodology and process improvement. A problem-solving methodology is a systematic approach to solving a particular problem. It assumes that there is an issue, concern, or problem with which the customer (internal or external) is not pleased and it is the responsibility of whoever has that issue, concern, or problem to fix it. On the other hand, process improvement deals with the insatiable desire to make the process better than what it is now. However, it is imperative that the process improvement be systematic and structured, as well. The reason for this is because it

- Provides a more tightly focused process than the problem-solving cycle (PSC), concentrating on improving the quality of a single process output (product or service)
- Lends itself especially well to customer issues because it builds and strengthens ties between customer and supplier
- Provides a catalyst for never-ending improvement through its continuous refocusing on customer needs and expectations
- Affords a prime opportunity to utilize statistical process control (SPC) methods for process analysis, control, and improvement
- Focuses improvement efforts on both product quality and waste reduction
- Provides improvement teams with an enhanced motivational environment for their improvement efforts

So, what is this structured process for improvement? Fundamentally, it is a 14-step process, shown in Figure 13.4. A clarification and requirements of each step follows.

Step 1. Identify the Output

- Prepare a statement that answers the question, "What is to be done?" and consists of two parts: a tangible, specific noun followed by a verb describing what you do to produce the output.
- The statement should be neither too broad nor too specific, if it is to be useful. Examples: shaft machined, doughnut fried, muffler installed, bicycle assembled, forecast developed, order taken, machine repaired, order shipped, report typed, and so on.

Step 2. Identify the Customer

- The (primary) customer is the next person in the work process, the person who will receive your output as input and act on it.
- The customer may be either external or internal.
- Secondary customers and end-users may be identified if their requirements are of significance.

Step 3. Identify the Customer Requirements

- What does the customer want, need, or expect of the output?
- Customer requirements may be general or very specific.

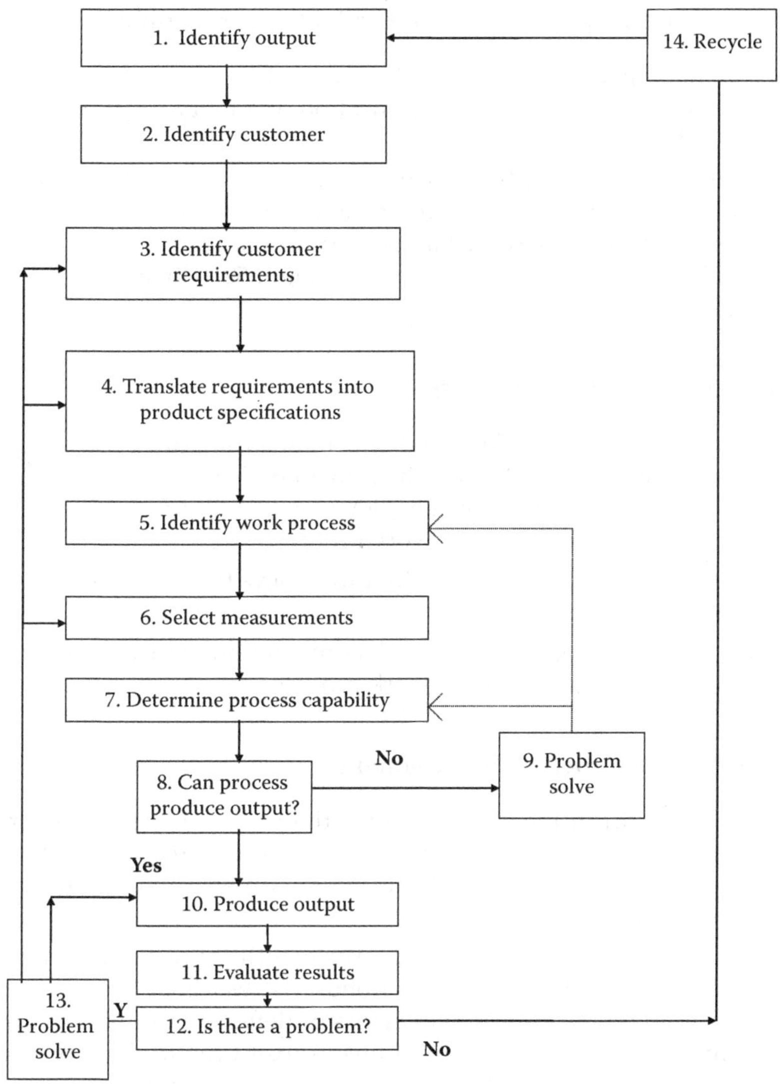

FIGURE 13.4
The 14-step process improvement cycle model.

- Customer requirements often fall into such categories as cost, accuracy, timeliness, quantity, completeness, dimension, performance, appearance, and so on.
- Supplier-customer interaction is essential. Both customer and supplier specify, negotiate, and evaluate information before reaching an agreement. Examples: report needed ASAP, expense cannot exceed the budget, heat shield must withstand operating temperatures, colors must match, quick response time, and so on.

Step 4. Translate Requirements into Product Specifications

- Customers' needs are put into language that reflects the supplier's active participation. (A good application of quality function deployment [QFD].)
- Specifications should be measurable, realistic, and achievable, and should be reviewed with the customer to be certain both parties understand and agree on what the output should be. Examples: need by 4:00 pm Friday, OD 1.625 +/–.005", weight 5200 lb maximum, type double-spaced, and so on.

Step 5. Identify the Work Process

- What are the process elements? A fishbone diagram used as a process analysis map (PAM) may help to identify the machines (equipment), measurement, labor (people), material (input), methods, and environmental factors in the work process.
- List step-by-step what must be done to achieve the output. Flowcharts are especially helpful here.
- Identify or develop the needed documentation, which describes or provides instructions for the work process.

Step 6. Select the Process Measurements

- Select the critical measurements for the *process output*. These measurements should be derived from customer requirements and product specifications and should permit an objective evaluation of the quality of output.
- Measurements should provide early identification of potential as well as actual problems, where emphasis (stress) is on the prevention rather than on the detection of errors or defects. Examples: cylinder OD in inches, cake weight in grams, transit time in days, response time in seconds, number of cycles before failure, circuit resistance in ohms, and so on.
- Also, select measurements for key *process inputs*, such as raw materials, incoming parts, energy, supplies, information, and so on. *In-process factors* (intermediate process outputs) include in-process conditions that impact the process output, such as temperature, pressure, time, and so on. *Process performance criteria* include scrap, rework, downtime, yield, delays, efficiency, and so on.

These measurements become the basis for process analysis, control, and improvement.

Step 7. Determine the Process Capability

- This step helps determine if the existing process can produce output that consistently meets the agreed on product specifications with minimal waste.
- The process is operated according to standard practices; output measurement data are collected and analyzed to provide a statistical evaluation of process control (consistency) and process capability (ability to meet specifications).
- This step is an important activity of SPC.

Step 8. Decide If the Process Is Capable

- If the evaluation indicates that specifications are *not* being met consistently and that excessive waste (scrap, rework, delays, errors, etc.) is experienced, the work process must be revamped or improved. This requires problem solving; proceed to Step 9.
- If the process capability assessment of Step 7 determines that the existing process *can* produce output that consistently meets specification requirements and does so with minimal waste, proceed to Step 10 to produce the output.

Step 9. Problem Solved to Improve the Work Process

- Use the 5-step problem-solving cycle (PSC) to revamp or improve the existing process.
- If the process capability problem indicated in Step 7 is a lack of process control (instability or inconsistency), indicated by sporadic deviations of the process average from the target, status quo, or historic level, the solution is one of identifying the destabilizing (special/assignable) causes and eliminating them. After the *special cause* has been identified, corrected, and is completed, loop back to Steps 7 and 8 (reassess process control). If the problem is solved, proceed to Step 10; if not, continue to repeat the PSC at Step 9 as needed, until process control is achieved.
- If the process capability problem indicated in Step 7 is an *inherent* inability to meet specification requirements, the process system must be fundamentally changed. This requires identifying those key process factors (common or random causes) that have a major impact on the quality of process output and modifying them and their effects. After *common cause* has been identified, corrected, and is completed, loop back to Step 5 to redefine the revised work process and proceed. If at Steps 7 and 8 the process capability problem is solved, proceed to Step 10; if not, continue to repeat the PSC at Step 9 as needed, until

process capability is achieved. (A very good methodology to use for special cause is the DMAIC model of the Six Sigma methodology.)

- If repeated attempts still leave you with an incapable process, talk with the customer, who needs to know what to expect. A renegotiation of the requirements may be possible. If not, the customer may want to find another supplier.

Step 10. Produce the Output

- Produce the output, continuing to follow the standard practices (or the newly established practices, which should be standardized for the new, revamped process).

Step 11. Evaluate the Results

- This step is an evaluation of the *results*. Evaluation of results must be based on the product specifications on which the customer and the supplier agreed as part of Step 4. Those specifications are a *template* against which results are compared.
- There is a major difference between this step and Step 7 (Determine Process Capability). Here you are evaluating how well you actually *did* (not how well you are *capable* of doing). The emphasis is on results rather than on process.
- There are two potential types of problems at this point: (1) the product output does not meet specification requirements, and (2) the product output meets specification requirements, but the customer is dissatisfied with the product.

Step 12. Decide If There Is a Problem

- If the evaluation of Step 11 indicates that there is a problem of either Type I or Type II, problem solving must be carried out. Proceed to Step 13.
- If no problem exists, continue production and proceed to Step 14.

Step 13. Solve the Problem to Improve Results

- Use the 5-step PSC to *troubleshoot* the problem.
- If the problem with the process output is of Type I, that is, the process output does not meet specification requirements, the cause of the problem is one of nonconformance to standard practices (since the process was shown to be capable in Step 7). The solution to this type of problem lies in identifying the nonconforming practices and

restoring operations according to established standard practices at Step 10.

- If the problem with the process output is of Type II, that is, the product output meets specification requirements, but the customer is dissatisfied with the product, the cause of the problem is usually one of (a) improper understanding of customer requirements, (b) improper translation of requirements into specifications, or (c) improper process monitoring (measurement). Any of these causes requires looping back to Steps 3, 4, or 6 for corrective action.

Step 14. Recycle

- At this point, there should be no evident problems of process capability or results.
- However, there may be opportunities for further quality improvement of the process output, for example, in terms of reduced variability. This could result in not merely meeting, but exceeding customer expectations.
- Of course, customer needs and expectations are rarely static. They are likely to change, especially in terms of becoming more restrictive.
- These opportunities for continuous process improvement require recycling back into the PDCA model.
- Under any circumstances, continue to monitor the work process and the results in order to maintain the required level of quality.

Comments on Problem Solving versus Process Improvement

One will notice that both the 5-step PSC and the 14-step PIC are proven, systematic approaches for continuous improvement. Both are extremely well suited for team-oriented activity. However, there is always confusion as to what method to use. To be sure, it is not always obvious which of the two methods to use to tackle a particular work issue. It takes some skill and experience to make the better selection. The PSC/PIC selection chart in Table 13.3 compares the two methods in detail and provides help for making this decision. Fortunately, no matter which method you select, once you start using it you are likely to learn quickly whether it is helping you to accomplish your objective.

In general, use the PIC when you need to improve the quality of a particular, currently existing output or you are about to produce a new output. Use the PSC when there is a gap between what is currently happening and what you want to happen. Note that the PSC is an integral and vital part of the PIC (at Steps 9 and 13).

TABLE 13.3

Problem Solving versus Process Improvement Selection Chart

Problem Solving Cycle	Process Improvement Cycle
A general method for making a change in	A tightly focused method for ensuring conformance of
Systems	A specific product
Results	A specific service
Conditions	
Work process	
Management process	
This method fosters	This method fosters
Definition of problems	Elimination of unneeded work
Analysis of data	Prevention of problems
Understanding of causes	Shared responsibility
Creative ideas	Strong customer/supplier communication lines
More alternatives	Evaluation of work processes
Teamwork	Confidence in results
Commitment	
Use this method when	Use this method when
There is a gap between what is happening and what you want to happen	You need to improve the quality of a particular, currently existing output
You want to move from a vague dissatisfaction to a solvable, clearly defined problem	You do not have agreed on customer requirements for an output
You are not sure how to approach an issue	You are about to produce a new output, the need for which has recently been determined
Switch to the PIC when	Switch to the PSC when
The problem you identify is a lack of quality or an inability to assess quality	Evaluation of process capability shows that the current work process cannot produce a conforming output
The recommended solution involves producing a specific output	Evaluation of results shows that the work process did not produce a quality output

We mentioned that both approaches are well suited for team-oriented activities. The question then becomes why is it necessary to use teams? When we deal with problems, all of us are better than any one of us. In fact, if we look at the word *team* as an acronym, we will find the answer. TEAM = **T**ogether **E**veryone **A**ccomplishes **M**ore. This is the essence of *synergy* and *consensus*. Specifically, management has recognized that the goal of any employee involvement (EI) is to give people increased influence over and responsibility for their own work. In recent years, one of the major routes to achieving this goal has been the creation of problem-solving teams: groups of employees who meet on a regular basis to identify and solve work-related problems. Organizations are indeed realizing that they have a vast, untapped potential in their employees' minds—especially for their own processes. To remain competitive, it is essential to make effective use of all the talents that our labor force has to offer, both physical and mental. Successful companies worldwide credit much of their success to the widespread use of employee teams to generate improvements in all departments and at all levels.

Quality improvement is hard work and takes a long time. It is all too easy for one person's commitment and enthusiasm to flag during a long project. The synergy that comes from people working together toward a common objective is usually enough to sustain the enthusiasm and support. Therefore,

- We need to provide employees with more challenging activities to keep their active minds fully utilized. Most employees want to participate in the decision making and problem solving that affect them. All of the people affected by a problem should be a part of the solution. This assures buy-in to the corrective action or solution put into place.
- We must recognize that no *one* individual has all the process and product knowledge plus the special skills and experience required for optimal problem solving.
- The most effective, proven problem-solving and process improvement methods and tools lend themselves best to a team approach.
- A team will invariably generate more problems, more causes, and more solutions than any individual can.

As important as the teams are, we must also recognize that there are two types of teams. The first is the local and the second is the cross-functional team.

Local or department improvement teams (DITs) are comprised of all the members of a department. These employees typically work in close proximity, experience common problems, and form a natural work group. Their purpose is to provide a focus and a means for all employees to contribute to an ongoing activity aimed at improving the quality and productivity levels of the department.

On the other hand, cross-functional or process improvement teams (PITs) are created to continuously improve quality, reduce waste, and improve productivity of a process that crosses many departmental lines. The PIT is made up of experienced, skilled problem solvers from all departments involved in and affected by the process. A typical PIT is the Six Sigma team that addresses major problems for the organization.

For either team to be effective, they require at least the following:

- A team charter, which specifies the team's purpose and enumerates the team's duties and responsibilities
- Selection of the proper team makeup

- Selection of an effective team leader and team coordinator
- Knowledge of the organization's mission, goals, and objectives
- Learning to work together as a team (team building)
- Adequate training in problem-solving and process improvement methods and tools
- Guidelines and ground rules for holding effective meetings and for making decisions
- An adequate meeting place with needed support services, such as typing, copying, information, and so on
- Use of meeting agendas and minutes (recaps)
- Liberal credit and recognition for team successes

So, what does it take to solve a problem? We believe that two essential attributes must be present. The first is an appropriate problem-solving methodology that supports the organizational goals (preferably standardized to reflect organizational needs) and the second is an appropriate team (members must have appropriate skills, ownership, and knowledge) for the task. The team must be part of an organizational culture that encourages

1. A climate of trust
2. A commitment to common goals
3. Honest and open communications
4. Common agreement on high expectations for the team
5. Assumed responsibility for work that must be done
6. A general feeling that one can influence what happens
7. Common access to information
8. A win-win approach to conflict management
9. Support for decisions that are made
10. A focus on process as well as results

References

Barry, K., E. Domb, and M.S. Slocum. (2010). TRIZ—What is TRIZ? *The Triz Journal*. http://www.triz-journal.com/archives/what_is_triz/. Retrieved July 25, 2014.

Blanchard-Fields, F. (2007). Everyday problem solving and emotion: An adult developmental perspective. *Current Directions in Psychological Science* 16(1): 26–31.

Bransford, J. and B. Stein. (1993). *The Ideal Problem Solver: A Guide to Improving Thinking, Learning, and Creativity*, 2nd ed. Duffield, UK: Worth Publishers.

Hua, Z., J. Yang, S. Coulibaly, and B. Zhang. (2006). Integration TRIZ with problem-solving tools: A literature review from 1995 to 2006. *International Journal of Business Innovation and Research* 1(1–2): 111–128.
Offord, P. (2011). *RPR: A Problem Diagnosis Method for IT Professionals*. Dunmow, UK: Advance Seven Limited.
Osinga, F. (2006). *Science Strategy and War, the Strategic Theory of John Boyd*. Abingdon, UK: Routledge.
Polya, G. (1945). *How to Solve It*. Princeton, NJ: Princeton University Press.
Sheng, I.L.S. and T. Kok-Soo. (2010). Eco-efficient product design using theory of inventive problem solving (TRIZ) principles. *American Journal of Applied Sciences* 7(6): 852–858.

Selected Bibliography

Novak, S. (2006). *The Small Manufacturer's Toolkit*. New York: Auerbach Publications.
QIP/PO Systems. (1991). *Improvement Tools*. Miamisburg, OH: QIP/PO Systems.
Stamatis, D. (2003). *Six Sigma and Beyond: Problem Solving*. Vol. II. Boca Raton, FL: St. Lucie Press.

14

Internal Audits

Overview

The general definition of an *audit* is a planned and documented activity performed by qualified personnel to determine by investigation, examination, or evaluation of objective evidence the adequacy and compliance with established procedures, or applicable documents, and the effectiveness of implementation (Stamatis 2002). The term may refer to audits in accounting, internal controls, quality management, project management, water management, safety, energy conservation, and so on.

On the other hand, *auditing* is defined as a systematic and independent examination of data, statements, records, operations, and performances (financial or otherwise) of an enterprise for a stated purpose. In any auditing, the auditor (the person who does the audit) perceives and recognizes the propositions before him or her for examination, collects evidence, evaluates the same, and on this basis formulates his or her judgment, which is communicated through an audit report. The purpose is then to give an opinion on the adequacy of controls (financial and otherwise) within an audited environment, and to evaluate and improve the effectiveness of risk management, control, and governance processes. Generally, it is system oriented.

Specifically, *quality audits* are performed to verify conformance to standards through review of objective evidence. A system of quality audits may verify the effectiveness of a quality management system. This is part of certifications such as ISO 9001. Quality audits are essential to verify the existence of objective evidence showing conformance to required processes, to assess how successfully processes have been implemented, and to judge the effectiveness of achieving any defined target levels. Quality audits are also necessary to provide evidence concerning reduction and elimination of problem areas, and they are a hands-on management tool for achieving continual improvement in an organization (Stamatis 1996, 2004).

To benefit the organization, quality auditing should not only report nonconformance and corrective actions, but also highlight areas of good practice and provide evidence of conformance. In this way, other departments may share information and amend their working practices as a result, also enhancing continual improvement (Stamatis 2012).

Types of Audits

There are many types of audits for different industries. However, all audits are divided into three categories. They are

1. First-party audit. An organization audits itself. The objectives usually are to validate
   a. A process: the process may be an activity that adds value to the product or service. Identify any gaps of what the process must do as opposed to what it is doing.
   b. Project implementation activities based on past deficiencies or current action plans for nonconformities.
   c. Product characteristics or performance requirements.
   d. Nonconformities have been addressed.
   e. Training, equipment capabilities, and process settings.
2. Second-party audit. A customer audits the supplier. The objectives are to validate
   a. Supplier organization processes used to provide a product or service.
   b. Supplier product or service characteristics or performance requirements.
   c. Supplier process capabilities, including capacity.
   d. Conformance to contract requirements.
   e. Material sources and traceability.
   f. Defects and nonconformities have been addressed.
   g. A QOS exists, is reviewed, and implemented.
   h. An internal audit is conducted and appropriate corrective actions are taken for any deficiencies.
3. Third-party audit. An independent certification body conducts the audit. This type of audit is certified usually for three years. At the end of this time frame, it needs to be conducted again. Some certified bodies require an external audit to be conducted once a year. The objectives are to approve/disapprove
   a. Process for certification.
   b. Product or service for certification.

In all three categories of auditing, it is imperative to identify the scope of the audit up front. The second imperative of all audits is the need to understand the process and product/service the auditor is going to audit. Reviewing the procedure, specifications, and records is a good starting point. If there is no procedure, you may need to ask the auditee to provide a description of the process or processes (a process flowchart may help). Ask questions and talk to people to get the information you need.

Several tools are available to help you to understand the process. They include

- Process flow diagrams, flowcharts, or process mapping
- Cause and effect diagrams, turtle diagrams, 3 × 5 Whys
- Tree or functional diagrams
- Failure mode and effect analysis (FMEA), design failure mode and effect analysis (DFMEA), and process failure mode and effect analysis (PFMEA) results
- Training documents (employee matrix)
- Inspection checklists
- Procedures and instructions
- Bill of materials, quantities, and specifications
- Relevant standards

In addition to the tools, the auditor must not underestimate the value of understanding the process input and output criteria for good performance. Appropriate criteria may include

- Specifications, lists
- Drawings, pictures, diagrams
- Planned arrangements for process approval
- Approved equipment and qualification/certification of personnel
- Test procedures
- Inspection methods (sampling, frequency)
- First article inspections
- Contract or regulatory requirements

In any audit, regardless of type, the history of the process plays a major role in identifying trends and gaps in corrective actions. Therefore, the auditor must review at least the following as part of his or her planning:

- Nonconformance reports and trend analysis
- Internal and field failures (how they happen and what was done)
- Corrective actions (check for appropriateness and timeliness)

- Process/product changes, date, and nature of changes (how change was handled)
- Operator/technician changes (are these changes completed with appropriate personnel?)
- Revalidation history (review history for continual improvement in the organization)
- Customer complaints (How are they handled? Is there a standardized problem-solving methodology? Are they completed in a timely fashion?)

Performing an Audit

All audits have a base of comparison. Usually, they start with a standard and then add the specific requirements of the customer as well as the organizational standards included in procedures and instructions of the organization itself. Details of how to conduct audits may be found in Arter (2003), Arter et al. (2012), and Stamatis (1996). Generally, auditors collect information by using open-ended questions to gather data about process inputs, outputs, and the process elements (people, environment, equipment, material, measuring, and method). They report not only weaknesses but also strengths in the organization based on objective evidence. Objective evidence is considered to be written documentation, personal interviews with more than one person, documented observation with witnesses, or a combination of these. An example of a typical preparation for an audit may be found in Appendix IV. The example is not an exhaustive audit, but hopefully the reader will see the flow of some requirements, the criteria for the requirements, and the items for which the auditor should look.

If the audit is conducted to validate conformance to a standard such as the ISO/TS 16949, then the auditor compares what he or she finds in the organization against the elements or clause of the standard. This same strategy is used in product/service audits to determine if the product conforms to specified requirements that are in a clause or element of a standard, specification, or condition document (this is very important if the customer has specific requirements). A product audit can include validation of product characteristics as well as performance requirements.

Verification audits are generally conducted to validate error-proofing actions for specific processes and are generally conducted in reverse order because the verifiable links are more obvious in the corrected process. In the auditing world, when the focus is on quality management audits, there are four approaches to identify gaps, correct (close the gaps), and monitor the improvements. They are

1. Product audit
2. Process audit
3. Layered audit
4. Error proofing

Product Audit

Product audit is a planned and systematic study of finished products prior to delivery, which must be regarded as a normal part of the quality assurance activities. In some cases, the customer may initiate such a product audit to be carried out jointly with the supplier such as in situations of a start-up of just-in-time deliveries.

The audit is to be performed with the help of written instructions and checklists, which have been drawn up by the supplier. Instructions should be based on the technical specification and other valid specified requirements.

A product audit has to be performed from the customer's perspective or, as commonly known, *with the eyes from the customer*. Measures must be duly taken whenever deviations have been discovered or when a negative trend is observed. Alarm and action limits must be set.

Sometimes, a product audit is called a *source inspection audit*. The differences are minor, if any, and most professionals in the quality field use them interchangeably. In any case, for the ones that see some differences, here is an overview.

Source inspection audit is the inspection of purchased products and process records at the supplier's premises to verify compliance with requirements before shipment to the client. It examines the potential customer's purchased product that is performed at the supplier's facility to verify product integrity and conformance to specified requirements prior to delivery. It involves the process of examination on a finished product before shipment in order to test the conformance of subassemblies, which may be obscured and thus impossible to inspect subsequently as the assembly is built up. In addition, it serves as the first inspection of a production piece that checks for compliance with engineering drawings. Typical applications are in tests castings, forgings, and machine parts to prove tooling before embarking on production runs, and electronic assemblies to prove the effectiveness of the production process before committing to order quantities. Precap inspection typically takes place prior to sealing or encapsulating high-reliability microelectronic components in cases where compliance with military, space, or international specifications is critical. A source inspection audit should be used when

- Companies and organizations see the need to examine a customer-purchased product at the supplier's facility with the intention of verifying product integrity and conformance to specified requirements prior to delivery.

- Companies and organizations need to inspect a finished product immediately before shipment.
- Companies and organizations need to examine and ensure that the conformance of subassemblies, which may be obscured and impossible to inspect, subsequently are thoroughly examined.
- Organizations need to inspect a first production piece to check for compliance with engineering drawings.
- Organizations intend to inspect castings, forgings, and machined parts in order to prove tooling before embarking on production runs.

Some of the benefits of source inspection audit are

- Enables the reviewing of the supplier's processing records to establish all plant manufacturing, inspection, and test operations have been carried out and properly recorded
- Enables the checking for evidence that any measuring equipment used in prior inspections or tests was under a system of calibrations control and within calibration at the time the inspections or tests were performed
- Ensures the verification of the product complies with pertinent engineering drawings
- Ensures that product identification marking or labeling requirements set out in drawings or associated specifications are properly observed

The reader should be very cognizant of the fact that product or source inspection audit is not very dependable and should be avoided whenever possible. The integrity of the product quality should be measured with statistical approaches including statistical process control (SPC). In any case, both of these approaches do have a business impact, which

- Determines the status of any related nonconformance report, waivers, or deviations, and how these bear upon the acceptability of the product
- Ensures that products inspected in the process conform to assembly drawings to the appropriate degree for the stage of manufacturing at which they are inspected

Process Audit

Process audit is a planned and systematic examination of the equipment, procedures, and activities according to the documented operations with respect to capability, application, and effectiveness. As a preparation, an audit plan should be drawn up incorporating the organization of the audit, the participants, the extent, and the basic information. The extent of the audit can be identified, for example, by

- Production area
- Department
- Workshop (floor) area
- Production line
- Special process

On the other hand, the basis for the audit can consist of

- Product specifications
- Process descriptions
- Written procedures
- Questionnaire/checklists
- Standards (international and/or specific customer requirements)

A process audit is generally used by customers to evaluate suppliers. However, the strength of the process audit is in the fact that organizations can use this type of audit to find nonconformances within their processes and correct them.

Arter (2003) has identified seven steps in the process approach to auditing. They are

1. Define the products
2. Define the processes used to make those products
3. Analyze and understand those processes
4. Develop objective evidence needed to explore the processes
5. Gather field data during the actual audit
6. Analyze the gathered data by sorting into piles
7. Present the information to the process owners

In order to use the process approach, the auditor must first understand how the business processes relate to the objectives of the organization. Then data are gathered to see how these processes are being controlled. Finally, conclusions must show how the identified strengths and weaknesses affect the operation of the business. A typical partial process audit form is shown in Figure 14.1.

In Figure 14.1, notice that the process audit uses questions to make sure that the procedures and processes that are agreed on during a selection audit

Process Audit										
Department Audited:						**Date:**				
Auditor's name:						**Shift:**				
***Note*: All questions are to be answered with a "Y" or "N" for yes and no respectively under the corresponding day's box. All nonconformances *must* have corrective action recorded under the specified column in this form with date** (MM/DD/YY).										
Item No.	**These Items Must Be Checked in Every Audit**	**Audit Area**	**M**	**T**	**W**	**T**	**F**	**S**	**S**	**Corrective Action**
Section 1: Operator Instructions										
1.	Are operator instructions accessible at the workstations and made available without disruption to the work being performed?									
2.	Do operators understand their work instructions and does the work being performed follow the work instructions step by step?									
3.	Do all operators performing containment activities have work instructions and have they been trained to the work instructions?									
4.	Do all operators have a clear understanding of their jobs? Do they know the right thing to do, are they doing it correctly, and can they demonstrate?									
5.	Are operators filling out required forms and documentation and turning them into their supervisors per their work instructions?									
Section 2: Product Identification, Housekeeping, and Workplace Organization										
1.	Is all product in the work area properly identified with product status clearly shown per work instructions and procedures (identify the actual number)?									
2.	Are all containment, quarantine, and product staging areas physically identified?									
3.	Are all work areas maintained in a state of order, well organized with good housekeeping?									
4.	Is suspect product properly identified per work instructions and procedures (identify the actual number)?									

FIGURE 14.1
Partial process audit.

process are followed. The complete document must identify nonconformances in the manufacturing process, engineering change process, invoicing process, quality process, and the supplier/shipment process. These audits are analyses that (1) identify internal nonconformances to standard, procedures, instructions, and (2) are done to document the relationship between different companies in order to verify compliance of a supplier's products and processes.

A process audit should be conducted when

- Companies need a supplier audit to verify their compliance to their supplier's products or processes
- Companies need to improve their return on auditing (ROA)
- Companies want to manage their supply quality management system so they know that their suppliers are in compliance
- Companies struggle with reallocating their resources such that they cannot keep current projects on schedule while their engineers are out in the field conducting supplier audits
- Businesses need to conduct mutual and beneficial audits for their company and suppliers
- Companies or businesses want to create an integrated auditing process aligned with internal goals and supplier commitments
- Companies require flexibility and consistently structured supplier audit services so that they can execute a value-adding audit program
- Companies want to improve their quality level

Some of the benefits of this type of audit are

- Recognition of implementing an overall quality system
- Recognition of establishing documentation control
- Recognition of the need to control business operations and records
- Recognition of establishing a measuring system for customer satisfaction
- Implementation of a business management review for key processes and indicators
- Implementation of communication policies to correct nonconforming products
- Implementing corrective actions in a timely manner

Just like everything else that is done in any organization, the process audits have business impacts. Some of them are

- Information and product improvements can be shared for common gain among departments, divisions, and different geographical facilities.

- Accomplishments of organizational measurement that can take place, risk mitigation exercises both proactive and reactive.
- When there is an existing stand-alone supplier quality management function, strong relationships with suppliers can be created from a customer's perspective.
- Knowledge can be integrated with deliverance of profit-generating opportunities for both organizations and the exploration of additional, above and beyond, contrasting business opportunities. This can be achieved through creation of a supplier quality management function for all those involved such as account managers, supply chain consultants, and supplier performance managers. (This is what supplier development is all about!)

Layered Audit

Layered process auditing (LPA) is an increasingly popular quality tool developed for manufacturing management, especially in the automotive industry. When utilized properly, LPA will drive cultural change throughout an organization to improve quality, reduce scrap and rework, and reduce customer rejections. However, what exactly are LPAs?

LPAs represent a formalized system focusing time and energy toward insuring high-risk processes and error-proofing devices do not exhibit a lot of variation and are working properly. An LPA system is comprised of three critical elements:

1. A set of audits focused on high-risk processes. An example of an audit form for high-risk processes is shown in Figure 14.2.
2. Layers of auditors from all areas of management who perform audits. A typical format for this type of a layered audit is shown in Figure 14.3.
3. A system of reporting and follow up to ensure containment as needed and drive Improvement. A typical form of a layered audit is shown in Figure 14.4, indicating the documentation and progress from shift to quarterly improvement.

AIAG's document CQI-8 provides the details of a generic layered audit. Fundamentally, the CQI-8 describes that the audit may be scheduled on a daily, weekly, monthly, or quarterly basis to make sure that the critical to quality (CTQ) characteristics are met from the customer's perspective. No matter what the frequency, each audit should last anywhere from 10 to 30 min and they should have the appropriate level of management. Of course, each level of management will be investigating/evaluating the items that are within that

Part name:							
Part Number:				**Shift:**			
Customer:				**Week Ending:**			
Department:				**Other:**			
Operation 10	Broach	Mon	Tue	Wed	Thu	Fr	Sat
	Is runout within specs?						
	Has perpendicularity been gauged? Spec is 20 lines.						
	Is there any broach tear? Check one part per broach.						
Operation 20	Olofssons						
	Is the ID on size?						
	Is the ID runout within specifications?						
	Is the ID taper within specifications?						
Operation 30	Hones						
	Has the ID-micro been checked and the micro log completed?						
Entire department	Are operators' verification sheets completed daily by all shifts?						
	Are the tagging procedures being followed?						
	Are there any temporary alert notices posted that have expired?						
Operation A	Assembly						
	Are the builders checking rotation and marking parts (touch point)?						
	Have the 8-hole sprags been inspected for the presence of 8 holes and identified with an orange marker?						
	Are bar-coding labels attached to shipping container in correct location?						
	Are shipping containers checked for correct count?						
	Machining operations total nonconforming items						
	Assembly total nonconforming items						
List any quality or manufacturing concerns in any operation:							

FIGURE 14.2
Typical format for layer audit focusing on high-risk processes.

Typical Layered Format					
Level	**Assigned Management Category**	**Assigned Management Personnel**	**Audit Assignment**	**Process A**	**Etc.**
First layer of management	Supervisors	Process A manager	Own department	1 per day	1 per day
		Process B manager	Own department	1 per day	1 per day
		Assembly supervising manager	Own department	1 per day	1 per day
		Quality supervising manager	Own department	1 per day	1 per day
Second layer of management	Plant staff and plant manager	Quality manager	Rotate departments		
		Maintenance manager	Rotate departments		
		HR manager	Rotate departments		
		Six Sigma/Lean training coordinator	Rotate departments		
		Assistant plant manager	Rotate departments		
		Assembly manager	Rotate departments		
		Process A manager	Rotate departments		
		Process B manager	Rotate departments		
		Shipping manager	Rotate departments		
Third layer of management	Corporate Management	Vice president of quality	Rotate departments		
		Vice president of operations	Rotate departments		
		Systems engineer	Rotate departments		

FIGURE 14.3
Typical form of management layered audit. (*Note*: If there is no supervisor, then the area manager shall be considered the firstlayer. Of course, other management representatives may participate as appropriate.)

layer's span of control. The higher the layer, the less tactical and more strategic the items reviewed.

It is imperative to understand that all layered audits must focus on areas in processes where deviation (variation) may cause a product to be a high risk for the customer, producing organization, or both.

Item	Shift		Daily		Weekly			Monthly			Quarterly		
	Op	Insp	TLead	Sup	P Mng	QA Mng	OP Mng	EX Mng	Cor Mng	P Mng (optional)	Pres	P Mng	QA Mng
Part/Product													
Error-proofing verification													
First piece inspection													
Last piece inspection													
Standard work instruction present													
Operator training tracking sheet													
Safety issues													
Process													
Setup													
SPC compliance													
Tooling approval (Sign-off)													
Quality gate data													
System													
Preventive maintenance													
Calibration													
Lot traceability													
Housekeeping													
Voice of customer													
6-panel posted													
Action plans updated and on time													
Customer delivery performance													
Other relative info													

FIGURE 14.4
Typical layered audit indicating documentation and progress from shift to quarterly improvement. (*Note*: Op = operator; Insp = QA inspector; TLead = team leader; Sup = supervisor; P Mng = plant manager; QA Mng = quality manager; OP Mng = peration manager; EX. Mng = executive manager; Cor Mng = corporate manager; Pres = president.)

An LPA should be conducted when the organization wants to ensure high-risk processes and error-proofing devices do not exhibit a lot of variation and are working properly. Typical areas are

- Assembly area
- Manufacturing operations
- Shipping/receiving
- Repair/rework area
- All operations and other support functions

Some of the benefits of using an LPA are it

- Minimizes variation in processes and error-proofing systems and makes significant progress toward single-digit parts per million (ppm) and even meets the goal of zero defects to your customers
- Reduces variation in production
- Improves quality
- Improves and maintains discipline
- Reduces scrap and eliminates waste
- Reduces customer rejections
- Improves discipline and communication
- Increases employee participation
- Improves overall quality and cuts costs
- Stops production problems from becoming rejections

The business impact of LPA is reduction of variation that is waste, which translates into profitability. An LPA is an ongoing chain of simple verification checks that ensure a defined process is followed correctly. It is a powerful management tool that can improve safety, quality, and cost savings by amplifying problem-solving systems and making continuous improvement almost routine. Through observation, evaluation, and conversations on the manufacturing floor, these checks ensure key work steps are performed properly.

Furthermore, LPA interactions are also an excellent way for managers to show respect for frontline workers. Some essential characteristics of LPA that have a direct business impact are

- Audits are done on the Gemba (source). There are no filters to management.
- Management must take ownership of the LPA process.

- Auditors must identify and ask the right questions.
- All management layers, top to bottom, must participate.
- There is immediate containment of found nonconformances.
- Continual improvement must be included in the process.
- Audits are scheduled and performed per shift, daily, or weekly. Survey data shows excessive process input variation is the single largest cause of manufacturing quality problems. This often results from a failure to reinforce process corrective actions.

Error-Proofing Audit

Error-proofing audits are generally confirmation audits. This means a nonconformance was identified, a corrective action was recommended, and now the auditor checks whether the action taken was effective. These audits are simply an organized group of questions designed to examine a device or process after the process has been error proofed. Therefore, they focus only on areas in the manufacturing process where deviation represents a high risk for producing defective products. For an error-proofing audit to be effective, it must integrate action, analysis, and improvements. If an auditor finds a nonconformance while performing this type of an audit, that auditor not only should record his or her finding, but also take immediate initial corrective action to ensure defective products do not go out the door. Information about the finding should be recorded and readily available to management for later analysis. With a good system for recording and reporting audit information, the error-proofing audit records provide an excellent tool for troubleshooting problem areas and identifying other similar places within the organization for improvement. A typical partial form of an error-proofing audit is shown in Figure 14.5.

General Comments about the Audits

By far the most important item in any audit is the element of surprise. Observations (walk the line) allow auditors to observe employees completing specific tasks in the company's work environment. Auditors can best use this process by showing up unannounced and conducting an observational review of the company's operations. This allows auditors to avoid managers and employees who will operate under the company's procedures for the planned audit day. Auditors need a true and unfiltered audit observation to ensure that employees properly fill out paperwork and follow all operational standards.

Error-Proofing Audit										
Only qualified personnel can perform error-proofing verification audits										
Department Audited:							**Date:**			
Auditor's name:							**Shift:**			
Note: All questions are to be answered with a "Y" or "N" for yes and no, respectively, under the corresponding day's box. All nonconformances must have corrective action recorded under the specified column in this form with the date (MM/DD/YY).										
Item No.	**These Items Must Be Checked Every Shift**	**Audit Area**	**M**	**T**	**W**	**T**	**F**	**S**	**S**	**Corrective Action**
Section 1: General										
1.	Are operators aware of the function and importance of any error proofing devices at their station?									
2.	Are all error-proofing verification parts identified and maintained with due care?									
3.	Are all error-proofing devices verified per work instructions? Are the verification results recorded?									
4.	When error-proofing devices fail verification checks, are the proper employees notified? Is the contact documented? Is there any follow up?									
5.	Is there a reaction plan to direct operators when error-proofing devices are not functioning? That is, how will the known defect or potential defect be detected in order to protect the customer in the absence of automatic error detection?									
6.	Are all operators performing error-proofing verification checks trained to do so?									
Section 2: Work Cell Specific										
1.	This area of verification will be specific and unique for every work cell containing error proofing.									

FIGURE 14.5
Typical partial error-proofing layered audit.

Many organizations require employees to fill out documentation of their activities in the working environment. Auditors review these forms to see which employees are filling out paperwork and the times or dates of the paperwork. This holds employees accountable for their actions, especially operational managers who must uphold the company's safety or standard operating procedures. While it can be difficult to determine if the paperwork is fraudulent, combining this process with employee interviews can help in the review of this information.

Consistency is also important. Audits must be carried out on a regular basis. While layered process audits seem to use up human resources because a large number of employees have to engage in audits, they ultimately reduce company waste by identifying processes that are wasteful or a threat to the overall success of a project. However, the most valuable aspect especially for an LPA is the fact that it allows management to

- Verify critical processes for conformance to requirements
- Have a reason to be on the floor
- Ensure problem-solving solutions (corrective actions) stay in place and are monitored
- Identify problems before issue reaches the customer

Yet another item necessary for any audit is the checklist. During any audit preparation, auditors fill out checklists to acquire the data necessary to fully understand the processes within the company. Depending on the audit, many levels of the company must assess the audit checklist to ensure that the checklist is perfected. In addition, checklists should be clear and specific, so that different levels of management will come to the same conclusions when answering compliance checklists. Typical things to remember when constructing a checklist are

- Management must own the process, so ask appropriate and applicable questions.
- Identify and ask the right questions for the process being audited. The fact that management is involved establishes accountability and encourages operator feedback at the source (Gemba). Of course, the starting point should be the highest priority for either a quality issue or risk for the customer
- Participation spans all management layers. Make sure the questions are focused for the appropriate level. As the level increases, the questions become strategic and policy oriented.
- Nonconformances lead to immediate containment. Make sure that there is a corrective action for each nonconformance. Quarantine alone is not a complete action. A typical form for keeping track of nonconformances in an LPA is shown in Figure 14.6.

Layered audit results						
Items *not* in compliance						
Item	Number of Occurrences	M	T	W	T	F

FIGURE 14.6
Typical form for tracking nonconformances in an LPA.

- Include a process for continuous improvement. Make sure that continuous improvement is the result of your findings, not busy work.
- Schedule and perform audits regularly. This is a very important consideration, especially for the layered audit. Because the upper level management staff that conducts audits does not always have a lot of time on its hands, audits usually need to be quick and simple. The management occasionally observes employees to see if they are complying with policies and procedures. The upper-level management audits can be somewhat unpredictable in when they perform audits, but lower-level auditors typically audit the same series of items in their department every day. Typical participation may be as follows:
    - Plant manager once a week
    - Area managers once a week
    - Production supervisor once a shift

References

Arter, D. (2003). *Quality Audits for Improved Performance*, 3rd ed. Milwaukee, WI: Quality Press.

Arter, D., C. Cianfrani, and J. West. (2012). *How to Audit the Process-Based QMS*, 2nd ed. Milwaukee, WI: Quality Press.

Ishikawa, K. (1992). *Guide to Quality Control*. White Plains, NY: Quality Resources.

Stamatis, D. (1996). *Documenting and Auditing for ISO 9000 and QS 900*. Chicago, IL: Irwin Professional Publishing.

Stamatis, D. (2002). *Six Sigma and Beyond: The Implementation Process*, Vol. VII. Boca Raton, FL: St. Lucie Press.

Stamatis, D. (2004). *Integrating ISO 9001:2000 with ISO/TS 16949 and AS9100*. Milwaukee, WI: Quality Press.

Stamatis, D. (2012). *10 Essentials for High Performance Quality in the 21st Century*. Boca Raton, FL: CRC Press.

Selected Bibliography

http://www.ehow.com/list_6911857_layered-audit-procedures.html#ixzz2nrlKipy2.

http://www.ehow.com/list_6935021_layered-process-audit-checklist.html#ixzz2nriLnvux.

http://www.sustainingedge.com/sustain-with-layered-process-audits/#sthash.VEjrqiJX.dpuf.

Lakhal, L. (2014). The relationship between ISO 9000 certification, TQM practices and organizational performance. *Quality Management Journal*. 21(3): 38–48.

Mills, C. (1989). *The Quality Audit: A Management Evaluation Tool*. Milwaukee, WI: Quality Press.

Parsowith, B. (1996). *Fundamentals to Quality Auditing*. Milwaukee, WI: ASQ Quality Press.

Phillips, A. (2009). *ISO 9001:2008 Internal Audits Made Easy: Tools, Techniques, and Step-By-Step Guidelines for Successful Internal Audits*, 3rd ed. Milwaukee, WI: Quality Press.

Russell, J. (Ed.). (2000). *The Quality Audit Handbook*, 2nd ed. Milwaukee, WI: ASQ Quality Press.

15

Poka-Yoke

Overview

The term *poka-yoke* was applied by Shigeo Shingo in the 1960s to industrial processes designed to prevent human errors (Robertson 1997). Shingo redesigned a process in which factory workers would often forget to insert the required spring under one of the switch buttons while assembling a small switch. In the redesigned process, the worker would perform the task in two steps—first preparing the two required springs and placing them in a placeholder, then inserting the springs from the placeholder into the switch. When a spring remained in the placeholder, the workers knew that they had forgotten to insert it and could correct the mistake effortlessly (Shingo 1987).

Shingo distinguished between the concepts of inevitable human mistakes and defects in the production. Defects occur when the mistakes are allowed to reach the customer. The aim of poka-yoke is to design the process so that mistakes can be detected and corrected immediately, eliminating defects at the source.

Historical Perspective

Poka-yoke can be implemented at any step of a manufacturing process where something can go wrong or an error can be made (Quality Portal 2013). For example, a jig that holds pieces for processing might be modified to only allow pieces to be held in the correct orientation, or a digital counter might track the number of spot welds on each piece to ensure that the worker executes the correct number of welds (Shimbun 1988).

Shingo recognized three types of poka-yoke for detecting and preventing errors in a mass production system (Shingo and Dillon 1989, Quality Portal, 2013). They are

1. The contact method, which identifies product defects by testing the product's shape, size, color, or other physical attributes
2. The fixed-value (or constant number) method, which alerts the operator if a certain number of movements are not made
3. The motion-step (or sequence) method, which determines whether the prescribed steps of the process have been followed

One can see that from the three options, either the operator is alerted when a mistake is about to be made or the poka-yoke device actually prevents the mistake from being made. In Shingo's lexicon, the former implementation would be called a *warning* poka-yoke, while the latter would be referred to as a *control* poka-yoke (Shingo and Dillon, 1989).

Shingo argued that errors are inevitable in any manufacturing process, but that if appropriate poka-yokes are implemented, then mistakes can be caught quickly and be prevented from resulting in defects. By eliminating defects at the source, the cost of mistakes within a company is reduced. This is because defects are a form of waste. Waste is not value-added and therefore, if present, productivity and profitability are decreased. A methodic approach to build up poka-yoke countermeasures has been proposed by the applied problem-solving (APS) methodology (Fantin 2014), which consists of a three-step analysis of the risks to be managed:

1. Identification of the need
2. Identification of possible mistakes
3. Management of mistakes before satisfying the need

This approach can be used to emphasize the technical aspect of finding effective solutions during brainstorming sessions.

Strategy of Mistake Proofing

The strategy of mistake proofing is to establish a *team approach* to *mistake-proof systems* that will focus on both *internal* and *external* customer concerns. This will include quality indicators such as on line inspection and probe studies. Appendix III provides several items of interest in the area of signals, forms, and rationale of defects and mistakes. A strategy of mistake proofing involves the following:

- Concentrating on the things that can be changed rather than on the things that are perceived as having to be changed to improve process performance

- Developing the training required to prepare team members
- Involving all the appropriate people in the mistake-proof systems process
- Tracking quality improvements using in-plant and external data collection systems (before/after data)
- Developing a *core team* to administer the mistake-proof systems process. This core team will be responsible for tracking the status of the mistake-proof systems throughout the implementation stages.
- Creating a communication system for keeping plant management, the local union committee, and the joint quality committee informed of all progress as applicable
- Developing a process for sharing the information with all other departments and/or plants as applicable
- Establishing the *mission statement* for each team and *objectives* that will identify the philosophy of mistake-proof systems as a means to improve quality
- Developing timing for completion of each phase of the process
- Establishing cross-functional team involvement with your customers

As any strategy must have a mission, so does the mistake-proofing approach. A typical mission statement may read: To protect our customers by developing mistake-proofing systems that will detect or eliminate defects while continuing to pursue variation reduction within the process. Typical objectives to support the mission may be to

- Become more aware of quality issues that affect the customer
- Focus the efforts on eliminating these quality issues from the production process
- Expose the conditions that cause mistakes
- Understand source investigation and recognize its role in preventing defects
- Understand the concepts and principles that drive mistake prevention
- Recognize the three functional levels of mistake-proofing systems
- Be knowledgeable of the relationships between mistake-proofing systems devices and defects
- Recognize the key mistake-proof system devices
- Share the mistake-proof system knowledge with all other facilities within the organization

Why must we have a strategy? Because many things can and often go wrong in our ever-changing and increasingly complex work environment.

Opportunities for mistakes are plentiful and often lead to defective product. Defects are not only wasteful, but also result in customer dissatisfaction if not detected before shipment. The philosophy behind mistake-proof systems suggests that if we are going to be competitive and remain competitive in a world market, we cannot accept any number of defects as satisfactory!

In essence, not even one defect can be tolerated. The method of mistake-proof systems is simple for making this philosophy a daily practice. Simple concepts and methods are used to accomplish this objective.

Mistake-Proof Systems

Humans tend to be forgetful and as a result make mistakes. In a system where *blame is* practiced and people are held accountable for their mistakes and mistakes within the process, the worker is discouraged and the morale of the individual is low. However, the problem continues and remains unsolved. To remedy this, the mistake-proof system is implemented, which is a *technique for avoiding errors in the workplace.*

The concept of error-proof systems has been in existence for a long time, only we have not attempted to turn it into a formalized process. It has often been referred to as idiot proofing, goof proofing, fool proofing, and so on. These often have a negative connotation that appear to attack the intelligence of the individual involved, and therefore are not used in today's work environment. For this reason, we have selected the terms *mistake-proof systems.* The idea behind error-proof systems is *to reduce the opportunity for human error by taking over tasks that are repetitive, or actions that depend solely upon memory or attention.* With this approach, we allow the workers to maintain dignity and self-esteem without the negative connotation that the individual is an idiot, goof, or fool.

There are many reasons why mistakes happen! However, almost all of the mistakes can be prevented if we diligently expend the time and effort to identify the basic conditions that allow them to occur, such as when they happen and why they happen. When we perform a diligent study for a given defect, we can then determine what steps are needed to prevent these mistakes from recurring—*permanently.* The mistake-proof system approach and the methods used give us an opportunity to prevent mistakes and errors from occurring.

In the case of human mistakes, we must understand the system that allows the individual to make the mistake and the conditions that an individual is operating under to do a particular task. In the following list, we have identified several conditions where one may be prone to create a mistake and, ultimately, a defect.

1. Forgetfulness: There are times when we forget things, especially when we are not fully concentrating or focusing. An example that can result in serious consequences is the failure to lock out

a piece of equipment or machine on which we are working. To preclude this, precautionary measures can be taken: post lock-out instructions at every piece of equipment and/or machine and have an ongoing program to continuously alert operators of the danger.

2. Mistakes of misunderstanding: Jumping to conclusions before we are familiar with the situation often leads to mistakes. For example, visual aids are often made by engineers who are thoroughly familiar with the operation or process. Since the aid is completely clear from their perspective, they may make the assumption (and often do) that the operator fully understands them as well. This may not be true. To preclude this, we may
   a. Test this hypothesis before we create an aid
   b. Provide training/education
   c. Standardize work methods and procedures
3. Identification mistakes: Situations are often misjudged because we view them too quickly or from too far away to clearly see them. For example, misreading the identification code on a component of a piece of equipment, and replacing that component with the wrong part. To prevent these, we might
   a. Improve legibility of the data/information
   b. Provide training
   c. Improve the environment (lighting)
   d. Reduce boredom on the job, thus increasing vigilance and attentiveness
4. Amateur errors: Lack of experience often leads to mistakes. A newly hired worker will not know the sequence of operations to perform his or her task, and often due to inadequate training, will perform those tasks incorrectly. To prevent amateur errors
   a. Provide proper training
   b. Utilize skill-building techniques prior to job assignment
   c. Provide work standardization
5. Willful mistakes: Willful errors result when we choose to ignore the rules. For example, placing a rack of material outside the lines painted on the floor that clearly designates the proper location. The results can be damage to the part, the material, or perhaps an unsafe work condition. To prevent this situation
   a. Provide basic education and/or training for the task and safety
   b. Require strict adherence to the rules
6. Inadvertent mistakes: Sometimes we make mistakes without even being aware of them. A wrong part might be installed

because the operator was daydreaming. To minimize this from occurring, we may

a. Standardize the work
b. Create appropriate discipline, if necessary

7. Slowness mistakes: When our actions are slowed by delays in judgment, mistakes are often the result. For example, an operator unfamiliar with the operation of a forklift might pull the wrong lever and drop the load. Methods to prevent this might be
   a. Skill building
   b. Work standardization
8. Lack of standards mistakes: Mistakes will occur when there is a lack of suitable work standards or instructions that are not understood by the workers. For example, two inspectors performing the same inspection may have different views on what constitutes a reject. To prevent this
   a. Develop operational definitions of what the product is expected to be, which are clearly understood by all
   b. Utilize proper training and education
9. Surprise mistakes: When the function or operation of a piece of equipment suddenly or unexpectedly changes without warning, mistakes may occur. For example, power tools that are used to supply specific torque to a fastener will malfunction if an inadequate oil supply is not maintained in the reservoir. Errors such as these can often be prevented by
   a. Work standardization
   b. Having a total productive maintenance system in place. Typical basic items of concern should be in the areas of: (i) precision alignment, (ii) imbalance, (iii) lubrication, (iv) appropriate bearing installation, and (v) visual inspection
10. Intentional mistakes: Mistakes are sometimes made deliberately by some people. These fall in the category of sabotage. Disciplinary measures and basic education are the only deterrents to these types of mistakes.

Essentials of Poka-Yoke

The goal of error proofing in the context of *doing it right the first time* is to

- Assess error-proofing implementation at a supplier site
- Identify error-proofing opportunities during site assessments

- Recommend corrective action during failure mode and effect analysis (FMEA) and 8D evaluations of detection rankings and minimizing or eliminating human-related problems, respectively.

This means that the supplier has the responsibility to

1. correct any deficiencies that may be found during the site evaluation in such a way that they will *not* recur. In fact, the correction documentation must be available for a certain amount of time (this is defined by mutual agreement between customer and supplier).
2. recommend corrective actions that will improve the detection ranking in the FMEA.
3. correct deficiencies in the system where human behavior is depended upon, for example, inspection, orientation, and so on.

To appreciate error proofing, one must understand the assumptions that are based on two essential attitudes about human behavior. They are

- Mistakes/errors are inevitable.
- Defects can be eliminated or minimized.

An error is any deviation from a specified process. This means that all defects are created by errors. However, *not* all errors result in defects. The reason for this is that some errors are corrected in the process itself before they reach the next operation. An example of this is a 5-stand rolling mill. An error may be generated in the rolling process of the second rolling stand, but by the time it finishes the fifth rolling stand, the error has corrected itself and there is no defect. Examples of errors are

- The wrong option package is sequenced into the assembly.
- An oil sender unit is defective.
- A hose clamp is not positioned correctly during the assembly.
- A worn installation tool causes molding clips to be installed incorrectly.
- A door is left open on an assembly finishing line.

On the other hand, a defect is the result of any deviation from product specifications that may lead to customer dissatisfaction. To be a defect,

- The product must have deviated from specifications.
- The product does not meet customer expectations.
- The nonconforming product must have reached the customer.

TABLE 15.1

Comparison between Error and Defect

Error	Defect
The wrong option package was sequenced into assembly.	Vehicle equipped with wrong option package.
An oil sender unit is defective.	Engine oil/coolant leak.
A hose clamp was not positioned correctly during assembly.	Squeak, rattles, or loose parts.
A worn installation caused molding dips to be installed incorrectly.	The door is damaged due to hitting an object.
A door was left open on an assembly finishing line.	

Sometimes in this discussion (of error and defect), the notion of mistake comes into the conversation. A mistake is an error that occurs during processing. For our purposes, a mistake is viewed as an error. A comparison between an error and defect is shown in Table 15.1.

Therefore, in general terms, error proofing is a process improvement designed to prevent a specific defect from occurring. In some industries, the specific terminology of error proofing is reserved for design issues. Mistake proofing is reserved for process issues. The reader should notice that the Japanese word poka-yoke (ポカヨケ) does not make any distinction between the two. It simply means POKA—inadvertent mistake that anyone can make—and YOKE—to prevent or proof the process from any undesirable results. In literal terms, it means foolproof. However, in the United States foolproof is not acceptable because it is considered a disrespectful adjective for operators. Mistake proofing applies to

- The application of tools and devices applied to a process to reduce the possibility of errors occurring
- The application of tools and devices to a process to reduce the possibility of defects that have occurred from continuing to the customer
- The use of functional design features to reduce the possibility of parts being assembled incorrectly

Therefore, when one talks about error proofing, a distinction has to be made. That is, are we talking about prevention or detection? (See Figure 15.1.) The way this question is answered will depend on the timing of the project (design or process) as well as whether a deficiency or nonconformance is identified in the course of conducting an FMEA or an 8D. Where does this assessment take place? Generally,

- During normal site assessment. When the assessment is conducted, the following observations should be made for identifying possible nonconformities:

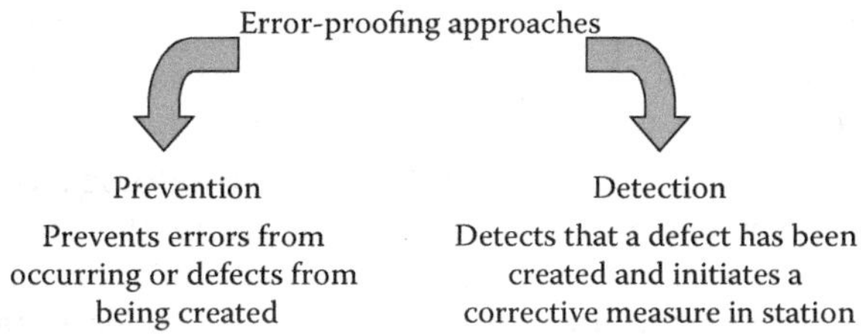

FIGURE 15.1
Comparison of prevention and detection approaches.

- All customer parts have documented design and process failure and effects analysis (DFMEA, PFMEA), and control plans (or dynamic control plans [DCPs]).
- All defect detection areas are reviewed and plans exist to move to defect prevention. There is evidence that defect prevention is replacing defect detection.
- Supplier has a maintenance system that contains reactive, preventive, and predictive maintenance. The maintenance system supports process capability improvement as well as overall equipment effectiveness (OEE) expectations.
- Supplier takes measures to minimize foreign material, chips, debris, contamination, excessive oil, and so on where part quality can be negatively affected.

- APQP reviews. When this review is conducted, the following should be evaluated with extra attention:
    - DFMEA. Consider process improvement techniques (design of experiments [DOE], robustness methodology, mistake proofing (poke-yoke), 8D with high emphasis on prevent recurrence, etc.) to develop preventative and corrective actions for severity ratings of 9–10, severity ratings of 5–8, with an occurrence of 4 or more, and high risk priority number (RPN) items. (Note that the RPN is the least desirable criteria for evaluating FMEAs. The priority should be first on high severity, second on severity and occurrence, and third on the RPN. It is important to mention here that some customers evaluate FMEAs based on the value of the RPN. We do not recommend this approach.)
    - PFMEA. Address the impact of failures as applicable to each of the following items: each part, subsequent operation, system, customer wants, government regulations, and operator safety.
    - Facilities/tools/gauges. Ensure that tooling and equipment design incorporates the recommend actions from the PFMEA.

- FMEA reviews. When reviewing the FMEA, critical items for review are severity, potential cause, design/process controls, and their effectiveness.
- 8D reviews. When a review for the 8D methodology is conducted, make sure that extra attention is on defining the root cause and escape point. Also, in Step D5, check for verification–effectiveness, and in Step D6 check for validation–date of implementation. Some items of concern should be the following. However, this is not an exhaustive list.
    - Critical features
    - Fit
    - Form
    - Function
    - Adherence to government regulations
    - Adherence to safety items
    - Customer satisfaction features
    - Customer concern issues
    - Things gone right/things gone wrong (TGR/TGW) reports
    - Service bulletins
    - Field actions and stop shipments

Therefore, based on what we have said thus far, one can surmise that mistake proofing is a method/approach process improvement designed to prevent a specific defect from occurring. The goal is *zero defects*. This goal is based on three attributes. They are

1. Physical: Install mistake-proofing devices
2. Operational: Ensure correct procedure, sequence, or execution
3. Philosophical: Empowerment of workforce

Because of this method, benefits occur and some of them may be the following:

- Enforces operational procedures or sequences
- Signals or stops a process if an error occurs or a defect is created
- Eliminates choices leading to incorrect actions
- Prevents product damage
- Prevents machine damage
- Prevents personal injury
- Reduces time to work on defective parts

- Reduces cost of all materials used, including those used for rework
- Reduces cost of time spent in rework
- Reduces cost of time spent inspecting defective parts
- Reduces lost production time spent on defective parts or rework
- Increases customer satisfaction
- Increases worker morale

All these benefits are actually materialized because mistake proofing focuses on increasing value-added activities of products/processes as opposed to non-value-added activities. Value-added is defined as any operation that changes, converts, or transforms material into the product that is sold to the customer and for which the customer is willing to pay. Satisfaction is achieved! On the other hand, non-value-added activity is defined as any operation that does not transform in any way, shape, or form the product that is sold to the customer. Examples of non-value-added activities are

- Inspection
- Storage/inventory
- Downtime (including changeovers)
- Defects/spoilage
- Waiting of any kind
- Rework of any kind
- Over/under production
- Early/late production

To take advantage of the poka-yoke methodology, improvement opportunities may come about when the following nine steps are followed:

1. Identify problem
2. List possible errors
3. Determine most likely error (verify that this is the error)
4. Propose multiple solutions
5. Evaluate solutions' effectiveness, cost, and complexity
6. Determine best solution (verify that this solution will resolve the problem)
7. Develop implementation plan
8. Analyze preliminary benefits
9. Develop plan for long-term measure of benefits

Zero Defect Concept in Mistake Proofing

When one talks about zero defects, it is accepted as a goal of zero defects. Everyone knows that in the real world there is no such thing as perfection. The zero defect concept, more often than not, is a definitional number. It is a quantitative statistic that everyone can support. A classic example is the parts per million (ppm) defect number. Of course, zero is the perfect number. However, in most manufacturing facilities, both customer and supplier are happy if the ppm is around 50 ppm defects. In order for the zero defect system to be effective, the following must be present:

- Source inspections
- Mistake-proofing devices
- 100% inspection (however, even 100% inspection is only 79% effective)
- Immediate feedback and action

Therefore, how do we practice the zero defect cycle? The answer is very simple. We have to follow a cycle similar to the improvement cycle of events. The reason for this is that zero defects is part of continual improvement attitude/philosophy/culture in any organization. A typical comparison is shown in Table 15.2. Table 15.3 shows some of the specific questions that may be asked in five specific categories for demonstrating the zero defect process.

Poka-Yoke Steps and Application

As previously mentioned, poka-yoke, as foreseen by Mr. Shingo, was to improve quality. As such, the fundamental principles for mistake proofing to which we all must adhere are defined in eight steps. They are

1. Build quality into the process.
2. Understand that inadvertent mistakes and defects can be eliminated.
3. Stop doing it wrong, start doing it right!
4. No excuses; think about how to do it right.
5. Anything over 50% chance of success is good enough. Implement! The reason for anything over 50% is that statistically one has a better than 50% chance of improvement from the status quo. Therefore, it is worth pursuing.

TABLE 15.2

Comparison of Improvement Cycle and Zero Defect Flow and Cycle

Improvement Cycle	Zero Defect Cycle
Adherence: The cycle begins with this stage. Its essence is to begin by following the current procedures, or the way things are done now.	Recognize defective work: An item has been identified as a mistake.
Analysis: Analyze what is being done now and the results we are getting.	Conduct problem analysis: The item is analyzed for a root cause and escape point.
	Improve: Improvement is made based on a data-driven analysis.
Improvement: Based on the results of the analysis, we take improvement action.	Stabilize: Repeatability and consistency are assured.
Standardization: Standardize the successful improvements and continue the cycle.	Adhere: Assurance of doing exactly what was identified under stability.
	Zero defects: Process is assured to be at zero defects as long as there is stability and as long as adherence is followed.

The flow and rationale of the zero defect cycle. It is imperative that when you are committed to a zero defects mentality, you must recognize that zero defects is a change in mindset. It is a philosophy that will take time for employee buy in. Zero defects involves using a systematic approach to completing the task, inspecting the process, and achieving zero mistakes, zero defects, zero problems. In essence, zero defects means a mistake-less environment with no exceptions. Absolutely no mistakes leave the work environment. The cascading process is

Zero mistakes
Zero defects
Zero problems
Zero flaws
Zero errors
Zero injuries
Zero waste
Zero downtime

The result of this flow will result in

Stability → Continuous Flow → Synchronized Production→ Pull System → Level Production

6. Mistakes and defects can be reduced to zero by working together and understanding both the task and process involved.
7. Build on synergy. Ten heads are better than one.
8. Seek out the true root cause and escape point of the defect.

These principles are very important. However, just like any system of improvement, with poka-yoke there must be a standardized process of implementation to successfully improve the process. A general generic approach is the 12-step approach. The steps are

1. Identify and describe the defect.
2. Show the defect rate by charting the defect occurrence over time.

TABLE 15.3

Zero Defect Process Overview

1. Defect	What is the product? What is the defect?
2. Mistake	Where was the defect found? Where was the defect made? What happens to the part in between? What is the most likely cause of the defect?
3. Root cause of mistake	Why? Why? Why? Why? Why?
4. Is short-term solution feasible?	If short-term action is necessary, then how can I stop the mistake from causing the defect now? If not, then the action is moved to long term.
5. Long-term action	How can I ensure 100% inspection so this mistake will never occur again?

3. Identify where the defect was found.
4. Identify where the defect was made.
5. Describe the current process where the defect was made by detailing the standard procedures used in the operation.
6. Identify any errors or deviations from process standards where the defect was made.
7. Use the 5 Whys or 3 × 5 Why problem-solving technique to identify the root cause of the defect or error. If you can use the 3 × 5 Why approach, it is even better.
8. Develop ideas for improving the process using the 5 Whys or 3 × 5 Why tools to eliminate or detect the error.
9. Improve the process by creating an error-proofing device.
10. Measure/document results of error proofing.
11. Standardize the improvement.
12. Ask where else could this improvement be used?

Obviously, in some industries these 12 steps may be modified to reflect their own needs. For example, Table 15.4 shows different steps for a typical automotive industry and a service organization as well as the most common mistakes that poka-yoke may be used for best results.

Now that we know how to approach a poka-yoke opportunity, let us look at the optimum way that a process can improve. Table 15.5 shows a generic approach of selecting an appropriate level of control. Table 15.6 shows the levels that are the most efficient and at the right column shows

TABLE 15.4

Typical Comparison of Poka-Yoke Approaches

Nine Typical Mistake-Proofing Steps for an Automotive Organization	Seven Generic Steps for Service Organizations	Most Common Mistakes That May Be Improved with Poka-Yoke
1. Locate the defect and isolate the process that created it	1. Identify and describe the defect	1. Omissions
2. List all possible mistakes	2. Determine where the defect is discovered and made	2. Processing mistakes
3. Determine the most likely mistakes	3. Detail the current standard procedure	3. Setup mistakes
4. Propose multiple solutions	4. Identify deviations from standards	4. Assembly/process mistakes (missing parts/information)
5. Evaluate each solution	5. Identify the red flag conditions where the defect occurs	5. Wrong components (part, tool, etc.)
6. Choose the best solution	6. Identify the type of mistake-proofing device type required to prevent the mistake or defect	6. Preparation mistakes
7. Develop a plan for implementation	7. Create devices and test for effectiveness	7. Procedural mistakes
8. Measure results/analyze benefits		8. Adjustment mistakes
9. Update documentation affected by the mistake-proofing implementation		9. Orientation issues
		10. Operator-dependent processes

some examples of production conditions in which poka-yoke may be used effectively.

Approaches to Correcting Human Mistakes

It has been said that to "err is human." Indeed, humans do make mistakes and that is why poka- yoke is a very powerful method to eliminate or at least minimize that human error. Human mistakes are part of someone performing a job or task. People will, on occasion, make mistakes no matter how hard they try not to. Mistakes can lead to defects, and people must be supported by a system that says mistakes can be prevented. However, when the focus is on fixing the person (the *who*) we are "blaming someone" instead of fixing

TABLE 15.5

Generic Approach of Selecting Appropriate Level of Control

	Level 1	Level 2	Level 3	Level 4	Level 5
Type	Autonomic operation	Automatic control	Self regulation	Personal supervision, auditing, inspection	Information control
Use	Highly repetitive	Highly repetitive	Stable environment, established practices	Changing environment, cost of failure	Large area of responsibility
Example	Traffic light cycle	Temperature control via feedback loop—thermostat	Machinist, typist, assembly	100% inspection by Q/C function	Warranty feedback

TABLE 15.6

Effective System of Poka-Yoke That Works Best

Level	Process	Approach	Production Condition
Level 1	People	Job is standardized and stabilized.	Adjustments Tooling/tool changes
Level 2	System	Many defects are produced but detected.	Many parts/mixed parts Multiple steps
Level 3	Subsequent station	Several defects produced but detected.	Infrequent production Lack of effective standards
Level 4	At the station (where work is done)	Possibly one defect produced and stopped at station.	Symmetry—parts that are mirror images Asymmetry—parts that look alike but are not
Level 5	Process parameters (manufacturing)	Mistake corrected, defect not produced.	Rapid repetition High production volumes
Level 6	Design (product/service)	Error eliminated.	Environmental conditions

the mistake. The focus should be on identifying the facts and eliminating the mistake through the appropriate level as shown in Table 15.6.

To approach the defects, we must first track and evaluate their significance to the system. Typical traditional tracking is based on the flow of mistake → mistake not corrected → defect. Specifically, typical mistakes are identified by

- Quantity
- Kind
- Percentage

- Scrap rate
- Rejects per defect

With mistake proofing, the focus of tracking must be on (1) the point at which the defect is discovered and (2) the point at which the defect occurred (escape point). The flow is quite different and is based on source inspection. That is, mistake → mistake discovered → feedback and action → zero defects.

Therefore, how can we differentiate the traditional approach from the mistake-proofing approach? In the first, we depend on self-check, and in the second, we depend on source inspection. Self-check means that the operator performing work checks on an item before passing it to the next process. This is self-inspection by the operator through sensory and visual cues. On the other hand, source inspection is taking preventive action, at the mistake stage, to prevent mistakes from turning into defects, rather than taking corrective action after defects have occurred. It is a preventive approach as opposed to a detection approach.

Therefore, the focus for any improvement using poka-yoke methodology is to develop sound approaches for controlling

- Value-added: Value-added activities are those that add form or function to the product. Any activity that does not add form or function is considered non-value-added waste.
- Just-in-Time information: The right information, in an appropriate form, in the hands of the people who can act on it, precisely when it is needed.
- Source inspection: (a) how to discover mistakes in conditions that create defects, and (b) how to prevent mistakes from turning into defects.
- The 3 actuals: To be effective, process control, inspections, and information sharing must be understood in terms of the three actuals: (a) the actual *place* or location an operation occurs, (b) the actual *people* in that place, and (c) the actual *process* occurring in that place.
- A visual language: This is a way of communicating fast and accurate information by (a) recognizing problems and abnormalities, (b) taking quick corrective action, (c) ensuring necessary learning, and (d) preventing problems from recurring.

For the strategies to be effective, the first requirement is that there must be an environment/culture in the organization that does not assign blame to any one individual but pursues issues in design and/or process. A no-blame environment means that (1) problems are the fault of the process, (2) work occurs in a dynamic world, (3) communication is vital to success, and (4) everyone is responsible for *no blame*. Focus the questions on *why* it happened, *what* happened, or *how* it happened. Avoid asking *who* did it?

The second requirement is to cultivate the zero defect mindset throughout the organization. In order to maximize the results of using a zero defect system for achieving zero defects, the following steps must be included in the process:

1. Reduce the amount of time between the discovery of defects and determining where the error occurred
2. Use 100% successive checks
3. Use 100% self checks

As we have already mentioned, poka-yoke is a way to minimize or eliminate defects. However, there are two types of defects:

1. Isolated: They happen once.
2. Serial: They happen more than once.

Both types can be either: (1) processing defects–process dependent or (2) material defect–material dependent. In both conditions, they are created by either human conditions or production conditions. To identify them, more often than not we depend on inspection, even though we know there are many shortcomings. For example,

- Although it is necessary to have efficient inspection operations, they are of little value in the process. Even the most efficient inspection operations are merely efficient forms of waste.
- Inspection involves the examination of a product or process in order to detect mistakes, defects, or problems.
- Inspection only makes sense from the company's standpoint, not the customer's.
- The customer does not care if the item is inspected. The customer's only concern is that the item is flawless.
- To the customer, inspection is *non-value-added*.

Therefore, even though inspection comes with some concerns, we still depend on it because it provides significant information based on *some* judgment. However, depending on the situation, there are three types of inspection.

1. Information inspections: They are based on observation to (a) generate feedback regarding quality issues so that corrective actions can be taken, (b) help reduce defects, but not prevent them, and (c) ensure focus is still reactive by correcting the problems after they occur. (d) Information inspections take two forms: (i) pass or (ii) fail.
2. Judgment inspections: They occur after the mistake and/or the defect has been made. They are reactive in nature. In essence, a

source inspection is (a) based on knowledge of the standards and a comparison to those standards, (b) reactive by determining the problems after the fact, and (c) difficult to obtain consistency because the inspections are based on a subjective human judgment rather than objective measurements.

3. Source inspections are done at the source of the task. Therefore, the inspection is capable of detecting mistakes as they occur and eliminating them.

All three types of inspection are forms of a check. This check can be a self-check or a successive check. Regardless of what type of check is conducted, the intent is to discover defects or reduce future defects. Specifically, the inspection to discover defects is concerned with (1) products compared with standards and defective items are removed and (2) sampling inspection when 100% inspection is too much trouble. The inspection to reduce future defects is based on statistical quality control activities such as SPC (control charts) to anticipate and avoid defects based on the assumptions of the notion that some level of defects is inevitable (type one error) and the interest is on results rather than cause. A comparison of these two approaches is shown in Table 15.7. So, is there a clear way to identify mistakes and defects? Yes. The distinction will be easier if the following information is available.

- Product: What is the product?
- Defect: What is the defect?
- Mistake: What is the most likely cause?

TABLE 15.7

Comparison of Self and Successive Checks

	Self Checks	Successive Checks
Description	The inspections are performed by the individual who performs the task.	The inspections are performed by the individuals who performed the task.
Advantages	Represents 100% inspection	May detect mistakes overlooked during the previous operations.
	Feedback is immediate and therefore corrective.	May detect errors missed during the self check.
		Encourages teamwork.
Disadvantages	Conscious and/or unconscious compromises may be made on quality.	Feedback may be slow.
		Corrective action occurs after the mistake is made.
	The inspections may not occur due to forgetfulness, time, and/or motivation.	May foster blame.

- Detection location: Where was the defect discovered?
- Source location: Where was the mistake made?
- Duration between source location and detection: How long between mistake and defect (time, workstations, processes, and so on)?

A comparison of mistakes and defects is shown in Table 15.8.

In addition to inspection practices to improve quality in any poka-yoke method, one will also find the use of mistake-proofing devices. Mistake-proofing devices are devices made up of elements that when followed will cause a predictable and desirable result and when not followed will result in a predictable defect or waste. Mistake-proofing devices should be developed by a team, developed for specific application, and be simple, reliable, and inexpensive. They are used to

- Prevent mistakes from occurring
- Detect mistakes that have occurred and prevent defects
- Detect defects that have been made before the devices are installed
- Perform 100% inspection
- Designed to detect only one mistake
- Provide prompt feedback and action

There are many types of mistake-proofing devices for many applications. They may be very technical and complex, but also very simple. Typical devices are

- Guide/reference/interference rod or pin
- Template
- Limit switch/microswitch
- Counter
- Odd-part-out method

TABLE 15.8

Comparison between Defect and Mistake

Defect	Mistake
Quartz watch does not work.	Dead battery
No hole drilled.	Broken drill
Undersized hole.	Worn drill bit
Inaccurate expense report.	Typographical error
Computer application glitch.	Software error
Car won't start.	Transmission not in "park"
Peeling paint.	Improper paint preparation

- Sequence restriction
- Standardize and solve
- Critical condition indicator
- Detect delivery chute
- Stopper/gate
- Sensor
- Mistake proof your error-proof device. This is the most misunderstood item in device implementation. This addresses the lack of good maintenance that can render the most brilliant error-proofing device useless—or even dangerous. Dangerous because if precautions are not taken, you can assume that the device is watching your quality or safety when it is not. To avoid this, you must regularly maintain every non-mechanical error-proofing device. This includes sensors, limit switches, counters, gates and stoppers, and any other device that relies on electricity, temperature and pressure gauging, or tolerances. Your device is reliable only to the extent that these predetermined specifications are precisely maintained.
- Eliminate the condition
- Redesign for symmetry
- Redesign for asymmetry
- Sticky notes on components
- Electronic eyes
- Components that will only attach one way
- Limit switches
- Probes
- Proximity switches
- Profile plates
- Alignment tabs
- Broken tooling indicator
- Color striping
- Photo cell eyes

The typical functions of mistake-proofing devices are shown in Figure 15.2. Some of the devices are the result of regulatory functions, but all devices are of three types. They are

1. Warning method: Signals the need for feedback and action when abnormalities occur (least effective). When an abnormality occurs, feedback is provided by signals or visual controls, for example,

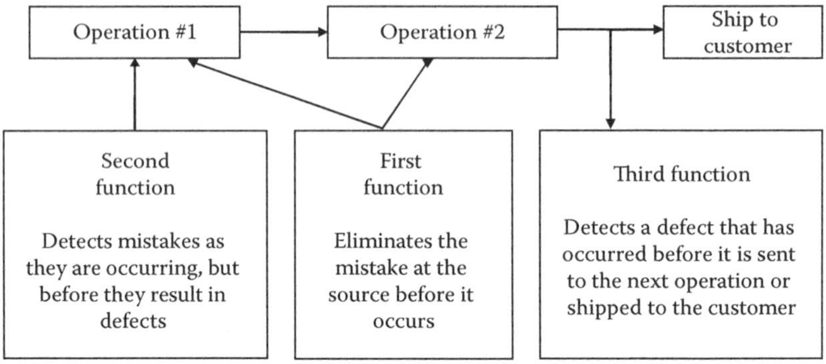

FIGURE 15.2
Functions of mistake-proofing devices.

lights and alarms. Once these devices are activated, the mistakes and/or defects will continue to be produced until an operator responds.

2. Shut down method: Shuts down or halts the operation when abnormalities occur. When an abnormality occurs, operation automatically shuts down. An operator needs to intervene before it starts up again.
3. Control method: Automatically corrects the mistake condition (most effective). When an abnormality occurs, the mistake-proofing device automatically corrects the mistakes.

Red Flag Conditions

A red flag condition is a condition in the manufacturing/service process that commonly provokes mistakes. Typical red flag conditions are

- Adjustments
- Symmetry
- Tooling/tooling changes
- Asymmetry
- Dimensions/specifications
- Critical conditions
- Rapid repetition
- High/extremely high volume
- Many/mixed parts
- Multiple steps

- Infrequent production
- Lack of an effective standard
- Environmental conditions:
    - Material/process handling
    - Foreign matter
    - Poor lighting
    - Housekeeping

How do you identify red flag conditions? There are many ways depending on the process. However, the following steps may help in identifying red flag conditions:

- Identify and describe defect
- Determine where the defect is discovered or made
- Detail the current standard or procedure
- Identify deviations from standard
- Identify red flag conditions where defect occurs
- Identify type of mistake-proof device required
- Create the device and test it for effectiveness

Corrective Action

Now that the defect has been identified, there must be an action developed to eliminate it. After all, taking corrective action means not using the same old solution or the solution is the result or accomplishment that will be better than what you had to start with. There are two types of corrective action. They are

1. Short-term corrective action: They aid in keeping things running but may not keep the problem or defect from recurring.
2. Long-term corrective action: They prevent the defect or problem from recurring.

To accomplish either one of the options, a strong rationale and mindset in the organization must be prevalent. This means that the organization must exercise the following on a daily basis and be on the lookout for any deviations:

1. Prevent defects from passing on to future product
2. Detect mistakes and give feedback
3. Correct or eliminate mistake

One of the *best* ways that the corrective action is maximized is by having appropriate and applicable support. That appropriate and applicable support is manifested through the team. Specifically, support means

- Having appropriate and applicable teams
- Having appropriate and applicable training
- Having a standardized methodology
- Having the commitment and support of management

Appropriate and applicable team means that all participants are working together for a common *fix* of a specific problem. Specifically, the team must be cohesive in the following areas:

- Treat each other with respect.
- Stay focused on the task.
- Offer suggestions, don't be afraid to speak up.
- Constructively criticize ideas, do not criticize people.
- Listen to others until they are finished.

This cohesion is very important because, more often than not, the composition of the team is very diverse in experience, education, and title of position. A typical composition may be made up of

- Support personnel
- Operators
- Skilled tradespeople
- Supervisors
- Engineers

The composition of the appropriate team members is of profound importance because the first decision has to be made whether the resolution of the problem at hand has to be with internal or external activities. Internal activities must be performed while the machine is shut down. On the other hand, external activities can be performed while the machine is running. Obviously, the best option is to do work while the process is operating. One way to analyze the situation is by brainstorming and specifically

- Analyzing each element
- Asking the 5 Whys or the modified 3×5 Whys
- Resolving if the element can be made externally

If brainstorming is used, then

- Record all ideas
- Do not emphasize only high-tech solutions
- Inexpensive solutions are often viable
- Always maintain a questioning attitude
- Zero changeover time is possible

Do not forget to mention the issues of scheduling. For example,

- Poor schedules from the scheduling activity
- Production ignores schedules from the scheduling activity
- Equipment breakdowns
- Material and/or personnel shortages
- Quality issues
- Product complexity

Once the team has evaluated the area of concern and disposition of a corrective action is about to be identified and recommended, it is imperative that the action taken *must* be based on the principles of employee empowerment (i.e., team members must have been given authority and responsibility for their situations) and the final decision *must* be consensus driven. Empowerment is critical because it allows the mistake-proofing team to

- Make changes to their work environment
- Take ownership in the product they manufacture
- Support continual improvement in the workplace

On the other hand, consensus is reached through a diligent discussion without fear of ridicule and/or intimidation by the other team members and/or management. Specifically, consensus is reached through the process of

- Discussing thoroughly each idea
- Eliminating duplicate ideas
- Working as a group to pick the best idea
    - Do not let one person dictate the outcome
    - All team members should agree on the name
- Decision is reached by agreement, not voting or majority for the best idea

Presenting Findings

When the team finishes studying the problem and has indeed figured out a solution to the specific defect, they present the results to the management team, which is the final judge as to whether the solution is acceptable. This means that the mistake-proofing team must do some preparations, starting with

- Deciding what information you are going to present
- Deciding how the information will be presented
- Deciding who will present (may be more than one person)
- Deciding if there is a need for practicing the presentation

Within these four categories, the predominant factors are the consequences of the *defect* (the reason for the study) and the *specific fix* with the associated potential benefits. The first one is addressed (at least the minimum information) by focusing on

- Mistake-proofing location
- Potential defect (escape and detecting)
- Mistake that could cause this defect
- How is this mistake prevented?
- Appropriate recommendation (prevention, detection or correction)

The second one is addressed by summarizing the benefits with quantifiable data. Without data, the mistake-proofing work will not be accepted. The benefits for any mistake-proofing undertaking are generally traditional costs and savings based on the number of defects eliminated. Do not include unquantifiable costs and undesirable situations as part of the presentation just because they sound good. They will distract from the focus and waste time. Typical unquantifiable costs and undesirable situations are

- Business costs: Cost of additional energy and equipment required to manage a poor-quality production environment.
- Liability costs: Cost of losing business due to poor product quality or inefficient compliance to government regulations and safety.
- Cultural costs: Reduction of worker morale and work standards. In some cases, the community acceptance may play a role in the decision.

Next Steps

The knowledge and skills you gained from conducting the study should be used (1) as a regular part of your work, and (2) diffused throughout the organization as applicable. Remember that success begets success! Remember also that mistake-proofing should come from everyone. Therefore mistake-proofing applications should be *simple*. Recommended next steps are

- Follow through on mistake-proofing projects.
- Use a mistake-proofing opportunity and tracking form to develop other mistake-proofing opportunities. A typical format for this is to
  - Identify the problem
  - List possible errors
  - Determine the most likely error
  - Propose multiple solutions
  - Evaluate solutions effectiveness, cost, and complexity
  - Determine the best solution
  - Develop implementation plan
  - Analyze preliminary benefits
  - Develop plan for long-term measure of benefits

Typical tracking items may be in the form of

- Percentage of errors proofed
- Percentage of errors detected versus percentage of errors prevented
- Percentage of errors attributed to design
- Percentage of errors attributed to process

Bring your knowledge and skills of error proofing back to your work group.

References

Fantin, I. (2014). *Applied Problem Solving. Method, Applications, Root Causes, Countermeasures, Poka-Yoke and A3. How to Make Things Happen to Solve Problems.* Milan: Createspace.

Quality Portal. (May 2013). Poka yoke or mistake proofing: Overview. *The Quality Portal.* http://thequalityportal.com/pokayoke.htm. Retrieved July 19, 2014.

Robinson, H. (1997). Using poka-yoke techniques for early defect detection. http://facultyweb.berry.edu/jgrout/pokasoft.html. Retrieved on July 19, 2014.

Shimbun, N.K. (1988). *Poka-Yoke: Improving Product Quality by Preventing Defects*. Portland, OR: Productivity Press, 111.

Shingo, S. (1987). *The Sayings of Shigeo Shingo: Key Strategies for Plant Improvement*. Portland, OR: Productivity Press, 145.

Shingo, S. and A. Dillon. (1989). *A Study of the Toyota Production System from an Industrial Engineering Viewpoint*. Portland, OR: Productivity Press.

Selected Bibliography

Hinckley, C. M. and P. Barkan. (1995). The role of variation, mistakes, and complexity in producing nonconformities. *Journal of Quality Technology* 27(3): 242–249.

Shingo, S. (1986). *Zero Quality Control: Source Inspection and the Poka-Yoke System*. Portland, OR: Productivity Press.

16

Measurement System Analysis (MSA)

Overview

Keep in mind the following truism: good outgoing quality depends on good incoming quality. The incoming quality for any customer depends on their supplier's quality. Therefore, the supplier's quality is an issue of both strategy and execution. Strategy is fundamental because it will define the necessary needs for excellence, market intelligence, good timing policies, risk management, and transparent collaboration of the customer–supplier relationship. On the other hand, execution is critical in understanding and implementing (or creating, in some cases) the analytics of the appropriate and applicable *data* that have been defined and collected.

Between strategy and execution, the supplier has the responsibility to focus on the *real data* of the organization with the intent to either modify and/or change drastically the way it is doing business for improved or continual improvement. This desire of being better has been accentuated with the latest recalls in the automotive industry. In fact, Walsh (2014, p. 11) reports, "the GM recall scandal has manufacturers scrambling to review documents, and practices to protect against potential liability."

We all have heard the old saying "what is measured gets done." Therefore, for the past 20 years, the quality profession has focused on the issue of data. If it is worth measuring, then appropriate and applicable data must be available, retrievable, and analyzed. However, to have data it is not a difficult thing. What is difficult is to have consistent and comparable data. This is of profound necessity because unless the data of what you are counting as *real* is real, the answer you will get, at best, is *not sure*. In the case of the supplier, measurements are a key item of concern in many of their processes and operations. So, where do we start? The following may help:

1. Start with a simple unit of measurement you can use as a basis for rationalizing all your data. This can be optimized by thinking in terms of what the customer wants. Next, coordinate the process

steps so that they follow a clear sequence tied to this unit of measure in real time. Finally, if possible, automate the process so that the measurements are consistent.

2. Keep it relevant. All data and measurements should focus on identifying key characteristics as to where they are (base line) and in what direction they are moving (monitoring).
3. Automate wherever possible. This will provide real-time measurements of real data but in addition to the increased speed, it will provide accurate, consistent, and reliable data for your analysis.
4. Make sure the instruments (methods) used are accurate, reliable, and repeatable. Remember that the instrument on which the measurement is being taken must be at least 10 times more sensitive than the measurement recorded.

In essence, data and measurement must contribute to some form of decision to fix or improve a process. From a supplier's perspective, the following items may be considered as possible sources for data:

- System complexity: The use of multiple suppliers in multiple locations and attendant inventory storage and transportation waste. Examples of measurement may be the cost of the system or delays created by complexity.
- Process efficiency: Optimizing process characteristics to control quality. Examples of measurement may be key input and output characteristics especially for special characteristics that the customer and supplier have identified as important to be controlled and monitored.
- Lead time: Generated by purchasing when negotiating lead time with supplier or delays built into the transit process. Examples of measurement may be the cost to the process or system for delays and waste of excess inventory or processing.
- Transportation: Wasted effort to ship product or wasted distance in the transportation process. Examples of measurement may be the cost of transportation, especially premium freight.
- Space (facilities): The space needed to transport or store products, prior to use. Examples of measurement may be facility (or cell) square footage and the cost associated with that space.
- Inventory: Inventory beyond what is needed to serve customers and satisfy the process. Examples of measurement may be the cost of inventory and the cost of maintaining it.
- Human effort: Wasted movement and motion of an operator or losses due to accidents. Examples of measurement may be wasted time and operator's compensation insurance costs, as well as effort for rework.

- Packaging: The costs associated with over- or underpackaging resulting in waste or product damage in transit. Examples of measurement may be the cost of premium freight, repeat shipping, and/or product replacement.

After reviewing these areas, specific considerations for quality improvements should be considered with the following characteristics in mind:

- Is it the right product? If not, how can we improve it?
- Is it the right quantity? If not, what is missing? Can we make it right?
- Is the right service available? Is the right service provided? If not, what can we do to make sure that the appropriate service is applied to satisfy the customer?
- Is it the right condition to collect data? If not, what can we do to make sure that the conditions are appropriate and applicable for appropriate data selection?
- Is it the right place and time to collect data? If not, what are the conditions to make sure that the place and time are appropriate and applicable for collection of the data?
- Is the frequency and source appropriate for representative data to be collected? If not, what can we do to mend this situation?
- Is the cost (price) right for what we are doing? If not, what did we miss? How can we recover data to reflect the right cost or price?
- Is safety an issue? If yes, what can we do to minimize the risk of accidents?
- Is the data taken from a confidential space? If yes, can we arrange the data to be drawn from an alternative source or method?
- Are the operators trained in sampling and measurement techniques? If not, what would it take to be appropriately trained?

To be sure, these questions are not one-to-one relationships with the items mentioned previously. However, they are important and they should be viewed and investigated because they could very well be seen as root causes of poor performance as long as the *right* data are collected.

Types of Data

There are two types of data with which to be concerned. The first type is attribute and the second type is measurable.

Attribute data is generally qualitative. Typical examples are yes/no situations, go/no go, good/bad, and so on. In analyzing this type of data, there are limitations. However, in some cases, that is the only data that we can have. Consequently, we have to be very careful how to use it and what statistics are applicable for such situations. Typical analyses that we can use are percentages and nonparametric statistics. A quality engineer should be consulted for specific recommendations or consult with a statistician for appropriate selection and evaluation of specific tests. A good overview for selecting some specific tools is Stamatis (2012) and Appendix V.

Variable data is generally quantitative. Typical examples are weight, temperature, pressure, dimensional specifications, and so on. In analyzing this type of data, we are very flexible as to the action that can be taken. There are many types of statistical tests that can be taken. However, due to the plethora of choices, one may be confused as to what is appropriate and applicable for a specific situation. Therefore, it is strongly suggested that a quality engineer or a statistician be consulted for the appropriate selection of specific tests and evaluation of results. A good overview for selecting some specific tools is Stamatis (2012) and Appendix V.

Using Minitab®

In today's world, computerization is everywhere. This is true in the quality field as well, especially in the measurement domain. Many software companies have specific programs dealing with all kinds of statistical tests. However, it seems that most companies have chosen the path of using Minitab®. It is very easy and powerful software and indeed provides many applications that most organizations can use. In addition to basic and advanced statistical analyses, Minitab® can plot many statistical process control (SPC) charts and perform gauge repeatability and reproducibility (R&R) studies. In this chapter, we are going to show a simple measurement system analysis (MSA) dealing with both attribute and measurable data.

MSA: Gauge R&R

MSA is a methodology used to understand measurement error and its effect on process and part quality. The collection of operations, procedures, gauge, software, and personnel are used to assign a number to the characteristic being measured. In other words, it is the complete process used to obtain measurements. Typical items of concern in an MSA are devices such as

- Micrometer
- Go/no-go snap gauge
- Standard plate inspection equipment
- Micro-hardness testers
- Infrarometers
- Vibration analyzers
- Coordinate measuring machine
- Automatic sorting gauge
- Air gauge

The reason for doing the MSA is to

- Provide information about what is happening in the process that produced the part
- Ensure good measurement data = good decisions
- Ensure accuracy and precision
- Eliminate false readings that indicate something is wrong in the process
- Reduce error in rejection of a null hypothesis that is actually true; something is bad when it is actually good (Type I error, producer risk, or alpha error)
- Reduce error when we do not reject a null hypothesis that is false; calling good parts bad (Type II error, consumer risk, or beta error)

At this point, one may ask, "Why do we need to measure anyway?" The reason is to understand whether a decision

- Meets standards and specifications
- Is detection/reaction oriented
- Produces short-term results

In addition, to stimulate continual improvement, ask the following:

- Where to improve?
- How much to improve?
- Is improvement cost-effective?
- Is this a prevention-oriented strategy?
- Is this a long-term strategy?

A good starting point—once you have determined the data—for a thorough MSA is

- Before you
  - Make adjustments
  - Implement solutions
  - Run an experiment
  - Perform a complex statistical analysis
- You should
  - Validate your measurement system or measurement process (preferably use the ANOVA method)
  - Validate data and data collection systems; always remember that variation may be in any of the following individually or in combination: machine, method, material, measurement, environment, and manpower (people)

So, what is an MSA? Specifically, one can say that MSA is

- A scientific and objective method of analyzing the validity of a measurement system
- A tool that quantifies
  - Repeatability; sometimes called equipment variation
  - Reproducibility; operator variation
  - The total variation of a measurement system
- Not just calibration
- Not just gauge R&R); it also addresses variation, which can be characterized by (a) location of the process, (b) width or spread, and (c) discrimination. When dealing with location, the following items are of concern:
  - Accuracy/bias: Difference between the observed average value of measurements and the master value. Remember that a master value is an accepted, traceable reference standard. If accuracy is suspected, then
    - Calibrate when needed/scheduled
    - Use operating instructions
    - Review operating specifications
    - Review software logic (drill downs)
    - Create operational definitions
  - Stability (consistency): Measurements remain constant and predictable over time for both mean and standard deviation. This

means, no drifts, sudden shifts, cycles, and so on (a good way to detect any movement is to evaluate an SPC control chart). If stability is suspect, then

- Change/adjust components
- Establish "life" timeframe
- Use control charts
- Use/update current standard operation procedures (SOP)

- Linearity: Measurement is not consistent across the range of the *gauge*. If linearity is suspect, then
  - Use only in restricted range (consider where the specification limits are)
  - Rebuild
  - Use with correction factor/table/curve
  - Sophisticated study required; contact either the quality engineer or a statistician

The second category deals with the width or spread (precision). This precision includes repeatability and reproducibility. Key points to remember:

1. Although most gauge repeatability and reproducibility (GR&R) studies are designed to study the significance of differences between appraisers, they can also be used to study the differences between multiple test units/end of line testers, multiple setups, multiple plants measuring the same outputs, and so on.
2. Gauges are sensitive instruments and their accuracy is adversely affected by abuse. Therefore, all users of gauges must properly use and care for them. When an instrument is inadvertently mishandled, it must be turned in to the calibration laboratory for checking and recalibration. Use the instrument only for its intended purposes. Here are some simple guidelines to keep tools and instruments in good condition:
   a. Never use any manipulative device or force the instrument
   b. Clean and remove burrs from the piece or product prior to use
   c. Never expose an instrument directly to heat
   d. Avoid exposure to dust, moisture, and grease
   e. Lightly oil instrument and store in a protected case or cabinet
   f. Return all instruments to storage case or cabinet after use

When dealing with repeatability, we are concerned with variation that occurs when repeated measurements are made of the same item under identical conditions. This means the same operator, the same setup (usually changes between measurements), the same units, and the same environmental conditions. It also implies short-term or within-operator variation. If repeatability is suspect, then

- Repair, replace, and adjust equipment
- Review/update SOP (i.e., clamping sequence, method, force)
- Gauge maintenance may be needed

When dealing with reproducibility, we are concerned with variation that results when different conditions are used to make the measurements. This means different operators (most common), different setups, different test units (end of line [EOL] testers), different environmental conditions, and different locations. It also implies variation between different operators when measuring the same part, as well as long-term or between-operator variation. If reproducibility is suspect, then

- Operators need more training in gauge usage and data collection
- SOP (standardized work) are needed
- The gathering-data device graduations, part location points, and data collection locations all need to be more clearly defined and identified
- Better operational definitions are needed

The third category deals with resolution/discrimination, which is the ability to measure small differences in an output. Poor resolution is a common issue. However, it is one of the simplest measurement system problems. Its impact is rarely recognized and/or addressed, even though it is easily detected. No special studies are necessary and no known standards are needed. When resolution issues are suspected, then

- Measure to as many decimal places as possible
- Use a device that can measure smaller units
- Live with it, but document that the problem exists
- Larger sample size may overcome problem
- Priorities may need to involve other considerations:
  - Engineering tolerance
  - Process capability
  - Cost and difficulty in replacing device
- MSA means walking the process (go see the gauging process, destructive testing, reduce the error in the process)
- MSA means having identified the correct operational definitions

Attribute MSA: Using Minitab®

The goal of an attribute MSA is to have 100% matches within and between operators to the known/correct attribute. We can measure this in two ways. The first one is a simple percentage of agreement and the second is a calculation of the Kappa value. To figure out the percentage, recall that the process average for a p chart is $\bar{p}$ and its formula is

$$\bar{p} = \frac{\sum np}{\sum n}$$

where

$\sum np$ = sum of the numbers of nonconforming units in all subgroups
$\sum n$ = sum of the numbers inspected in all subgroups

We use the process average ($\bar{p}$) to calculate the average capability as follows:

$$\text{Average Capability} = \left(1 - \bar{p}\right) \times 100\%$$

For the Kappa evaluation, we do the following. If Kappa = 1, there is perfect agreement; if Kappa = 0, then the agreement is the same as would be expected by random chance. For practical reasons, a good value is considered .75 or higher. (In some cases for *nominal* data [go/no go], use Fleiss' Kappa statistic. Avoid the use of Cohen's Kappa statistics because you can only have two appraisers and a single trial per appraiser. For *ordinal* [higher numbers represent higher values] data, use Kendall's Coefficient of Concordance. Nominal data are categories or groups. This means that the information is qualitative, not quantitative; e.g. go/no go, yes/no, marital status, religious preference, race, gender, and so on. On the other hand, ordinal data are higher numbers that represent higher values, but intervals between numbers are not necessarily equal. The zero point is chosen arbitrarily, e.g., casting porosity rating [ratings of 1 to 5], race finish, opinion poll response [difference between rating of 2 and 3 may not be the same as the difference between ratings of 4 and 5] and so on.) To calculate the Kappa value, we use the following equation:

$$\text{Kappa} = \frac{P_{\text{observed}} - P_{\text{chance}}}{1 - P_{\text{chance}}}$$

where

$$P_{\text{chance}} = \left(P_{\text{inspect 1 good}}\right)\left(P_{\text{inspect 2 good}}\right) + \left(P_{\text{inspect 1 bad}}\right)\left(P_{\text{inspect 2 bad}}\right)$$

that is, the proportion of the judge's agreement is the $P_{observed}$, and the proportion expected to occur by chance is P_{chance}. In everyday language, this means

$$\frac{\text{Proportion of times that the appraisers agree (corrected for chance)}}{\text{Maximum proportion of times that the appraisers could agree (corrected for chance)}}$$

In both cases, the parts should be independently measured with a variable gauge so that the physical measure of each part is known. When measuring a true attribute that cannot be measured with a variable gauge, use other means, such as experts, to predetermine which samples are good, bad, or marginal (borderline) defective. In essence then, an attribute MSA must

- Allow for R&R analysis within and between appraisers
- Test for effectiveness against standard
- Allow nominal data with two levels
- Allow for ordinal data with more than two levels (i.e., survey data; low, medium, high categories)

Remember that for nominal data, the Kappa coefficient provides a measure of relative agreement between appraisers, and for ordinal data, Kendall's coefficient of concordance provides a measure of relative agreement between appraisers. Therefore, the question is how many Kappa values are there in an analysis and which Kappa value must be higher than .75? The answer to the first question is that there are four possibilities. They are

- Each appraiser to itself (within)
- Between appraiser
- Each appraiser to the standard
- All appraisers to the standard

The answer to the second question is that all the values must be higher than .75. Finally, if the experimenter/evaluator wants to use a different statistic, talk to a quality engineer, a statistician, or a Master Black Belt.

The actual analysis is based on (1) the use of at least 50 samples—more is better, (2) the use of at least three operators—any operator that will use the gauge should complete the gauge R&R study, and (3) each operator measuring the parts at least three times.

To start the analysis, we begin with the truth table. This means that based on independent measurement, the true assessment is defined in the truth column. This indicates whether the part should pass or fail when measured by the attribute gauge. The truth table in Minitab® will look like the one shown in Figure 16.1.

	A	B	C	D	E	F	G
1	Part Number	Measurement	Truth	Appraiser 1 Trial 1	Appraiser 1 Trial 2	Appraiser 1 Trial 3	Appraiser 2 Trial 1
2	1	2.60	Pass	Pass	Pass	Pass	Pass
3	2	1.99	Fail	Fail	Pass	Fail	Fail
4	3	2.44	Pass	Pass	Pass	Pass	Pass
5	4	3.01	Fail	Fail	Fail	Pass	Fail
6	5	2.47	Pass	Pass	Pass	Pass	Pass
7	6	2.01	Pass	Fail	Pass	Pass	Pass
8	7	2.42	Pass	Pass	Pass	Pass	Pass
9	8	2.46	Pass	Pass	Pass	Pass	Pass
10	9	1.98	Fail	Pass	Pass	Pass	Pass
11	10	2.02	Pass	Pass	Pass	Pass	Pass
12	11	2.01	Pass	Pass	Fail	Pass	Pass
13	12	1.90	Fail	Fail	Fail	Fail	Fail

FIGURE 16.1
Typical Minitab® truth table.

In Minitab®, the column order is very important. Here we have transferred the data from the Excel sheet to the Minitab® worksheet. To avoid mistakes and confusion, it is strongly recommended that the data be entered into the Minitab® worksheet, thereby bypassing the Excel sheet altogether. An example of this is shown in Figure 16.2.

Once the data is entered, go to the tool bar and select: stat → quality tools → attribute GR&R study. The screen will look like Figure 16.3.

Once selected (by clicking Attribute Gage R&R Study), the screen will look like Figure 16.4.

> If data is stacked, select one column. If data is unstacked, then chose multiple columns.
>
> Select the number of appraisers and trials,
>
> Enter the column of *known* if you know it.
>
> Select OK and the following screen (Figure 16.4) will appear showing the variation between appraisers and standard.

↓	C1-T	C2-T	C3-T	C4-T	C5-T	C6-T	C7-T
	Attribute	Appraiser 1 Trial 1	Appraiser 1 Trial 2	Appraiser 2 Trial 1	Appraiser 2 Trial 2	Appraiser 3 Trial 1	Appraiser 3 Trial 2
1	Pass	Pass	Pass	Pass	Pass	Pass	Pass
2	Pass	Pass	Pass	Pass	Pass	Pass	Pass
3	Pass	Pass	Pass	Pass	Pass	Pass	Pass
4	Pass	Pass	Pass	Pass	Pass	Fail	Pass
5	Fail	Fail	Fail	Fail	Fail	Pass	Fail
6	Fail	Pass	Pass	Pass	Pass	Pass	Pass
7	Pass	Pass	Pass	Pass	Pass	Pass	Pass
8	Pass	Pass	Pass	Pass	Pass	Pass	Pass
9	Fail	Fail	Fail	Fail	Fail	Fail	Fail
10	Pass	Pass	Pass	Pass	Pass	Pass	Pass
11	Pass	Pass	Pass	Pass	Pass	Pass	Pass
12	Pass	Pass	Pass	Pass	Pass	Pass	Pass
13	Pass	Pass	Pass	Pass	Pass	Pass	Pass

FIGURE 16.2
Typical worksheet where the data is entered.

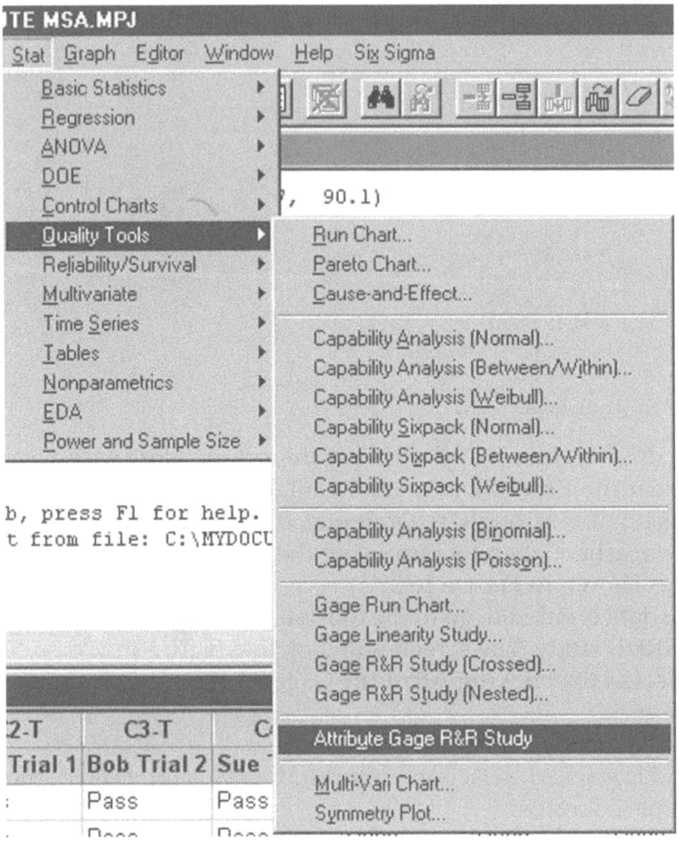

FIGURE 16.3
Minitab® screen showing the attribute gauge R&R.

Based on the selections, a screen like that shown in Figure 16.5 will be displayed. A visual interpretation shows that the graph on the left shows the within-appraiser agreement—how well each appraiser agrees with itself. The graph on the right shows the appraiser agreement to the standard. How well does each appraiser agree with the Truth (C1)? You can see that John has a lower agreement to himself and to the appraiser.

A numerical evaluation of the same data may be selected for evaluation and is shown in Table 16.1. CI is the confidence interval. The numerical analysis of the agreement shows the individual versus standard, disagreement assessment between appraisers, within appraisers, agreement within appraisers, and all appraisers versus standard. Specifically,

- Agreement assessment between individual versus standard (John had the lowest number matched)
- Disagreement assessment within appraisers (repeatability)

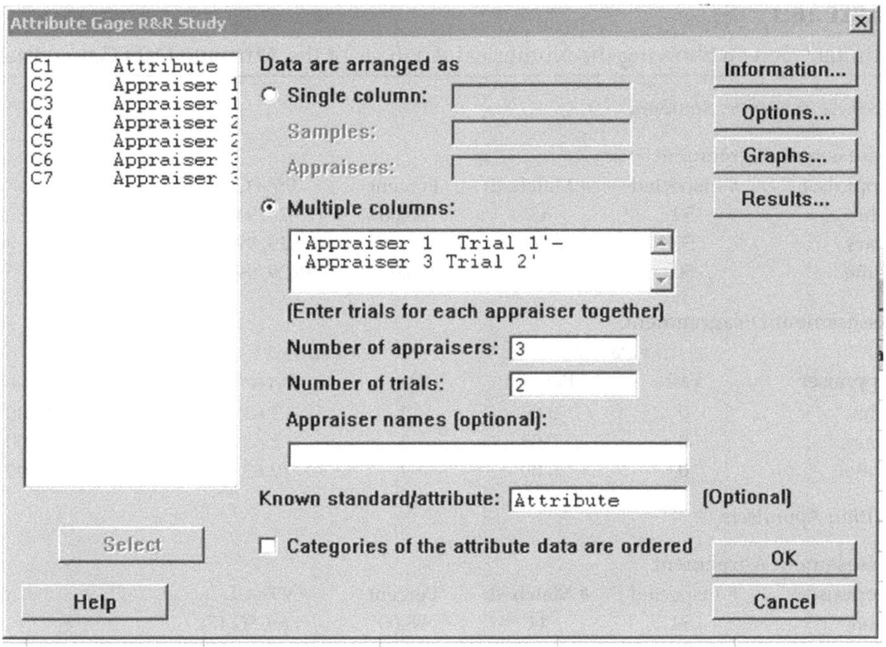

FIGURE 16.4
Minitab® screen showing the selection options for attribute gauge R&R.

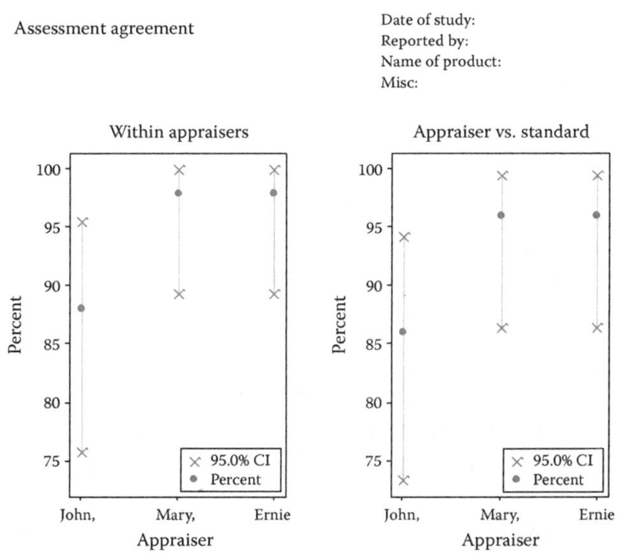

FIGURE 16.5
Minitab® screen showing the assessment agreement for an attribute gauge R&R.

TABLE 16.1

Minitab® Screen Showing the Numerical Analysis of the Attribute Data Gauge R&R

Each Appraiser vs. Standard

Assessment Agreement

Appraiser	# Inspected	# Matched	Percent	95% CI
John,	50	43	86.00	(73.26, 94.18)
Mary,	50	48	96.00	(86.29, 99.51)
Ernie	50	48	96.00	(86.29, 99.51)

Assessment Disagreement

Appraiser	# Pass / Fail	Percent	# Fail / Pass	Percent	# Mixed	Percent
John,	0	0.00	1	2.63	6	12.00
Mary,	0	0.00	1	2.63	1	2.00
Ernie	0	0.00	1	2.63	1	2.00

Within Appraisers

Assessment Agreement

Appraiser	# Inspected	# Matched	Percent	95%CI
John,	50	44	88.00	(75.69, 95.47)
Mary,	50	49	98.00	(89.35, 99.95)
Ernie	50	49	98.00	(89.35, 99.95)

Matched: Appraiser agrees with him/herself across trials.

Between Appraisers

Assessment Agreement

# Inspected	# Matched	Percent	95%CI
50	43	86.00	(73.26, 94.18)

Matched: All appraisers' assessments agree with each other.

All Appraisers vs. Standard

Assessment Agreement

# Inspected	# Matched	Percent	95%CI
50	43	86.00	(73.26, 94.18)

- Agreement assessment within appraisers (repeatability)
- Agreement assessment between appraisers (reproducibility)
- Agreement assessment: all appraisers versus standard (against known)

Therefore, to do a good attribute MSA, the following must be adhered to:

- Some parts need to be selected as good, bad, and close to the specification limits (areas of risk).

- The goal is to have 100% match within and between operators to the known/correct attribute.
- The gauge should reject defective parts.
- All Kappa values should be greater than 0.75.
- There should be a minimum of 3 operators, 3 measurements each, 50 parts. All people who use the gauge should complete MSA.
- A very accurate and precise gauge (e.g., coordinate measuring machine [CMM]) should be used to determine the exact value of each part. This is used for the truth table.
- When measuring a true attribute that cannot be measured with a variable gauge, use other means such as experts to predetermine which samples are good or defective.

Variable MSA: Using Minitab®

The supplier must do some prework before the variable MSA is conducted. At least seven steps are required. They are

Step 1: Establish the need for the study and information desired from it before any data is collected (%Study, %Tolerance).

Step 2: Determine the approach to be used and operational definitions.

Step 3: Determine the number of operators, number of sample parts, and the number of repeat readings in advance.

Criticality of the dimension/function

Part configuration

Level of confidence and precision that is desirable for the gauge system error estimate

Step 4: Whenever possible, select the operators that normally operate the gauge.

Step 5: Choose sample parts to represent the entire range of variability of the process (not tolerance; do not manipulate or doctor up parts).

Step 6: Ensure that the gauge has graduations that are at least one-tenth of the tolerance or process variation of the characteristic being measured (depends on how the gauge is being used—to check for good/bad parts or for controlling the process).

Step 7: Ensure the measuring method and gauge are both measuring the desired characteristic.

Conducting the Study

Once the prework is finished, then the experimenter is ready to conduct the study. Generally, there are four steps. They are

Step 1: Take all measurements in a random order to ensure that any drift or changes that occur will be spread randomly throughout the study.

Step 2: Estimate all readings to the nearest number possible. At a minimum, all readings should be taken at ½ of the smallest gradation.

Step 3: If operator calibration is suspected to be a major influence on variation, the gauge should be recalibrated by the operator before each trial.

Step 4: The study should be observed by someone who understands the importance of the precautions required to conduct a reliable study.

Trials and Data Collection

The actual process begins by making sure that

- There are generally two to three operators.
- There are generally five to ten process outputs to measure.
- Each process output is measured two to three times (replicated) by each operator.

To use the variable data MSA, one must begin with the worksheet of the software (Figure 16.6). Here we enter the raw data. Once the data is entered, then we follow the rubrics for analysis in the manner shown in Figures 16.7 and 16.8.

Manipulate the Data

Once the data is in the work sheet, now we go to the commends to *stack* the data. In Minitab®, this is a very important step. If stacking is not selected, the analysis may not be correct.

10 process outputs

3 operators

2 replicates

- Have Operator 1 measure all samples once (as shown in the outlined block)
- Then, have Operator 2 measure all samples once
- Continue until all operators have measured samples once (this is Replicate 1)
- Repeat these steps for the required number of replicates (parts in random order)
- Enter data into Minitab® columns as shown

Replicate 1			Replicate 2 (randomized order)		
Part Trial 1	Operator Trial 1	Response Trial 1	Parts Trial 2	Operator Trial 2	Response Trial 2
1	1	0.60000	8	1	0.85000
2	1	1.00000	2	1	1.00000
3	1	0.80000	10	1	0.60000
4	1	0.95000	6	1	1.00000
5	1	0.45000	5	1	0.55000
6	1	1.00000	4	1	0.85000
7	1	0.95000	7	1	0.95000
8	1	0.80000	1	1	0.65000
9	1	1.00000	9	1	1.00000
10	1	0.70000	3	1	0.85000
1	2	0.65000	1	2	0.55000
2	2	0.95000	2	2	1.05000
3	2	0.75000	3	2	0.80000
4	2	0.75000	4	2	0.80000
5	2	0.40000	5	2	0.40000
6	2	1.05000	6	2	1.00000
7	2	0.90000	7	2	0.95000
8	2	0.70000	8	2	0.75000
9	2	0.95000	9	2	1.00000
10	2	0.60000	10	2	0.65000
1	3	0.55000	1	3	0.60000
2	3	1.00000	2	3	1.05000
3	3	0.80000	3	3	0.80000
4	3	0.80000	4	3	0.80000
5	3	0.50000	5	3	0.45000
6	3	1.05000	6	3	1.00000
7	3	0.95000	7	3	0.95000
8	3	0.80000	8	3	0.80000

FIGURE 16.6
Data in worksheet.

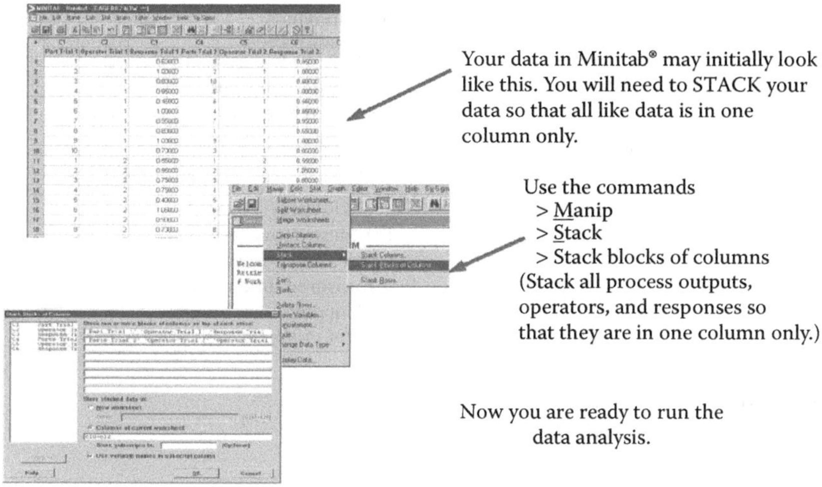

FIGURE 16.7
Process for stacking the data.

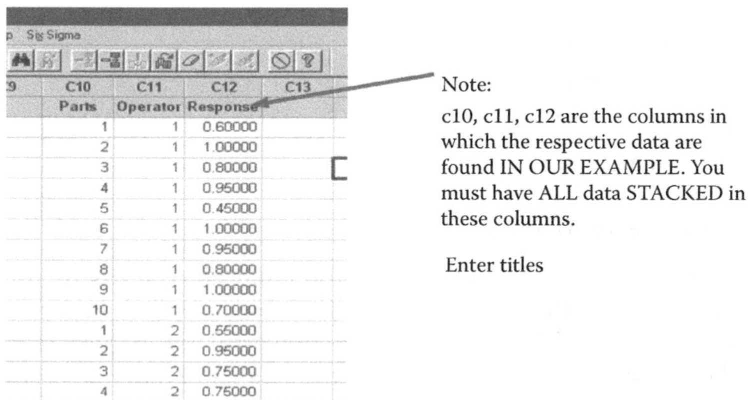

Parts	Operator	Response
1	1	0.60000
2	1	1.00000
3	1	0.80000
4	1	0.95000
5	1	0.45000
6	1	1.00000
7	1	0.95000
8	1	0.80000
9	1	1.00000
10	1	0.70000
1	2	0.55000
2	2	0.95000
3	2	0.75000
4	2	0.75000

FIGURE 16.8
Data stacked and ready for analysis.

Prepare the Analysis

Now that the data are appropriately stacked, we are ready to select our analysis. Figure 16.9 shows the selection process. There are multiple opportunities to measure crossed gauge R&R, so make sure, for example, that the appropriate and applicable method of analysis is chosen and that the error or measurement can be estimated.

On the other hand, if a single opportunity measurement (nested gauge R&R) is needed, make sure that the characteristics of the item being tested or the components of the measurement process are no longer the same as

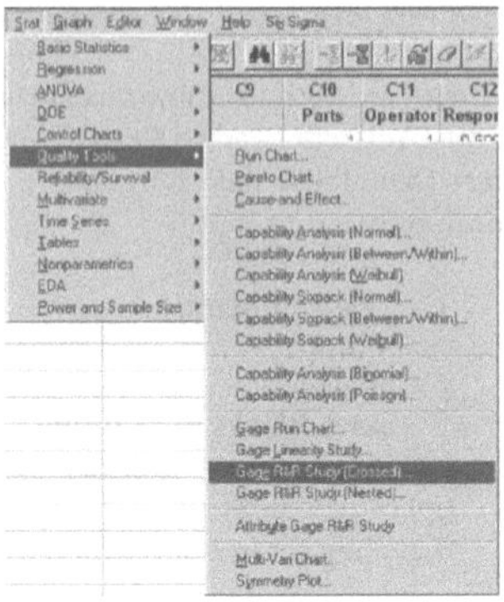

FIGURE 16.9
Selection process for crossed gauge R&R.

when the test began. It is important here to note that there is a need to collect a batch of *like* parts (minimize within variation) for the measurement system study. Some examples are

- Destructive testing
- Hardness testing
- Aging
- Chemical analysis

Choose a Method of Analysis

Generally, there are two approaches to the analysis. The first is to use the ANOVA method and the second is to use the Xbar and Rbar. Both are acceptable. However, the ANOVA is much more powerful and it gives more information. Figure 16.10 shows the selection process and Figure 16.11 shows the selection process if the tolerance is necessary.

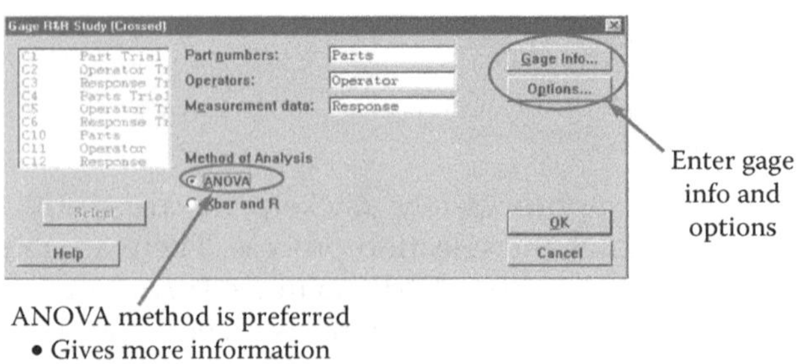

FIGURE 16.10
Selection process for method of analysis.

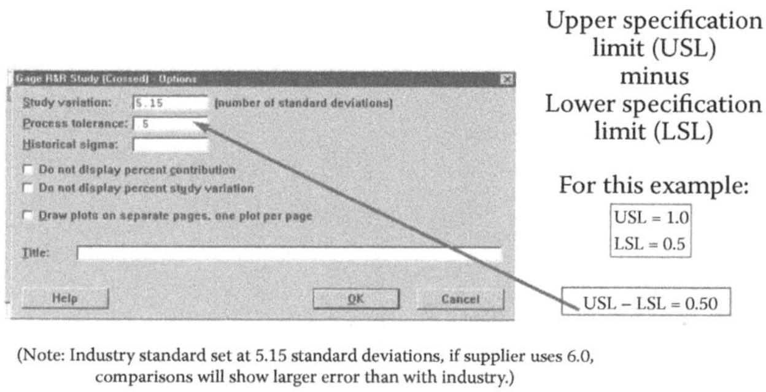

FIGURE 16.11
Selection process for adding tolerance (optional).

MSA Graphical Six Panel

The typical output for a variable MSA is both numerical and graphical. Both are shown in Figure 16.12.

Minitab® Graphical Output

The graphical output of Minitab® for a variable MSA is shown in Figure 16.13 with minimal comments.

What We Look for in These Charts

An output of software is useless unless one knows what one is looking at and for. The following is an overview of the generic concerns that should be examined.

1. Total gauge R&R < 30% (% study variation), and part-to-part to be the biggest contributor (similar to that shown).
2. R chart must be in-control. An out-of-control range chart indicates poor repeatability. In addition, you want to see five or more levels (look across the points) of range within the control limits.
3. Xbar chart must be 50% out-of-control or more (indicating the measurement system can tell a good part from a bad part) and similar patterns between operators.
4. Spread of the 10 MSA parts should represent the actual process variation.
5. The operator means should be the same (straight line).
6. Lines should be parallel and close to each other. This indicates good reproducibility.

Session window

Two-way ANOVA table with interaction

Source	DF	SS	MS	F	P
Part	9	2.05871	0.228745	39.7178	0.00000
Operator	2	0.04800	0.024000	4.1672	0.03256
Operator*Part	18	0.10367	0.005759	4.4588	0.00016
Repeatability	30	0.03875	0.001292		
Total	59	2.24912			

Gage R&R

Source	VarComp	%Contribution (of VarComp)
Total Gage R&R	0.004437	10.67
Repeatability	0.001292	3.10
Reproducibility	0.003146	7.56
Operator	0.000912	2.19
Operator*Part	0.002234	5.37
Part-To-Part	0.037164	89.33
Total Variation	0.041602	100.00

Source	StdDev (SD)	Study Var (5.15*SD)	%Study Var (%SV)	%Tolerance (SV/Toler)
Total Gage R&R	0.066615	0.34306	32.66	68.61
Repeatability	0.035940	0.18509	17.62	37.02
Reproducibility	0.056088	0.28885	27.50	57.77
Operator	0.030200	0.15553	14.81	31.11
Operator*Part	0.047263	0.24340	23.17	48.68
Part-To-Part	0.192781	0.99282	94.52	198.56
Total Variation	0.203965	1.05042	100.00	210.08

Number of Distinct Categories = 4

Graphs

Gage R&R (ANOVA) for response

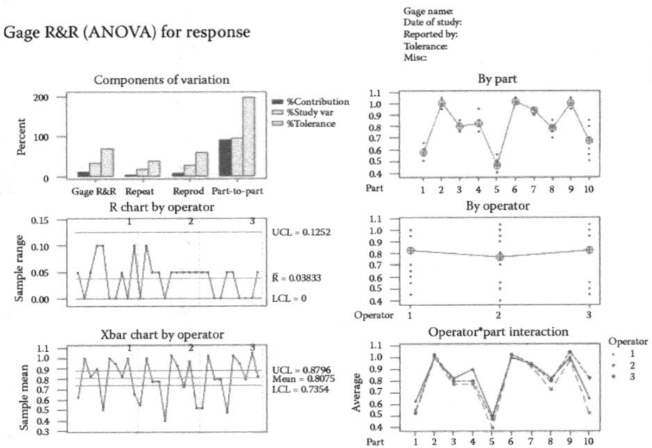

What does all this mean?

FIGURE 16.12
Minitab® output for variable MSA output.

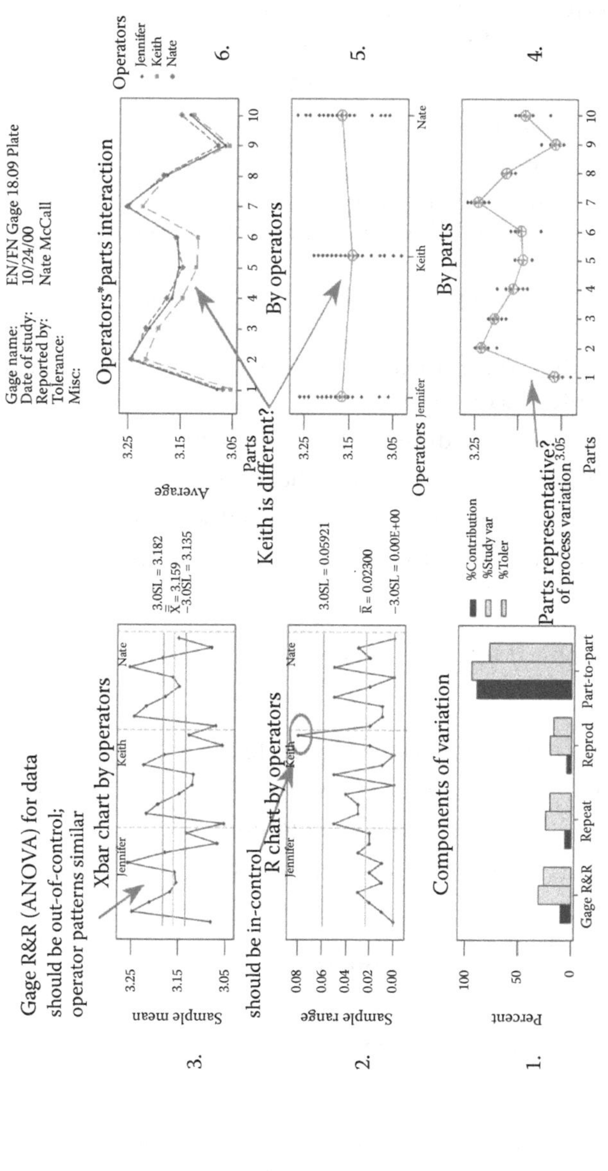

FIGURE 16.13
Typical graphical output of Minitab® for variable data MSA.

Graphical Output: Six Graphs in All

The six items that we just mentioned might be separated into two categories (see Figure 16.14). They are the health side and the troubleshooting side.

Destructive Test

When we are conducting a destructive testing, we are interested in a nested approach of gauge R&R MSA. The output for a nested gauge R&R is similar to the normal output and is shown in Figure 16.15.

Chart Output Interpretation

Now that we have identified the process of entering data and selection of the appropriate analysis, and we have shown the typical outputs of the Minitab® software in reference to the MSA, let us attempt to interpret the results.

- Tall bar charts: Distinguishes the components of variation in percentages (%)
  - Repeatability, reproducibility, and parts (want low-gauge R&R, high part-to-part variation)
- R chart: Helps identify unusual measurements
  - Repeatability/resolution (no outliers permitted)
- Xbar chart: Shows sampled process output variety
  - Reproducibility/sensitivity (want similar patterns for each operator)

The leading graphical indicators with which the evaluator should be concerned the most are

- Bar charts for components (see Figure 16.16)
- Look at the R chart first, then Xbar chart (see Figure 16.17)
- More R chart indicators (see Figure 16.18)
- ANOVA tabular output % (see Figure 16.19)

Explanation of Some of the % Numbers

$$\%\text{ Study Variation} = \frac{\sigma_{R\&R}}{\sigma_{TOTAL}} * 100$$

If the gauge does not pass %Study, it cannot perform the job of process control (it will not be able to adequately distinguish one part from another). Anything less than 10% is very good. Anything less than 30% is acceptable.

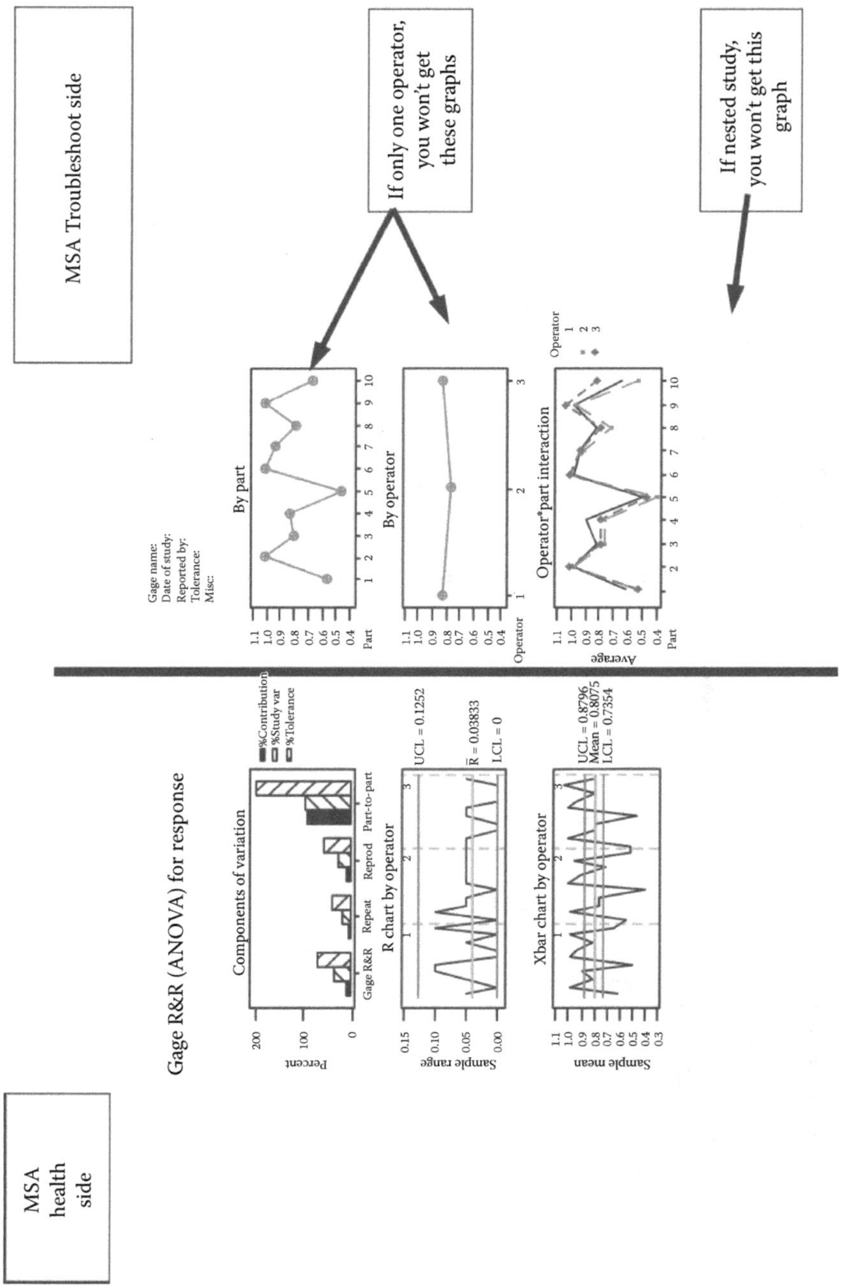

FIGURE 16.14
Typical output of the health and troubleshooting side of the graphical output.

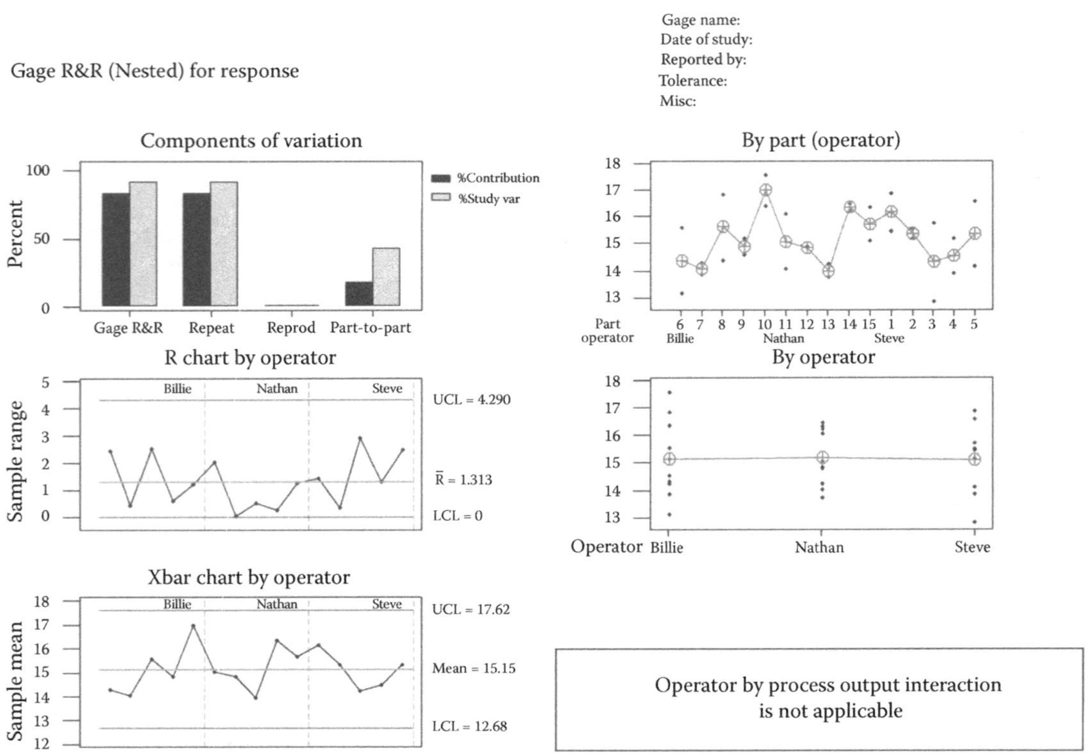

FIGURE 16.15
Nested gauge R&R.

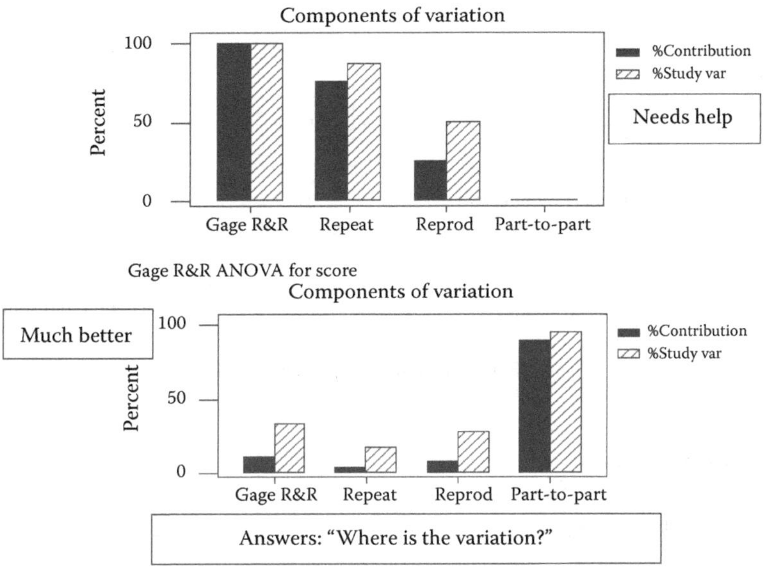

FIGURE 16.16
Bar chart for components.

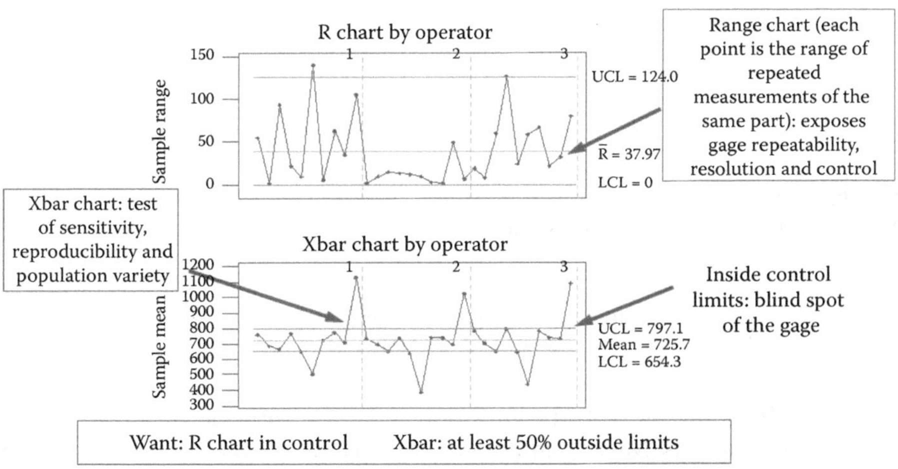

FIGURE 16.17
Evaluation of the R chart and the Xbar chart by operator.

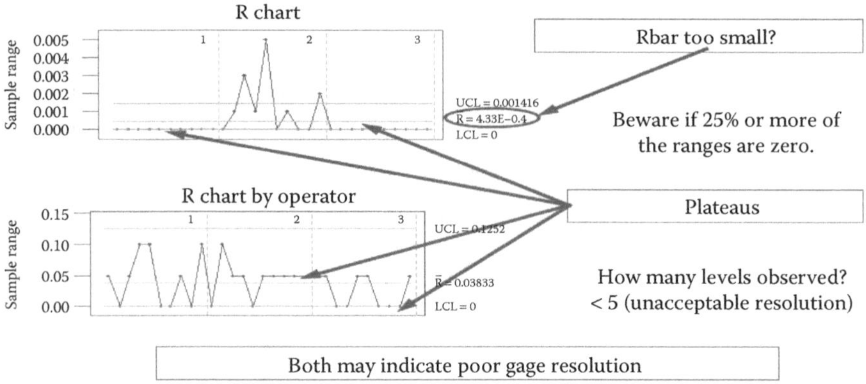

FIGURE 16.18
More R chart indicators.

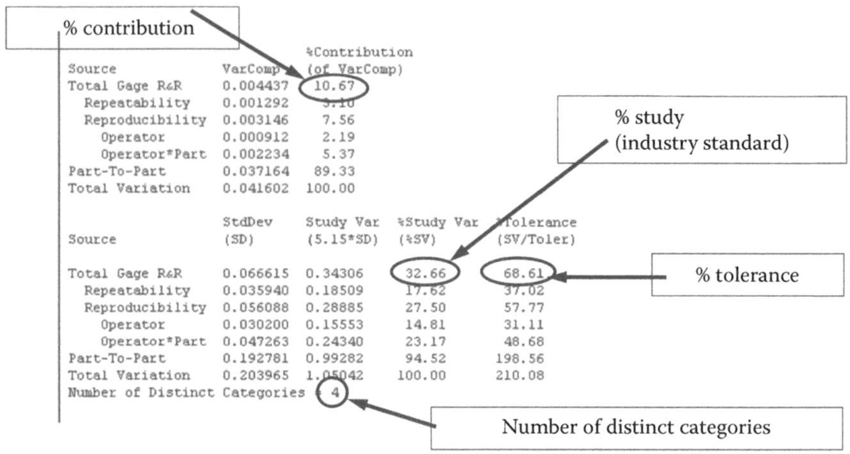


```
                                  %Contribution
Source               VarComp      (of VarComp)
Total Gage R&R       0.004437       10.67
  Repeatability      0.001292        3.10
  Reproducibility    0.003146        7.56
    Operator         0.000912        2.19
    Operator*Part    0.002234        5.37
Part-To-Part         0.037164       89.33
Total Variation      0.041602      100.00

                     StdDev     Study Var  %Study Var  %Tolerance
Source               (SD)       (5.15*SD)  (%SV)       (SV/Toler)

Total Gage R&R       0.066615   0.34306     32.66       68.61
  Repeatability      0.035940   0.18509     17.62       37.02
  Reproducibility    0.056088   0.28885     27.50       57.77
    Operator         0.030200   0.15553     14.81       31.11
    Operator*Part    0.047263   0.24340     23.17       48.68
Part-To-Part         0.192781   0.99282     94.52      198.56
Total Variation      0.203965   1.05042    100.00      210.08
Number of Distinct Categories = 4
```

FIGURE 16.19
Minitab® output of an ANOVA MSA analysis.

$$\% \text{ Contribution} = \frac{\sigma^2_{R\&R}}{\sigma^2_{TOTAL}} * 100$$

This is the same as %Study Variation, except a ratio of variances versus standard deviations. Measurement system variation (R&R) is a percentage of total observed process variation. Anything less than 9% is considered good.

Precision to Tolerance P/T

$$\% \text{ Tolerance} = \frac{5.15 * \sigma_{R\&R}}{\text{Tolerance}} * 100$$

This measurement error is a percentage of blueprint specification tolerance. Less than 10% is very good; less than 30% is acceptable. Product control includes both repeatability and reproducibility. 5.15 Study Variation = 99%

$$\text{Number of Distinct Categories} = \sqrt{2 * \left(\frac{\sigma^2_{\text{Process Output}}}{\sigma^2_{\text{R\&R}}} \right)}$$

The number of divisions that the measurement system can accurately measure across the process variation is effective resolution, not gauge resolution. This statistic tells us how well a measurement process can detect process output variation, process shifts, and improvement. Less than 5 indicates attribute conditions.

What to Look For

Repeatability Issues (within Operator)

- No gauging instructions or operational definitions
- Lack of skill in reading vernier scales
- Clamping sequence differs from trial to trial
- Loose gauge features/gauge not rigid enough
- Measuring location on part differs from trial to trial
- Semi-destructive testing changes the part (squeezing soft parts)
- Not enough gauge resolution
- Rounding away (up and down) gauge resolution
- Instrument needs maintenance

Reproducibility Issues (between Operators)

- No gauging instructions or operational definitions
- Differences of skill in reading vernier scales
- Clamping sequence differs from operator to operator
- Different (operator-owned) measurement devices
- Ergonomics: operator size, height, strength, and so on
- Different rounding methods, operator to operator
- Temperature of gauge over time, that is, expansion

Key to Successful MSA

- Define and validate measurement process
- Identify known elements of the measurement process (operators, gauges, SOP, setup, etc.)

- Clarify purpose and strategy for evaluation
- Set acceptance criteria: operational definitions
- Implement preventive/corrective action procedures
- Establish ongoing assessment criteria and schedules

Possible Causes for Bias

- Instrument needs calibration
- Worn instrument, equipment, or fixture
- Worn or damaged master, error in master
- Improper calibration or use of the setting master
- Poor-quality instrument: design or conformance
- Linearity error
- Wrong gauge for the application
- Different measurement method: setup, loading, clamping, technique
- Measuring the wrong characteristic
- Distortion (gauge or part)
- Environment: temperature, humidity, vibration, cleanliness
- Violation of an assumption, error in an applied constant
- Application: part size, position, fatigue, observed error (readability, parallax)

Possible Issues with Linearity

- Instrument needs calibration, reduce the calibration interval
- Worn instrument, equipment, or fixture
- Poor maintenance: air, power, hydraulic, filters, corrosion, rust, cleanliness
- Worn or damaged master, error in master: minimum/maximum

One of the benefits of using Minitab® is the fact that it has one of the best help menus. The help menu can be consulted at any time by the user for either help in selecting a particular test or help in the interpretation.

References

Stamatis, D. (2012). *Essential Statistical Concepts for the Quality Professional*. Boca Raton, FL: CRC Press.

Walsh, D. (2014). Automakers, suppliers buckle up. *Crain's Detroit Business*. 11–12.

Selected Bibliography

AIAG. (2010). *Measurement System Analysis (MSA)*, 4th ed. Southfield, MI: Chrysler Company, Ford Motor Company, and General Motors. Automotive Industry Action Group. (AIAG).

Arendall, C., A. Tiger, and K. Westbrook. (2014). Evaluate the potential of a process establishing process control. *Quality Progress*. 28–32.

Conover, W. (1980). *Practical Nonparametric Statistics*. New York: John Wiley & Sons.

Cox, H. (2014). Calibration: Doing your own thing. *Quality*. 19.

Gleason, S. (2014). Bending the stream of big data. *Quirk's Marketing Research Review*. 50–54.

Grachanen, C. (2014). Calculating uncertainty: Understanding test accuracy and uncertainty ratios. *Quality Progress*. 44–45.

Grachanen, C. (2014). Pass or fail: Understanding calibration content and requirements. *Quality Progress*. 44–45.

Johnson, L. and M. Deener. (2014). Necessary measures: Expanded gage R&R to detect and control measurement system variation. *Quality Progress*. 34–38.

Marascuilo, L. and J. Levin. (1970). Appropriate post hoc comparisons for interaction and nested hypotheses in analysis of variance designs: The elimination of Type-IV errors. *American Educational Research Journal*. 7(3):397–421.

Peck, R. and J. Devore (2011). *Statistics: The Exploration and Analysis of Data*. Stamford, CT: Cengage Learning.

Rocheleau, E. (August 2014). Calibration explained. *Quality*. 31–32.

Rohde, J. (2014). Asking more of the data. *Quirk's Marketing Research Review*. 62–67.

Schuetz, G. (2014). Custom gages to go: When you want to have it your way. *Quality*. 24–30.

Seaman, J. and E. Allen. (2014). Don't be discrete: Caution when categorizing quantitative variables during data analysis. *Quality Progress*. 41–43.

Sheskin, D. (2004). *Handbook of Parametric and Nonparametric Statistical Procedures*. Boca Raton, FL: CRC Press.

Stamatis, D. (2012). *10 Essentials for High Performance Quality in the 21st Century*. Boca Raton, FL: CRC Press.

17

Supplier's Perspective of APQP

Overview

As we mentioned in Chapter 1, the principles of advanced product quality planning (APQP) are very important from a customer's perspective because they assure timing, capacity, and quality characteristics being within the necessary requirements. In this chapter, we will focus on the requirements of APQP from the supplier's perspective.

Traditionally, APQP has been misunderstood and taken seriously very late in the process of launching. Suppliers do not take advantage of the positive benefits that a system such as APQP can deliver and the many value-added goals that it presents when it is viewed as a process rather than a single tool to be checked in a list of multiple activities just because "the customer says so." APQP is a structured method of defining and establishing the steps necessary to assure that a product satisfies the customer. The goal then is to facilitate communication with everyone involved to make sure that all required steps are completed on time. It is imperative that any form of quality planning in any organization *not* be viewed as a system to be used by certain individuals and for specific situations but rather as a *process* that encompasses all aspects of a program, that is, from conception to completion. This commitment to planning for quality applies to everyone involved—from management to operators.

Effective advanced quality planning depends on a company's top management commitment to achieving customer satisfaction. In essence, this effectiveness is a culture change and a mindset that calls for continual improvement and waste reduction.

For a typical supplier, the benefits of APQP are

- Promotion of early identification of required changes
- Avoidance of late changes
- Provision of a quality product on time at acceptable cost to satisfy the customer

- Provision of uniformity and standardization in the process of communicating and reporting any issues or problems to the customer

APQP Initiatives

The APQP initiative is important in that it demonstrates the continuing efforts of both customer and supplier to become quality leaders in their respective industries. It does this by providing value-added goals explicitly focused on all pertinent activities to avoid concerns rather than focusing on the results in the product or service throughout all the phases, until delivery of the product or service to the customer.

It is this reason that the deliverables discussed earlier must be clear and understood by all concerned. In fact, to make the point, we suggest reviewing Appendix I. The appendix is a typical detailed *expectation list* of a site assessment by the customer to the supplier's facility. By understanding the value-added goals, the supplier will benefit immensely because they add to better quality, lower cost, productivity increases, and timely delivery. Some of the value-added goals might become instrumental to

- Reinforce the organization's focus on CI in quality, cost, and delivery
- Provide the ability to look at an entirely new program as a single unit
- Be prepared for every step in the creation of the product
- Identify where the most amount of effort must be centered
- Create a new product with efficiency and quality
- Provide a better method for balancing the targets for quality, cost, and timing
- Deploy targets that use detailed practical deliverables within specific timing schedule requirements
- Provide a tool (methodology) for all concerned to follow-up all planning processes
- Provide consistency and stabilization results in cost improvement opportunities including reduction of special sample test trials

Understanding the program requirements makes it easier for the supplier to fulfill the expectations of the customer, but it also provides the following advantages to the customer.

- Clarifies the expectations early on in the relationship
- Focuses on the deadlines to make sure they are met

- Tracks special characteristics (critical and significant)
- Evaluates and avoids quality, cost, and timing risks
- Clarifies any bottlenecks and concerns for all appropriate product specifications using a common control plan

APQP Reporting

It is the responsibility of the supplier to do the APQP and appropriately report the status to the customer. The frequency of reporting depends on the program and the milestone. Early in the program, the reporting may be once a quarter, but as the time gets closer to Job#1, the frequency may be weekly.

The initiation process depends on the cross-functional team deciding that APQP will be used to track the program. Once this step is agreed on, APQP is initiated. The lead and support teams are made aware of their roles in completing each of the elements. The lead responsibility is required to establish the layout for each APQP element.

As the program moves along the timing and specific milestones approach, reviews are conducted to discuss the milestones and whether these milestones are met. Based on these reviews, a decision is made as to the status of the APQP process. This assessment is reflected in the coding of three colors, that is, green, yellow, and red. If a program is assessed at green, it proceeds to the next step. If a program is assessed at yellow or red, it proceeds to the next step, which is risk assessment. It is the lead's responsibility for each element to provide the appropriate and applicable documentation for all supporting activities.

The definition of the designations of green, yellow, and red is very specific and must be used accurately at all times. The classic definitions of the colors are

- Green is the best possible evaluation for an APQP element. A green or "G" is given before the program milestone date to indicate the element will meet the program milestone and will meet all quality expectations. A "G" rating given on the program milestone date indicates the element is complete and meets *all* quality expectations.
- A yellow or "Y" rating is given prior to the program milestone to indicate that an element might not meet the milestone or quality expectations. To be a "Y," a recovery plan must be in place for the element. A Y rating indicates a need for program management attention. A Y rating can *only* be given to an element prior to the program milestone date.

- Red is the worst possible evaluation for an APQP element. A red or "R" rating is given prior to the program milestone date to indicate that an element will not meet the program need date or quality expectations and has no recovery plan as to the specific milestone. R signifies the program is at risk and needs immediate management attention. Any element rated R at its program need date must carry the R rating through the remainder of the program. Completion of the element after the program need date does not change the timing status of the element. In other words, the element is late and must remain red. To reflect improvements of a red element status after missing the milestone date, progress to green will be shown in the next milestone review.

Fundamental Elements of APQP

In the automotive industry, there are several (generic) elements in the APQP process and they are published in AIAG (2008). Here we focus on key items of which the supplier must be aware and complete them.

We begin with the APQP focus, which is twofold:

- Up-front quality planning; the earlier the planning, the less the risk of failure
- Identifying key processes so that customers are satisfied by evaluating and accepting the output; determining appropriate and applicable support for continual improvement

To deliver on these two focus items, the APQP process is divided into five phases. They are

1. Plan and define program
2. Product design and development verification
3. Process design and development verification
4. Product and process validation and production feedback
5. Launch, assessment, and corrective action

Within these phases, there are five major activities:

1. Planning
2. Product design and development
3. Process design and development

4. Product and process validation
5. Production

Out of all the key items in the APQP, seven major elements stand out. They are

1. Understanding the needs of the customer
2. Having a proactive feedback and corrective action
3. Designing within the process capabilities
4. Analyzing and mitigating failure modes
5. Having appropriate and applicable verification and validation
6. Having design reviews
7. Having a process to control special (critical and significant) characteristics

By now, you may feel somewhat confused because of all the divisions we just covered. The reason for all these divisions is to make sure that the supplier understands why the APQP is so important for them to make sure they complete on time and without any glitches. To demonstrate further the importance of each element, we will provide specific definitions that split each element into six separate areas. These areas are

- Definition: Identifies the requirements for the element.
- Expectations: Defines the requirements for the element. This usually is given to the supplier by the customer.
- Lead responsibility: Defines and identifies the activities responsible for lead reporting. Activities that all others will support in completion of the element.
- Support function: Defines and identifies the support activities that will provide input to and assist the lead responsibility.
- Timing: Identifies the initial and final milestone timing for the element. It is worth mentioning here that although the customer initiates the boundary of the timing milestones, the supplier confirms the material delivery date (MDD). Therefore, it is imperative that the supplier understands and agrees to the timetable. Furthermore, it is the responsibility of the supplier to completely understand the timing plan for each element and the supplier should be able to identify the period of each phase. If the supplier disagrees with the MDD, let the customer be aware as soon as possible.
- Deliverables: Indicates the items that must be completed during three separate time frames for the element. The three time frames are

- Initial: Deliverables that must be completed upon initiation of the element or before the element enters the intermediate phase.
- Intermediate: Deliverables that must be completed during the process of the element. These deliverables may be started during the initial phase, but must be completed before the final phase.
- Final: Deliverables that must be completed before the process moves on to the next element.

Note two very critical issues in these elements. First, it is the supplier's responsibility to complete the elements of APQP accurately and on time. If there is a problem, it is also their responsibility to contact the customer as soon as possible and notify the appropriate customer representative of the pending problem. At that point, the customer representative may offer suggestions for fixing the issue and/or be personally involved in either training or helping the supplier to fix the problem. Second, the supplier must understand that all APQP elements are closely related and require correlation of inputs and outputs. For example, to successfully complete Element 15, Prelaunch control plan, one must be provided and utilize outputs from Element 3, Design FMEA; Element 5, Design verification plan; and Element 8, Prototype build control plan. For the identification of the numerical values, see Table 1.1.

Risk Assessment

The risk assessment step is used to identify the problems that have caused a yellow or red assessment. At this point, a corrective action to return the flow to a green assessment is created and implemented. The corrective action is then reassessed in the major review of the program. A typical risk assessment form is shown in Figure 17.1.

A summary of the aims, expectations, and requirements associated with documenting the individual elements of the APQP Status Report can be found in Schaeffler (2013a,b). A very cursory review is given in Table 17.1. Schaeffler provides a good overview of QAA/S 296001 in Parts 1–6 and several appendices. In Table 17.1, we have rearranged his information in a table form for convenience. In Appendix I, we provide a detailed guideline form for addressing expectations of each of the several elements of APQP primarily based on the automotive industry.

Program name:	Part name/#:
Model/year:	Element name:
Months before Job#1:	APQP status rating:
List all functional activities affected by the concern:	
Summarize the concern creating the risk. Quantify the concern as appropriate:	
Define the risk due to the concern. Include potential impact on quality, cost, and timing:	
Recommend appropriate action:	
Sign assessment by authorized representative to indicate approval of recommended actions:	

FIGURE 17.1
Typical risk assessment form.

Control Plans

In conjunction with APQP, there are also two significant additions as part of the methodology. The first one is the control plan and the second is the Supplement K. Many suppliers do not understand the significance of a control plan and therefore they dismiss its value by identifying simple and often insignificant controls for their processes so they can "move on" to issues that are more important. They fail to understand that most of the customer's problems are supplier-related issues. In fact, from my experience, I would estimate that the percentage of supplier-related issues is close to approximately 60%. Of this percentage, approximately 25% is design related and approximately 35% is manufacturing related.

The primary reason for the control plan's existence is to make sure that there is a plan in place to fix a deficiency if it occurs. It must be linked to the process failure mode and effect analysis (FMEA) and process flow chart.

Supplement K

Supplement K, on the other hand, is a formal form (worksheet) to facilitate all special process and product characteristics agreed by the customer and the supplier. The characteristics are numerically identified in a sequence so the supplier does not overlook them when the control plan is completed. Generally, this worksheet is a supplement to the control plan form and is viewed as an extension of it.

TABLE 17.1

Aims, Expectations, and Requirements Based on Schaeffler's Characterization of APQP

	Customer Order	*Customer Specifications*	*Contract Review*
	This is what the customer wants	Customer specifications include, e.g., the design briefs, drawings, or specifications that are required at the start of a project	Customer's expectations are reviewed and evaluated
Aim	Formal placing of an order by the customer so that investments can be effected on schedule by the supplier	To avoid misunderstandings by using clear specifications	To carry out a commercial and technical assessment of the documents provided by the customer to check that they are complete, up-to-date, and feasible
Expectations	The customer selects a supplier and informs the supplier of the decision	The supplier must be familiar with the requirements of the product/project, e.g., • Installation situation • Ambient conditions • Functional performance requirements • Dimensions • Weight • Material • Reliability (durability) • Guarantee objectives • Quality objectives for incoming parts (ppm rating, defect levels, and rejection quotas) • Capacity data or volumes • Milestones • Checking the program status • Final definition of the design (design freeze) • Prototype parts • Preproduction • Initial samples	Before entering into a delivery agreement with the customer, a contract review is carried out to ensure that • The product requirements have been suitably defined and documented (e.g., drawing, specification, design requirements) • Deviating requirements are clarified prior to submitting the quotation or closing the contract (e.g., drawing deviations, drawing modifications made by the customer, delivery dates, price) • The defined requirements can be met; The quotation is prepared following a cross-functional manufacturing feasibility analysis

Documentation with supplier	• Nomination letter • Individual order indicating project classification for APQP and submission level for sampling • Delivery schedule	• Design briefs, drawings, or specifications • List showing issue level of the individual documents	• Capacity confirmation • Feasibility confirmation
Submission to customer	• Status in APQP status report	• Status in APQP status report • List showing issue level of the individual documents	• Status in APQP status report • Feasibility confirmation • Capacity confirmation
	Craftsmanship: Appearance	*Design FMEA*	*Design Review*
		Only applicable to suppliers with personal responsibility for product development A design FMEA is a systematic procedure used to ensure that potential development and design defects and their respective causes have been taken into account and have been countered by preventive measures within a technical cross-functional team The design FMEA must cover all functions of the product and must take into account experiences and concerns	
Aim	To define characteristics relating to appearance, texture, handling ability, and acoustics	To prevent defects during product development	To prevent misunderstandings and problems, to monitor the progress of measures and to ensure that objectives are met

(*Continued*)

TABLE 17.1 (Continued)

Aims, Expectations, and Requirements Based on Schaeffler's Characterization of APQP

Expectations	The supplier should be familiar with and meet the requirements that relate to the above-mentioned characteristics. In order to check the characteristics, characteristics catalogues and reference samples and so on should be made available and agreed on with the customer	Problems relating to product layout are solved in good time so that all sampling and mass production dates can be observed Checks can be made on progress in respect to the design verification plan Unforeseen potential defect modes, which occur at the design review stage, must be incorporated into the design FMEA and the layout criteria modified in agreement with the customer Checks must be made on potentials for improvement regarding product reliability and manufacturing costs Checks must be made on progress in terms of achieving goals relating to reliability, quality, costs, and planning Influencing defect modes are described and assessed and, where necessary, corrective measures are introduced and monitored by the persons responsible	Problems relating to product layout are solved in good time so that all sampling and mass production dates can be observed Checks can be made on progress in respect to the design verification plan Unforeseen potential defect modes, which occur at the design review stage, must be incorporated into the design FMEA and the layout criteria modified in agreement with the customer Checks must be made on potential improvements regarding product reliability and manufacturing costs Checks must be made on progress in terms of achieving goals relating to reliability, quality, costs, and planning
Documentation with supplier	Characteristics catalogues and reference samples	Design FMEA	Meeting reports prepared by the supplier or the customer
Submission to customer	Status in APQP status report	Status in APQP status report Cover sheet to design FMEA including participants and issue level	Status in APQP status report

	Design Verification Plan (DVP)	*Quality Planning of Subcontractor*	*Equipment and Tools*
	Only applicable to suppliers with personal responsibility for product development	The suppliers must forward the APQP requirements to their subcontractors, check implementation, and give the results in summarized form in the APQP status report	
Aim	Systematic planning of all tests or calculations to check whether the product or design is suitable for the application	To provide a clear illustration of the progress that a subcontractor is making on a project for process/products associated with increased risks or at the customer's specific request	Capable/released operating materials
Expectations	The suitability of the product must be proven by means of planned tests or calculations Responsibility for tests on mass products must be defined	The supplier must carry out a risk assessment and specify the extent of its subcontractors' participation in the advance quality planning process The suppliers must check the progress of the project on a regular basis with their subcontractors, particularly if they deliver these products with special characteristics	Planning (monitoring of dates) and provision of all necessary operating materials Deadlines for the procurement or manufacture of equipment and tooling must be monitored The equipment and tooling should be tested prior to the production trial run
Documentation with supplier	Design verification plan Test reports, layout calculations, and tolerance studies	Subcontractor's status report on the progress of the project	Schedules, capacity plans, proof of capability established within the framework of the initial sampling process
Submission to customer	Status in APQP status report Status for design verification (current status of DVP)	Status in APQP status report Detailed schedules at the customer's request	Status in APQP status report Detailed schedules at the customer's request
	Inspection Methods and Inspection Equipment	**Inspection Plan/Control Plan for Prototype Parts**	**Production and Inspection of Prototype Parts**
Aim	To obtain inspection methods that have been coordinated with the customer and inspection equipment which is suitable for the measuring task	To secure the prototype quality	To despatch, on schedule, prototypes that meet cost and quality requirements

(Continued)

TABLE 17.1 (Continued)

Aims, Expectations, and Requirements Based on Schaeffler's Characterization of APQP

Expectations	The inspection methods should be coordinated between the supplier and the customer Deadlines for the procurement or manufacture of equipment and tooling must be monitored Proof of inspection equipment capability and, where necessary, inspection process suitability must be provided	The type and scope of the inspections and the corresponding inspection equipment for prototypes are defined and agreed on with the customer All special characteristics are included	Dates and quantities for the manufacture of prototypes must be planned, monitored, and met Inspection reports must be supplied with the prototypes Approval must be obtained from the customer for nonconforming prototypes prior to delivery
Documentation with supplier	Stipulation in control plan, proof of capability within the framework of the initial sampling process	Control plan for prototypes	Prototypes, inspection report
Submission to customer	Status in APQP status report	Status in APQP status report Control plans for prototype	Status in APQP status report Prototypes and inspection report
	Design Freeze (Drawings/Specifications)	*Confirmation of Manufacturing Feasibility*	*Process Flow Chart and Mass Production Layout*
	The term *drawings and specifications* includes all technical drawings, CAD data, material specifications, and technical specifications from the customer/supplier	A cross-functional team must assess the manufacturing feasibility of the intended product within the framework of the contract review. Even if the customer is responsible for the design, the supplier must assess the manufacturing feasibility of the products, both at the quotation/prototype phase and at the planning stage of the mass production process	The mass production process flow chart is a graphical representation of the planned sequence of operations

Aim	To provide, on schedule (design freeze) all drawings and specifications that are needed to observe the initial mass production sample date/start of production	To assess production feasibility (mass) relative to the intended design	To provide a foundation for investment planning, process FMEA, production plan, control plan, and visual aids
Expectations	The customer must be notified by the supplier of the latest possible date for making changes to serial drawings and specifications to ensure that the planned initial mass production date is met In case the supplier has the responsibility for product development, the drawings and specifications are agreed on with the customer at that time Any special characteristics that have been specified must be given appropriate consideration within the framework of the product and process planning process	The team must be convinced that the product is suitable for the intended application and can be produced, checked, packed, and delivered to the customer in sufficient quantities, at a competitive price, and to the required quality If subcontractors are involved and their activities have an influence on special characteristics, the supplier must decide whether to request confirmation of manufacturing feasibility from them	To ensure the sequence of all mass production stages and inspection stages from goods inward through to goods issue
Documentation with supplier	Drawings Specifications	Feasibility confirmation	Process flow diagram Machine setting plan
Submission to customer	Status in APQP status report	Status in APQP status report Feasibility confirmation	Status in APQP status report Process flow chart

(*Continued*)

TABLE 17.1 (Continued)

Aims, Expectations, and Requirements Based on Schaeffler's Characterization of APQP

	Process FMEA	*Inspection Equipment Capability*	*Preproduction/Prelaunch Control Plan*
	A process FMEA is a systematic procedure used to ensure that potential production defects and their respective causes have been taken into account and have been countered by preventive measures within a technical pan-divisional team The process FMEA must cover all production and inspection stages and must take into account experiences and concerns		
Aim	To prevent defects during process development	To evaluate the suitability of the intended inspection equipment by means of a measuring system analysis/capability study, e.g., using the MSA method of the AIAG	To ensure that process and product requirements are met during preproduction or prelaunch, e.g., through increased inspection frequency and additional inspection characteristics
Expectations	All production and inspection stages are listed and analyzed Potential defect modes are described and evaluated and, where necessary, corrective measures are introduced and monitored by the persons responsible Significant numbers are coordinated with the customer	Capability studies must be carried out in order to check the suitability of the inspection equipment The customer should be given the opportunity to check the results Where appropriate, inspection/measurement methods must be agreed on in good time with the customer If modifications are made to the inspection and measuring equipment, the capability studies must be repeated	The type and scope of the inspections and the corresponding inspection equipment for preproduction or prelaunch are defined and agreed on with the customer Reaction plans are defined for handling discrepancies All special characteristics are included

Documentation with supplier	Process FMEA Action plan for high risk priority number in the order of S, SxO, RPN	Proof of capability with individual values, e.g., printouts from the statistical program used	Control plan for preproduction/prelaunch
Submission to customer	Status in APQP status report Cover sheet to process FMEA including participants and issue level Pareto analysis of high-severity numbers or risk priority numbers (top 20 of RPN)	Status in APQP status report Proof of capability with individual values within the framework of the production process and product release procedure	Status in APQP status report Control plan for preproduction/prelaunch
	Process Instructions	*Logistics Concept*	*Packaging Materials*
	All instructions for production personnel, e.g., production plans, work instructions and inspection instructions, maintenance plans, defect catalogues, and process parameters		
Aim	To ensure quality and quantity	To ensure delivery capability in agreement with the relevant customer contact	To maintain product quality using suitable packaging materials
Expectations	Easily understood (available in the national language of the respective production site) and accessible instructions, at the workplace, to ensure that procedures are followed and requirements regarding the process or the product are implemented Descriptions are given of the procedures involved in controlling defective products All employees must have been trained or instructed in the tasks they are to perform. Proof of training must be kept and competences regulated	Call-off system, e.g., Web-EDI Definition of transport routes Customs arrangements (where applicable) Inventory, e.g., consignment warehouse, safety stock Consideration of special items required by either the customer or supplier	Suitable packaging materials have been defined for transport from or to the subcontractor, internal transport/storage, and despatch to customer Packaging specifications and corrosion protection for shipments to the customer have been agreed on with the customer Related customer packaging specifications must be observed It must be ensured that the quality of the product is not impaired when it is packed, despatched, put into storage, and removed from storage

(*Continued*)

TABLE 17.1 (Continued)

Aims, Expectations, and Requirements Based on Schaeffler's Characterization of APQP

Documentation with supplier	Process instructions Proof of training	Logistics agreement, minimum inventory planning	Defined packaging specifications and corrosion protection
Submission to customer	Status in APQP status report	Status in APQP status report	Status in APQP status report Packaging specification and corrosion protection agreed on with the customer
	Production Trial Run	*Mass Production Inspection Plan/Control Plan*	*Preliminary Process Capability Study* Process capability studies use statistical methods to prove that the product can be manufactured in line with the specification. Proof of the capabilities of specific characteristics and, where applicable, other characteristics defined with the customer within the framework of APQP must be provided
Aim	To check the effectiveness of the mass production process	To ensure that process and product requirements are met during mass production	To provide statistical proof of capable processes. With some customers, it is also important to document a preliminary capacity
Expectations	Use of mass production installations, machinery, tooling, inspection equipment, and conditions (including regular operating personnel), also at the subcontractors premises Use of mass production materials Verification of the required product quality and the planned nominal quantities (capacity)	The type and scope of the inspections and the corresponding inspection equipment for mass production are defined and agreed on with the customer Reaction plans are defined for handling discrepancies All special characteristics are included	Preliminary process capability under mass production conditions, e.g., at P_p/P_{pk} values of > 1.67 (at least 25×5 parts) and machine capability (50 parts) at C_m/C_{mk} values of > 1.67 In the event of incapable processes, suitable corrective measures must be introduced in order to achieve process capability. A 100% inspection must be carried out until process capability is achieved

	Verification of the mass production process The production quantity must consist of at least one production batch size, which is representative of the process (usually daily requirement from annual requirement) Taking of initial mass production samples from this batch Customer participation, if stipulated beforehand		
Documentation with supplier	Documentation with records and/or proof of capability	Documentation with supplier Control plan for mass production	Proof of capability and precapacity with individual values
Submission to customer	Status in APQP status report	Status in APQP status report Control plan for mass production	Status in APQP status report Proof of capability with individual values within the framework of the production process and product release procedure Action plan to achieve required process capabilities, where required
	Technical Tests on Mass Production Parts	*Initial Sampling Inspection*	
		Documented proof that the product manufactured under mass production conditions meets the customer's requirements. The production process and product release procedure is described in detail in Part 2 of QAA/S 296001	
Aim	To provide proof that the mass product meets customer requirements with the aid of a structured process verification plan (PVP)	To provide proof of product and process release	

(Continued)

TABLE 17.1 (Continued)

Aims, Expectations, and Requirements Based on Schaeffler's Characterization of APQP

Expectations	Technical tests with products from the production trial run, if defined in the design verification plan or in the specification	To produce, on schedule, the initial mass production samples; to prepare documentation for all elements requested by the customer in accordance with the production process and product release procedure To provide documentation on schedule; scope in accordance with defined submission level
Documentation with supplier	Records and/or examination reports	Initial sample inspection report, initial mass production samples Documentation for all requested elements in accordance with the production process and product release procedure
Submission to customer	Status in APQP status report Proof within the framework of the production process and product release procedure	Status in APQP status report Scope in accordance with defined submission level

Some customers have improved the Supplement K worksheet with a computerized version called Special Characteristics Communication and Agreement Form (SCCAF). The primary difference between SCCAF and Supplement K is that SCCAF includes where special characteristics are controlled at subtier suppliers, and customer approvals are recorded and communicated to the appropriate personnel.

References

AIAG. (2008). *Advanced Product Quality Planning and Control Plan*, 2nd ed. Southfield, MI: Chrysler Corporation, Ford Motor Company, General Motors. Automotive Industry Action Group (AIAG).

Schaeffler, S. (2013a). QAA/S 296001. Quality Assurance Agreement with Production Material Suppliers. Part 1. Advanced Product Quality Planning. http://www.schaeffler.com/remotemedien/media/_shared_media/12_suppliers/quality/quality_assurance_agreement__qaa_/history_of_revisions.pdf. Retrieved on August 27, 2014.

Schaeffler, S. (2013b). QAA/S29600. Quality Assurance Agreement with Production Material Suppliers. Part 2. Production Process and Product Release Procedure. http://www.schaeffler.com/remotemedien/media/_shared_media/12_suppliers/quality/quality_assurance_agreement__qaa_/history_of_revisions.pdf. Retrieved on August 27, 2014.

Stamatis, D. (1998). *Advanced Product Quality Planning*. Boca Raton, FL: St. Lucie Press.

Stamatis, D. (2004). *Integrating ISO 9001 with ISO/TS 16949 and AS 9000*. Milwaukee, WI: Quality Press.

18

Supplier's Perspective on the PPAP

Overview

The essence of a production part approval process (PPAP) is to make sure that the supplier knows and monitors its process to the point where the product going to the customer is *flawless*. The PPAP is product specific and *never* facility specific. Specifically, any PPAP focuses on six primary areas. They are

1. If a process owner exists
2. If a process is defined
3. If a process is documented
4. If linkages of a process are established
5. If a process is monitored, analyzed, and improved
6. If records are maintained

These 6 areas are embellished into 18 specific requirements. These are not new but repositioned requirements of the advanced product quality planning (APQP) process (AIAG, 2008). This means that if the supplier has done diligence work on time for all APQP requirements when they were due (milestone), then the PPAP process is not difficult or time consuming. The 18 elements of the full PPAP are

1. Design records: A printed copy of the drawing needs to be provided. If the customer is responsible for designing, this is a copy of the customer drawing that is sent together with the purchase order (PO). If the supplier is responsible for designing, this is a released drawing in the supplier's release system. Every feature must be *ballooned* or *road mapped* to correspond with the inspection results (including print notes, standard tolerance notes and specifications, and anything else relevant to the design of the part). Prints (if applicable) must be in the geometric and dimensioning tolerance (G&DT) format.
2. Engineering change documents: A document that shows the detailed description of the change. Usually this document is called engineering change notice, but it may be covered by the customer

PO or any other engineering authorization. If the customer initiates a change during the PPAP process, it is called an alert and is good for 60 days. If during these 60 days nothing happens, it reverts back to the original status. If the change is complete, then the appropriate documentation is updated and made part of the final PPAP. If it is work in process, it may be closed and a new one issued or the current one may be extended. (Obviously, the days may depend on the customer.)

3. Engineering approval: This approval is usually the engineering trial with production parts performed at the customer plant. A *temporary deviation* usually is required to send parts to the customer before PPAP. The customer may require other engineering approvals.
4. Design FMEA: A copy of the design failure mode and effect analysis (DFMEA) is reviewed and signed off by the supplier and the customer. If the customer is responsible for the design, the customer may not share this document with the supplier. However, the list of all special (critical and significant) or high-impact product characteristics should be shared with the supplier, so they can be addressed on the PFMEA and control plan. If the supplier is responsible for the design, then the FMEA must be shared with the customer. If the part is confidential, the customer may review it in the supplier's facility, but the customer cannot remove the document from the facility. It cannot be faxed, e-mailed, copied, or in any other way transferred unless properly authorized by the supplier.
5. Process flow diagrams: A copy of the process flow, indicating all steps and sequence in the manufacturing process, including incoming components, must be available for review.
6. Process FMEA: A copy of the process failure mode and effect analysis (PFMEA) is reviewed and signed off by supplier and customer. The PFMEA follows the process flow steps, and indicates what could go wrong during the manufacturing and assembly of each component. All special characteristics must be accounted for and linked to the control plan.
7. Control plan: A copy of the control plan is reviewed and signed off by the supplier and the customer. The control plan follows the PFMEA steps, and provides more details on how potential issues are checked in the incoming quality, assembly process, or during inspections of finished products.
8. Measurement system analysis (MSA) studies: MSA usually contains the gauge repeatability and reproducibility (R&R) for the special or high-impact characteristics and a confirmation that gauges used to measure these characteristics are appropriate and calibrated.

9. Dimensional results: A list of every dimension noted on the ballooned drawing. This list shows the product characteristic, specification, the measurement results, and the assessment showing if this dimension is "OK" or "not OK." Usually a minimum of six pieces is reported per product/process combination.
10. Records of material/performance test results: A summary of every test performed on the part. This summary is usually on a form of the design verification plan and report (DVP&R), which lists each individual test, when it was performed, the specification, results, and the assessment pass/fail. If there is an engineering specification (ES), usually it is noted on the print. The DVP&R shall be reviewed and signed off by both the customer and the supplier engineering groups. The quality engineer will look for a customer signature on this document. In addition, this section lists all material certifications (steel, plastics, plating, etc.) as specified on the print. The material certification shall show compliance to the specific call on the print.
11. Initial process studies: Usually this section shows all statistical process control (SPC) charts affecting the most critical characteristics. The intent is to demonstrate that critical processes have stable variability and that they are running near the intended nominal value. Sometimes in this element, an initial sample inspection report will be attached to identify the material samples that were initially inspected before the prototype stage.
12. Qualified laboratory documentation: A copy of all laboratory certifications (e.g., A2LA, TS) of the laboratories that performed the tests reported in Element 10. All laboratory tests must have traceability to the National Institute of Standards and Technology (NIST) or be approved by the customer.
13. Appearance approval report: A copy of the appearance approval inspection (AAI) form signed by the customer. This is applicable for components affecting appearance only.
14. Sample production parts: A sample from the same lot of initial production run. The PPAP package usually shows a picture of the sample and where it is kept (customer or supplier).
15. Master sample: A sample signed off by customer and supplier, this is usually used to train operators on subjective inspections such as visual or for noise attributes.
16. Checking aids: When there are special tools for checking parts, this section shows a picture of the tool and calibration records, including dimensional report of the tool.
17. Customer-specific requirements: Each customer may have specific requirements to be included on the PPAP package. It is a good

practice to ask the customer for PPAP expectations before even quoting for a job. For example, North American automakers original equipment manufacturer (OEM) requirements are listed on the International Automotive Task Force (IATF) website.

18. Production part approval process submission warrant (PSW): This form summarizes the whole PPAP package. This form shows the reason for submission (design change, annual revalidation, etc.) and the level of documents submitted to the customer. There is a section that asks for "results meeting all drawing and specification requirements: yes/no," which refers to the whole package. If there is any deviation, the supplier should note on the warrant or inform that PPAP cannot be submitted. A successful launch is where the ramp-up curve is met without missing parts or shortages and without the requirement to retrofit any component for quality, capacity, or engineering reasons. Here we must make a very strong point: Even though a PSW may be completed and approved, the customer must approve the shipping for production purposes. That approval comes after *functional testing* at the customer's production facility. If the functional testing passes, then production is commenced; if not, then the parts of the PPAP that fail must be redone.

The notion of the PSW is critical and many suppliers do not understand the ramifications. Therefore, let us look at the PSW in detail. We start with the scope of the PSW, which is to redefine both the interpretation and execution of the PSW and PPAP. Due to *compromised* interpretations because of timing pressures as well as parts that either do not meet specification or do not meet the required volume, launches and Job1 have often been disrupted and delayed. This is a very serious problem and every OEM is very concerned about it.

Therefore, to understand the importance of the PSW, we must understand and internalize what is and what is not a PSW. A PSW run is conducted at the intended point of production using the production machinery, tools, and facilities with the intended people at the intended production rate that is capable of meeting the customer's daily production rate (as specified on the request for quotation) and results in components that are correct to drawing and meet *all* customer's specifications. Typically, the volume produced will be sufficient to supply the customer with one full day's production.

On the other hand, the following list will indicate what is not a complete PSW. This list is not an exhaustive one, but it gives a direction that should be followed.

- Parts produced at the facility supplier and not at the production source

- Parts made from one impression tool but production intent is from four impression tools
- Parts produced meet the ramp-up volume but not the full laid down volume
- Parts are produced off one line but in production, three lines are required (e.g., wiring looms)

To avoid issues of the PSW, we recommend

- PSW runs should be conducted from a run of a minimum of one day's production (a day's production is identified from the request for quotation), run at the intended production rate. Any exceptions require an alert signed by the customer's engineering, purchasing, and manufacturing departments.
- Any part not meeting the required production rate (specified on the request for quotation) is not PSW and requires an alert signed by the customer's engineering, purchasing, and manufacturing departments.
- Any part that does not meet any of the quality or specification criteria is not PSW and requires an alert signed by the customer's engineering, purchasing, and manufacturing departments.
- Any part not meeting the statement for what is a good PSW is not a PSW and requires an alert signed by the customer's engineering, purchasing, and manufacturing departments.
- At a program review, whenever PSW data is shown it should always state % PSW and separately % PSW (interim) with approved alert.

In some cases, the supplier may be issued a conditional PSW *if and only if* everything is acceptable except the results of some testing that have been conducted but for some reason are not reported yet. Obviously, when the results are available, the final determination will be made.

In addition, some customers demand that part of the PSW must include the evaluation of capacity volume. If that is the case, the volume produced for the PSW run should be identified on the PSW form. In all cases, the PSW is always a *legal document*.

PPAP Submission Levels and Approvals

The evaluation of the PPAP takes the form of levels (AIAG 2009). This means that there are different criteria for reporting the PPAP. However, the requirements for completing the PPAP are the same for all levels. This is very important because some suppliers assume different interpretations on what the levels are about. There are five levels of submission. They are

Level 1: PSW only submitted to the customer. Supplier self-certifies PPAP documentation and supplier keeps on file for customer review as requested. (Very important note: Even though the supplier is self-certified, the requirements for Level 1 are exactly the same as Level 5. In other words, the only difference between Level 1 and Level 5 is the fact that with Level 1 there is generally no site visit.)

Level 2: PSW with product samples and limited supporting data.

Level 3: PSW with product samples and complete supporting data. Supplier submits PPAP documentation to customer representative for review and approval. Supplier keeps signed copy on file. There is also a facility audit on the supplier's premises. Generally, for most suppliers, this is the default level of submission.

Level 4: PSW and other requirements as defined by the customer.

Level 5: PSW with product samples and complete supporting data available for review at the supplier's manufacturing location. Customer's representative on-site review of site PPAP submission and accompanying documentation is required.

It is imperative to recognize that a supplier site may have several parts for a customer. Some parts may be Level 1, some Level 3, and some Level 5.

As previously mentioned, the PPAP level must be approved by the customer. It is also important to recognize that some customers require capacity planning and verification as part of the PPAP process. This means that a supplier is required to report demonstrated capacity as purchased part capacity based on average purchased part capacity (APPC) and maximum purchased part capacity (MPPC) in the customer's capacity planning systems, as recorded on the PPAP warrant.

Flow of Verification

All PPAPs are a summary of the APQP process. However, if one looks at the requirements for the first time, it becomes a very frightening experience due to the overwhelming requirements. Table 18.1 breaks down these

TABLE 18.1

Flow of PPAP Implementation

Flow of Verification* ⟶

Run-at-Rate	**Quality Verification**	**Production Verification**	**Capacity Verification**
1. Design records	Continue production Part Approval process (PPAP) using parts from run-at-rate.	Run-at-rate for all production streams (Element 1–8, 16, 17)	Capacity verification
2. Engineering change documents	Parts made during run-at-rate must meet all 18 PPAP requirements for one production stream	Quality verification for all production streams (Element 9–15)	Capacity analysis report
3. Customer engineering approval	PPAP warrant (Element 18)	PPAP warrant (Element 18)	Input purchased part capacity (PPC)
4. Design FMEA			PPAP warrant (Element 18)
5. Process flow diagrams			
6. Process FMEA			
7. Control plans			
8. Measurement system analysis studies			
16. Checking aids			
17. Customer-specific requirements capacity analysis report			

*Quite often, many organizations do the complete PPAP at one stage. However, whether it is completed in one stage or four stages (as shown), the requirements are the same and must be completed before the product is delivered to the customer.

requirements in a more comprehensive way so that the supplier can find the implementation and reporting of the PPAP much easier.

Capacity Analysis

Capacity is the principle of making sure that the supplier can produce enough product to satisfy the customer's production needs. It is important to recognize that the supplier may have an impeccable record of quality performance, but if that supplier cannot produce to the requirement of expected capacity, it will not be approved or it may be approved conditionally. A typical capacity analysis is made up of the following items:

- Capacity analysis report: The capacity analysis report is used for capacity planning, run-at-rate, production verification, and capacity verification. At this stage, the suppliers are required to use the report for capacity planning at the run-at-rate and all streams of production phases. This means that the supplier must have and substantiate accurate and appropriate capacity planning with historical and surrogate OEE summary (for a detailed analysis of OEE and capacity, see Stamatis 2010 and Chapter 19). It must also include a shared loading plan (if production will include other customers and a declarations page, including capacity planner training).
- Capacity planning: This planning must be completed prior to installation of facilities, equipment, tooling, and gauging at the appropriate milestone. Also during this phase, the use of historical and surrogate data is to determine historical and surrogate OEE. This form of planning will determine the likelihood of the supplier being able to meet average production weekly (APW) rates in 5 days and maximum production weekly (MPW) rates in 6 days for planned production operating patterns (shifts, breaks, cycle time, etc.). Quite often, as part of the requirement, there is also a build in factor of 10% capacity to cover special swings in production requirements.
- Run-at-rate: The supplier will produce at least 300 consecutive parts. The customer representative may grant exceptions; however, all efforts must be taken to make 300 parts. During the run-at-rate phase, the net ideal cycle time and other operating pattern factors must be at production level on *home line* to give early indication of meeting APW in 5 days and MPW in 6 days on the home line, not a tool shop or prototype supplier. In this stage, a capacity analysis report is required to be submitted per PPAP submission level and a preliminary P_{pk} capability (not C_{pk}) is required to be established.
- Capacity verification: The last item of the capacity analysis report is the verification of production level at the supplier's site. The supplier will produce a complete day of production. This means that if three shifts are planned, all three must work in sequence. If lines or shifts are shared with other customers, then the total amount of produced parts must reflect the total production requirement of *all* customers with shared loading plan completed and changeovers conducted as needed. In addition, the verification must include the same production operating pattern and assumptions that were agreed on by the customer and supplier when sourced. (A full day run must represent planned steady state long-term conditions, with all expected sources of manufacturing variation.)

Capacity Report

A typical capacity report is divided into three areas covering the capacity planning (A), run at rate (B) and capacity verification (C) sheets of the report. Each of these categories is divided into the following sections.

Category A provides information that deals with supplier/part, capacity requirements, and contacts. Specific items required are operating pattern, required parts/week, required OEE, and shared process allocation plan. In essence, this portion of the report is used to gather supplier site and part information, including part description, site and contact identification, and customer capacity requirements.

Category B provides information that deals with equipment availability, performance efficiency, quality rate, OEE, and process-specific weekly part estimates as well as observed cycle time. The supplier is required to complete the report with actual run data, such as

- Total run duration
- Actual downtime, changeover times
- Parts produced and parts rejected/reworked
- Actual cycle time
- Reasons for breakdowns

Obviously, for the planning sheet, surrogate/historical information may be used.

Category C provides information that deals with gap analysis, predicted good parts per week, and purchased part capacity (PPC) as well as appropriate approvals. Specifically, this section shows if the APW and MPW capacity requirements are met, overall predicted good parts per week and hence PPC are given, notes regarding the specific assumptions and conditions of the capacity demonstration if applicable, a graphical summary comparing requirements and results for each manufacturing process with an indication if requirements were met, and a record of supplier and customer personnel approvals of the analysis.

Misconceptions about PPAP

There are three key misunderstandings regarding PPAP. They are

1. The existence of an annual PPAP; this might have been a misinterpretation of the annual layout requirement or ongoing PV testing

2. The requirement for packaging approval as part of PPAP
3. The PSW is only a customer requirement

Actual Requirements

Appendix II identifies some of the key issues and concerns the supplier should evaluate for its PPAP process before it calls the customer for validation of the process. However, the following three comments are considered to be very high on the list of awareness.

- There is no requirement for an annual renewal of PPAP. PPAP approval is granted only for the reason for PPAP submission. Nothing should change if the process is constant.
- There is no connection between packaging approval and PPAP approval, so no special requirements are needed as a condition of PPAP approval.
- The PSW is a customer requirement but it is also a legal document.

Key Clarifications for PPAP

The culmination of any PPAP report is indeed the PSW. The significance of this is understood if we understand the types of PSW. There are three options and each has its own significance. They are

- Full PSW: This indicates that all parts/materials meet all technical and functional specifications and requirements. The supplier is authorized to ship production quantities. Mandatory: For some customers, attach OEE and capacity documentation to full PSW submission.
- Interim PSW: This indicates that material/part is not in line with specification and volume requirements, but supplier is authorized to ship product with approved alert. Mandatory: For some customers, attach OEE and capacity documentation to interim PSW submission. Typical options with an interim PSW are
    - For PPAP Level 2–5, the authorized customer representative ticks the "Approved" box and writes "Interim" above the "Approved" box and above the header of the PSW, if part/material conforms to the alert.

    - Alerts: An interim PSW always needs an alert and vice versa. A copy of the alert must be attached to the PSW. The alert number must be entered in the "Additional Engineering Changes" box and a brief one-line description in the "Explanation/Comments" box. For capacity issues, write the actual versus the daily production volume (DPV) needed (taken from the request for quote [RFQ]).
    - Submission results: An interim PSW always means the "No" box is ticked by the supplier on the PSW. The "Yes" box should be ticked for capacity reasons only because all specifications are met.
    - Format: An interim PSW must have the word INTERIM in the top right-hand corner of the PSW, followed in brackets by either CAPACITY for capacity-related deficiencies or TECHNICAL/QUALITY for all other nonconformancies, for example, dimensional, process, test, and so on.
    - Elect: Program buyers should enter only interim (capacity) PSWs into elect as "INTERIM." Interim (technical/quality) PSWs should not be entered into elect. Interim (technical/quality) PSWs have to be followed up by the authorized customer representative until full PSW approval.
- Rejected PSW: This indicates that the supplier is not allowed to submit parts/materials to the customer because the parts/materials do not meet specifications and an alert is not available. Mandatory: For some customers, attach OEE and capacity documentation to rejected PSW submission. Typical options with a rejected PSW are
    - For PPAP level 2–5, the authorized customer representative ticks the "Rejected" box and writes "Reject" on the top right-hand corner of PSW to make it clear to others it is not an Interim Approved or Full PSW.
    - Any PSW with a nonconformity that is not covered by an alert is rejected, that is, no parts can be shipped.

References

AIAG. (2008). *Advanced Product Quality Planning and Control Plan*, 2nd ed. Southfield, MI: Automotive Industry Action Group, Chrysler Corporation, Ford Motor Company, General Motors.

AIAG. (2009). *Production Part Approval Process (PPAP)*, 4th ed. Southfield, MI: Automotive Industry Action Group, Chrysler Corporation, Ford Motor Company, General Motors.

Stamatis, D. (2004). *Integrating ISO 9001 with ISO/TS 16949 and AS 9000*. Milwaukee, WI: Quality Press.

Stamatis, D. (2010). *The OEE Primer: Understanding Overall Equipment Effectiveness, Reliability and Maintainability*. Boca Raton, FL: CRC Press.

Selected Bibliography

DaimlerChrysler. (2010). *DaimlerChrysler Customer-Specific Requirements for Use with ISO/TS 16949*. http://www.ask.com/web?qsrc=1&o=11517&l=dir&q=DaimlerChrysler+specific+requirements&qo=serpSearchTopBox. Retrieved on September 29, 2014.

Ford Motor Company. (2010). *Ford Motor Company Customer-Specific Requirements For use with PPAP*, 4th ed. Dearborn, MI: Ford. http://www.iatfglobaloversight.org/docs/Ford_Specifics_for_PPAPJune2013.pdf. Retrieved on September 29, 2014.

GM. (2010). *GM Customer Specifics—ISO/TS 16949: Including GM Specific Instructions for PPAP*, 4th ed. GM. http://www.iatfglobaloversight.org/docs/Minimum%20Automotive%20Quality%20Management%20System%20Requirements%20for%20Sub-tier%20suppliers%20-%20AUG%2014.pdf. Retrieved on September 29, 2014.

19

Capacity Analysis

Overview

One of the areas that has been overlooked or misunderstood by suppliers is the issue of capacity. In fact, capacity problems are a concern about 30% of the time as production, delivery, and volume fall short of the expected performance. The simple definition of capacity is a study that is regularly conducted to get confirmation of supplier ability to support new model launch and/or manufacturing plan changes. Supplier capacity means that the supplier is capable of providing enough product to sustain the customer's requirements. It is performed under the following conditions. However, understand that the timetable given here is approximate and may vary depending on the product and/or customer requirements:

- New model confirm capacity studies are conducted at 2 weeks post final data judgment (FDJ) milestone at 17 months before Job 1 (MBJ1).
- Prelaunch readiness capacity studies are also conducted at 7.5 MBJ1 for new model programs. This should align with the tool trial (TT) build milestone.
- Capacity studies are conducted for planned capacity uplifts both within the new model launch timeline and post launch.
- Capacity studies are also conducted to evaluate alternative product offerings to customers (mix rate changes).
- There is no difference in supplier response expectations for both What If and confirm studies.

Goal of Capacity

In no uncertain terms, capacity must be understood as a process not a form. As such, data must be used to confirm that a supplier can produce the required *volume* of *quality parts* in the planned *operating pattern* at the planned *production rate.* The process starts with

- Planning: Planning and asking are the planned manufacturing equipment and processes projected to be capable of producing the required volume during planned operation patterns accurate?
- Verification. Do the demonstrated run-at-rate and capacity verification stages of the production part approval process (PPAP) results validate that the equipment and process can produce the required volume at the appropriate efficiency?
- Managing growth and shortages. Can the equipment and process supply additional volume? What can be done to increase production and efficiency? (The supplier must have at least a 10% buffer capacity beyond the requirement.)
- Communicating: Do suppliers cascade capacity study requirements and commitments to their affected plants? (Quite often, the requirements are assumed to be cascaded, but in reality they are not—especially to the subsuppliers.)
- Identifying wrong assumptions of past performance: Are suppliers misinterpreting new study as superseding prior studies? (In case of capacity shortages, communication with the customer must be immediate, clear, and with an action plan to be back on schedule. This communication cannot be emphasized enough because any failure in capacity volume will have serious consequences to the customer's production requirements. It is imperative that the communication is understood as crucial to mitigating supply risks to the plants. The earlier the notification of pending issues, concerns, or problems, the higher the likelihood of being averted.)
- Identifying the process of *when* and *how* to escalate: Is there an escalation process for reporting any type of deviation? If there is no appropriate action taken, escalate issues impacting supply to all concerned as soon as possible.
- Recognizing wrong capacity estimates: Are the capacity estimates appropriately calculated? Sometimes capacity reporting and data discrepancy are the causes for errors. The result will be an overstated or understated capacity reported to the customer and, of course, the data will not be aligned with the capacity verification. Here we must note that the most common issue is the overstated capacity. The reason for this is more often than not the appropriate allocation has not been accounted for or the correct overall equipment effectiveness (OEE) has not been recorded.
- Identifying insufficient information from the customer: Is there sufficient information for an appropriate decision? If not, the supplier

capacity uplift execution falls apart. Typical issues here are delay tooling investment until uplift volumes are cascaded in customer releases. This is a major concern especially when these delays are close to design freeze. Sometimes, this lack of information falls on the supplier itself by not communicating early enough potential issues via a supplier request for engineering approval (SREA). It is very important to note here that any supplier process and/or location changes to support capacity uplifts must be communicated to the customer.

- Reviewing the meeting capacity requirements: Are reviews conducted appropriately? This requirement is based on 5 days and maximum production weekly requirements in 6 days. This means that the supplier is to manage its tooling, equipment, and facilities such that, during a 7 calendar day week, the average production weekly (APW) capacity requirements are to be met by operating the tooling, equipment, and facilities based on a 5-day work week (unless otherwise dictated by government regulations. In some countries, the normal workweek is defined as 6 days). On the other hand, the maximum production weekly (MPW) capacity requirements are to be met by operating the tooling, equipment, and facilities based on a 6-day workweek. The remaining time during the week is reserved for completing the required tooling, equipment, and facility maintenance. If the supplier is unable to meet the APW based on a 5-day workweek, or the MPW based on a 6-day workweek, the supplier *must* contact its customer to develop a resolution plan to meet the capacity requirements.

Any exceptions to these requirements must be requested by the supplier and concurred in writing by the customer.

Environment for Capacity Planning

To validate a supplier's capacity plan, some type of capacity analysis report (CAR) form must be filled. (Do not confuse this report with the corrective action report, which is also designated as CAR.) There is no standard form for all customers. However, it is imperative that both customer and supplier agree on a CAR and it should be utilized to make a comparison between the supplier's planned manufacturing plan (required OEE for planned volumes) and its historical manufacturing performance. Relevant information on the CAR during capacity planning is and can be recorded in a tabular form:

- Capacity planning
- Historical manufacturing performance

- Shared loading plan
- Supplier declarations including subtier suppliers

The integrity of each tab is critical in completing a proper analysis—the selection of appropriate processes and parts for historical manufacturing performance and the inclusion of appropriate changeover times (if applicable) are required.

The intent of the capacity planning analysis is to provide validation of the supplier's capacity plan. One of the very basic questions is, does the supplier plan have an appropriate work pattern, with appropriate equipment cycle times, to meet the expected program volume adjusted for their historical manufacturing efficiency? Note that if historical data is not available, surrogate production data from a similar manufacturing process may be used.

To make sure that the basic question is answered, the supplier conducts the initial run-at-rate. This is a very important step because parts produced from a production stream (from a minimum of one production tool, line, process stream) at production feeds and speeds will give an indication whether the supplier can indeed produce what is expected. This initial stage provides an early indicator if a supplier can make future timing of delivery and production requirements.

In conjunction with the requirement to satisfy the appropriate requirements for producing parts at designed cycle times and achieving print specification requirements, a CAR form is required at run-at-rate to validate the supplier's ability to achieve all the run-at-rate requirements. This means typically a utilization of a short duration of production (~300 pieces, although according to AIAG [2009, p. 3] this can be changed with the approval of the customer representative). It is also strongly recommended that a capacity planning document with historical manufacturing performance data be reviewed—if it has not previously been completed. It also allows for the analysis of additional production streams as required for completion of the quality verification. By adjusting the planned net ideal cycle time to account for additional production streams, the overall capacity can be analyzed in support of the quality verification requirements.

The second issue of planning is the concern for quality verification. Here parts and processes from a minimum of one production stream (tool, line, facility, etc.) are evaluated for quality performance. Typical issues are dimensional, lab, and engineering specifications (ES) testing complete for this production stream (less appearance approvals). If the supplier is ready, this stage may be combined with the run-at-rate.

Production verification is the third stage of planning. Here the complete actual production stream (tool, line, facilities personnel, etc.) intended for this specific program/launch is in place and operational. Dimensional, lab, and ES testing is complete for all tools, cavities, molds, and production streams (including all appearance approvals). The supplier submits a part submission warrant (PSW). Quite often, this stage is combined with the second stage. It is only necessary if the supplier has multiple streams.

The last stage of planning is *the* capacity verification. Here, parts are produced from the complete actual production stream (tooling, equipment, facilities personnel). Capacity verification is demonstrated by yielding quality parts to meet a minimum or one day of the customer's production requirement (daily planning volume [DPV]). If three shifts are planned, the capacity verification study *must* reflect that pattern of production.

In conjunction with the requirement to satisfy the mass production (full production) requirements for producing parts at designed cycle times and achieving print specification requirements, a CAR form is required at the capacity verification stage to validate the supplier's ability to contain total program capacity. Again, as the capacity verification continues, so does the run-at-rate event continue to be a relatively short duration of production (~24 h). It is also strongly recommended that a capacity planning document with historical manufacturing performance data be reviewed if it has not previously been completed. At this stage, it is also assumed that all production equipment, tooling, personnel, and gauging are in place and that all production streams are capable of achieving the required production rate.

Overall Equipment Effectiveness (OEE)

One of the most underrated and misunderstood components of the capacity study is the OEE analysis. OEE combines key manufacturing metrics to state the overall health of the production process. It can be communicated as a minimum required OEE, or an effective/demonstrated OEE. The relationship between these two determines the feasibility of the manufacturing process to meet customer volume requirements. For a very detailed analysis of OEE, the reader is encouraged to see Stamatis (2010).

- (Minimum) required OEE: It is the minimum OEE a supplier must achieve to support the customer volume requirements. Warning: the traditional 85% initial OEE may not be accurate and may not be as good as one thinks. The best OEE is 100%; anything else and there is an opportunity for improvement. If there is more than 100%, the cycle time is miscalculated. For a detailed discussion on this, see Stamatis (2010).
- Demonstrated OEE: This is the OEE a supplier demonstrates through a PPAP event or historical/surrogate analysis.

Both types of OEE depend on the following three calculations:

Availability: Includes operating pattern, downtime, and allocation percentage. The actual formula is

$$\text{Availability} = (\text{Operating time})/(\text{Net available time})$$

Performance efficiency: Rate at which parts are produced compared to ideal rate. The actual formula is

$$\text{Performance} = (\text{Total parts run}) \times (\text{Net ideal cycle time})/(\text{Operating time})$$

Quality rate: Comparison of *good parts* to *all parts* produced. The actual formula is

$$\text{Quality rate} = (\text{Total parts run}) - (\text{Total defects})/(\text{Total parts run})$$

The relationship between these three components produces the minimum level of efficiency required to meet customer volume requirements, or required OEE. Using OEE as the unit of measure also allows assessment of a supplier's capacity for future production using current production performance data, surrogate OEE, or historical manufacturing performance.

Comparing required OEE and demonstrated OEE determines the supplier's capacity risk.

The reader hopefully has noticed that the OEE combines key manufacturing metrics to state the overall health of the production process. It can be communicated as either a minimum required OEE or an effective or demonstrated OEE. The relationship between these two determines the feasibility of the manufacturing process to meet customer volume requirements. This relationship is shown in the actual form of the OEE with the appropriate headings and calculations of (1) net ideal cycle time, (2) total good parts produced, and (3) net available time (Stamatis 2010). Remember that these are the same inputs used on the historical manufacturing performance sheet.

Evaluating the OEE

Once the calculations have been completed, then an analysis of both the data and results is appropriate. One of the most important questions that has to be asked is: Is demonstrated (surrogate) OEE greater or equal to (≥) required OEE? The answer to this question will be used to determine the capacity risk at all the appropriate milestones and final delivery date. When evaluating the OEE, one should remember that the demonstrated OEE is based on (1) surrogate data in capacity planning and (2) actual run data during run-at-rate and production verification stages

On the other hand, required OEE is the minimum level of efficiency required to meet customer demand, based on (1) the customer's capacity requirements and (2) the supplier's manufacturing plan.

If all processes have a favorable answer to the above question, there is no risk to capacity. Focus on continuous improvement and transferring lessons learned to new model processes. On the other hand, if *any* process has an *unfavorable* answer to the above question (i.e., demonstrated OEE < required OEE), then there is a risk to capacity. The capacity gap between demonstrated OEE and required OEE must be closed to meet customer volume requirements. Obviously, if there is a deficiency in capacity, the supplier has two possible alternatives. The first is to increase the demonstrated OEE and the second to decrease the required OEE. The best option is to work on improving the demonstrated OEE. (There is no threshold for the initial OEE. Whatever it is, make sure it is the right one for it serves as the benchmark for the future OEE. That means as the supplier progresses, the OEE should get better. If the initial OEE is below 85%, try to identify the reasons and correct them.)

Identifying potential capacity risks early in the time line cycle allows necessary time to implement improvements to minimize the risk to launch. Regardless of *when* the capacity gap is identified, the process of resolving the capacity gap is the same. This is true even when capacity gaps are identified after launch.

Some Specific Actions for Improving OEE

Improving Demonstrated OEE

By improving surrogate processes and duplicating lessons learned, capacity gaps can be eliminated and the overall *health* of the supplier is improved. Remember that the demonstrated OEE is the actual level of efficiency achieved by a supplier in historical or current production. For capacity planning, surrogate processes are used for demonstrated OEE; at run-at-rate and capacity verification, actual performance data is used. So, in capacity planning (forward-model), the demonstrated OEE is made of historically similar (surrogate) processes. At run-at-rate, the demonstrated OEE is made of actual production data. This is true at all times after this stage as well, whether before launch or in a capacity constraint situation. The approach to improving demonstrated OEE—whether surrogate or actual—is the same.

Some Specific Improvements on the Components of OEE

For Availability

- Track any Pareto losses to reduce unplanned downtime
- Identify special and common causes
- Analyze mean time between failure (MTBF), mean time to repair (MTTR), and so on
- Utilize rapid-response maintenance teams and hourly count production boards
- Prioritize equipment for maintenance activities

For Performance Efficiency

- Ensure performance to constraint ideal cycle time.
- Identify blocked and starved conditions; ensure cycle times support Takt time requirements. Make sure you distinguish between Tact (theoretical) and Takt (actual) time.
- Machine/equipment: Have an engineering study to identify gap, and assignment of appropriate technical resources for closure.
- Human resource: Address shortage of operators, imbalanced work, or lack of standardization.

For Quality Rate

- Track, Pareto, and reduce scrap and rework losses.
- Install poka-yoke/error-proofing devices.
- Communicate good versus bad parts (include boundary samples).
- Allocate problem-solving resources and process experts to launch.
- If the capacity analysis is performed early in the timing process, there is time to implement OEE.

Demonstrated OEE

Demonstrated OEE is generated from actual performance data for each process and is used to assess risk when compared to the required OEE for each process. At run-at-rate and capacity verification stages, the demonstrated OEE is generated from the performance data of the run event. Generally, there are no significant changes from the current capacity standard at both stages relative to the inputs required for completing the analysis. However, by moving the initial assessment of capacity earlier to, say, program approval (about 30 months before Job#1), actual performance data may not be available during capacity planning. As such, the demonstrated OEE is generated from performance data of *surrogate* processes.

When selecting a surrogate process, the best surrogate should be aligned with the planned process for the new part. Considerations include

- Part complexity.
- Technology used in the production.
- Manual or automated process.
- Greenfield (new) or brownfield (existing) site.
- Process layout comparison.
- Volume, operating pattern, and so on.
- Similar part size and cycle time.
- Potential cultural differences.
- Surrogate processes selected do not necessarily need to be from a common part or time (e.g., milling processes, welding processes). On the other hand, it is important to remember that when selecting surrogate processes, it is not required that the overall parts are exact. The goal is selecting *manufacturing processes* that are similar to the new part's *manufacturing processes.*

To obtain historical OEE performance data for each process, we need three known data inputs. They are

1. Net ideal cycle time (NICT): NICT should be verified through observations or design, *not* engineering standards. NICT must *not* include any baked-in efficiency losses. NICT is the best, sustainable cycle time achieved in production (steady state, not ramp-up).
2. Net available time (NAT): NAT is the amount of time that the process was in production for the specific part number. NAT also *excludes* planned downtime (lunches, breaks, etc.) but *includes* unplanned downtime. This means that if the process was shut down early on Friday because requirements were met, do not include the time that the process was shut down. For shared processes, NAT must also include changeover time.
3. Good parts produced (GPP): GPP includes only the good parts through the process—no scrap or rework parts (first time through).

Enablers to Improve the Demonstrated OEE

- With each action, the demonstrated OEE should be monitored to validate the OEE improvement.
- Improving the demonstrated OEE must be the first approach considered to close capacity gaps and eliminate risk.

Other Enablers

- Install strategic buffers to prevent starve conditions.
- Establish energy room to communicate metrics and progress to plan.
- Run simulations with multiple iterations—changing input variables (cycle times, downtime, quality rate, buffer sizes, MTBF, MTTR, etc.) to optimize throughput.

Analyze Results and Implement Actions

- Ensure cascade of lessons learned (TGW and TGR) from previous launches.
- For shared equipment, reduce changeover times.
- Review/update preventative maintenance schedules.
- Investigate incremental tooling and equipment potential changes.
- Ensure there is a method of verifying the results. Translate the results to OEE so that a common language is being used to track improvements.
- Make sure that action items are prioritized appropriately.

Required OEE

Required OEE is directly proportional to required good parts. However, if consideration of other losses (scrap, rework, changeover, etc.) exceed the maximum possible parts, the manufacturing plan is not feasible. This condition can occur *even when* the required OEE appears reasonable. Remember, required OEE is the *minimum level of efficiency* that is required to support the demand, based on the defined operating pattern, allocation percentage, and operating parameters. Once all potential losses are assessed, a process may not be feasible, even if the required OEE < 100%.

All *value streams* are required to be assessed for capacity risk. That is, all processes that are utilized for production require assessment for risk. This may require the use of more than one document, and is consistent with PPAP requirements. For end-item part numbers that have multiple *value streams*, the capacity planning analysis must begin by looking at a single *value stream*. This is due to considerations made to individual process scrap losses, which have impact on the individual process required OEEs. In effect, the capacity analysis is generally setup on a *part-specific* basis. However, depending on the part and manufacturing process, it *may* be appropriate to group families of parts or several manufacturing lines. If in doubt, check with the customer for approval.

Improving Required OEE

By adding operating time, reducing up-line scrap loss, or decreasing net ideal cycle time, capacity gaps can be eliminated. The minimum level of efficiency required by a supplier to meet customer's volume requirements is based on the supplier's planned operating pattern and manufacturing assumptions. The required OEE for each process must be understood and identified as early as possible by both customer and supplier. For each process, the required OEE is the *target*—capacity gaps can only be identified by comparing demonstrated OEE to this target. The basic question here is: Are the required OEEs for each process feasible? Specific items to reduce the required OEE with minimal investment that may be considered are to expand operating pattern, add extra shifts of expanded hours, and consider overtime. Another option is to reduce constraint cycle times by: (1) machine/equipment design, (2) coordination with equipment suppliers to reduce cycle time, (3) introduce faster processes when and if possible, and (4) track and optimize cycle times at ramp-up.

Shared Loading Analysis

Shared loaded analysis is the evaluation of how much time is allotted for production using a stream for the customer. Make sure that the allotted time is rated correctly and fulfills the requirements of the customer. Generally, there is only one exception to this rule where total allocation is calculated. The exception is for machining and outsourced e-coat. To complete a Shared Loading Plan for a process, the following information is needed:

1. APW and MPW volume information adjusted for downstream scrap losses, along with the NICT for all parts that are planned for production on the specific process; average demonstrated OEE from most

recent production performance data. The Shared Loading Plan validates the accuracy of the planned allocation percentage for the process, adjusted for the demonstrated OEE.

2. The equipment/process is not oversold considering the total book of business of the process.

Total Allocation

For machining, if the total percentage of allocation is 54.2% for both APW and MPW, this means that, adjusted for the supplier's demonstrated OEE, the operation can contain all business (customer and non-customer) without exceeding the acceptable operating patterns. However, for outsourced e-coat, if the total percentage allocation is *greater than 100%*, the process is oversold, that is, *capacity is at risk* (must follow-up with action plan). The total percentage allocation can be addressed in the same ways that gaps are addressed between required OEE and demonstrated OEE—by utilizing capacity gap closure to improve demonstrated OEE, by reducing required OEE (and, thus, the part-specific allocation percentage), or by off-loading some of the business.

Reaction and Plan Expectations

When all the data have been collected and the analysis has been completed, it is time to evaluate the results. This evaluation is based on (1) reaction to the data and (2) planning for the expectations. Both are dependent on

1. An assessment of current capacity and potential supply impact by
    a. Understanding the maximum possible output
    b. Examining alternate manufacturing options
    c. Establishing output tracking of constraint operations
    d. Reviewing the supply plan based on customer release requirements as far in the future as it is available
    e. Establishing supply mitigation if there is a supply shortage projected
    f. Establishing contact with appropriate customer representative if there is permanent degradation in output
2. Communicating with the customer immediately. This means contact the appropriate customer representative immediately and convey any issue, root cause, and supply status.

   a. Presenting a supply protection plan
   b. Executing the shipping and actions as agreed
3. Corrective action. Here the supplier is required to investigate process and system escape contributors to root causes by utilizing formal problem-solving techniques. Furthermore, the supplier must
   a. Establish permanent and systemic corrective actions
   b. Update preventive maintenance plans and/or process control plans with permanent corrective actions as applicable
   c. Capture systemic improvement actions in customer's manufacturing site assessment system
   d. Implement corrective actions per plan and validate results

Capacity Information Maintenance

As in any system, capacity has to be maintained on a regular basis. Typical appropriate maintenance approaches are

1. At a minimum, suppliers are required to update their capacity information in the customer's capacity planning systems on a quarterly basis or as required by the customer.
2. Suppliers must also update their capacity information in the customer's system (if available) as soon as there is a significant change in their operating output such that they are not able to support previously declared capacity in the customer's system. For significant drop in capacity, please notify the customer for immediate supply risk mitigation.
3. Besides maintaining capacity information in the customer's system, suppliers must monitor all customer's releases for significant amount of time (in some cases, for 6-month releases) and communicate any supply constraints to the appropriate customer representatives.

Some specific items of review for appropriate maintenance are in the areas of

1. Study
   a. Confusion and misinterpretation due to multiple studies
   b. Supplier response errors
   c. APW/MPW operational practices not consistent across supply base
   d. No subtier confirmation

2. Tool order
   a. Missed tool funding due to supplier response errors
   b. Supplier execution
      i. Delayed tooling kickoff through supply chain
      ii. Inadequate resources impacting timely execution
      iii. Poor subtier planning and execution
      iv. Does not believe increased releases are real
3. Capacity verification
   a. Overcommitted and overstated capacities
   b. Subtier not verified prior to capacity confirmation with customer
   c. Changeover times not accounted in shared lines planning
4. Sustainable manufacturing
   a. Domino effect from lack of quality of event upstream
   b. Lack of ability to recover from nonrobust plans
   c. Unplanned downtimes due to extended operating patterns (lack of preventive maintenance)

Thought Starters to Improve Supplier Capacity Volume

OEE improvement actions maximize the following: (1) equipment availability, (2) performance efficiency, and (3) quality rate. Specific questions may be

1. Are there hours, days, and shifts available in which the supplier can run extra parts during the week? Understand the operating patterns.
2. Is there any opportunity to improve the cycle of time constraint stations? Develop performance to Takt tools (Rational: performance efficiency improvement).
3. Does the supplier collect accurate data for downtime and scrap/rework?
4. What are the major downtime issues? Deep diving into the top 5 downtime issues. Are there any available Pareto charts on downtime for 4 weeks (Rational: equipment availability improvement)?
5. What are the major scrap/rework issues? Deep diving into the top 5 scrap/rework issues. Are there any available Pareto charts on downtime for 4 weeks? Parts loss to rework/scrap (Rational: quality rate/performance efficiency improvement).

6. Can the changeover process timing be improved? Review the changeover process at the supplier's site. Apply single minute exchange of dies (SMED) opportunity lean tools and principle. (Rational: quality rate/performance efficiency improvement.)
7. Are there any opportunities to improve the consumable changeover (C/O) timing?
8. Is the supplier doing too many changeovers?
9. Are there any opportunities to implement PM scheduling, cleaning during the changeover or breaks/lunches (parallel tasks)?
10. Are there any opportunities to run the rework/reject parts during the breaks/lunches at the final inspection cell?
11. Are there any machine start-up checklists available? Can the start-up checklist be performed during lunches/breaks?
12. Does the supplier have a complete spare part available (electrical/tooling, etc.)?
13. Does the supplier have adequate raw materials on-site for the increased production volume (especially parts that are shipped from Europe or Asia)?
14. Are there any opportunities to implement the tag relief during lunch and breaks?
15. Are there any stations or machines on bypass mode, which impacts the efficiency throughput?
16. Does the shift change occur in a timely manner with minimum impact to the equipment availability?
17. Is there any opportunity to optimize the start-up between each shift? Minimize/optimize production meeting between the shifts, and so on.
18. Do the production counts shift to shift have a large fluctuation? Deep dive into the root-cause analysis and standardize operation procedures.
19. Is there any opportunity to add stations to the constraint station/cell to sustain volume increase with minimum capital investment? Is there any opportunity to make more parts in the offline station? Is there any opportunity to outsource the parts?
20. What are the next constraint stations or cell beside the current cell?
21. Is there any opportunity to develop a short-term/long-term road map to capture the improvement job per hour/average production weekly (JPH/APW) Pareto to sustain the customer production demand?

References

AIAG. (2009), *Production Part Approval Process (PPAP)*, 4th ed. Southfield, MI: Chrysler Corporation, Ford Motor Company, General Motors, AIAG.

Stamatis, D. (2010). *The OEE Primer: Understanding Overall Equipment Effectiveness, Reliability and Maintainability.* Boca Raton, FL: CRC Press.

20

Geometric Dimensioning and Tolerancing (GD&T)

Overview

Engineers communicate with mathematical formulas and blueprints (drawings). Since the early days of humans, mathematical formulas have been improved and have stood the test of time. We have indeed learned how to communicate with numbers at a very sophisticated level. However, the second form of communication for engineers—blueprints (drawings)—has been drastically changed because of different interpretations that humans put on the pictures. The mathematical foundation of this change started in the late eighteenth century when Gaspard Monge solved the problem of "two-dimensional representation of three-dimensional objects," known as *descriptive geometry*. With this breakthrough, engineers were able to provide a method to create auxiliary views projected with right angles and work lines to yield three basic solutions:

- True lengths of lines
- Point view or end view of a line
- True shape of a plane or surface

Because of this breakthrough, in 1994 the manufacturing community decided to do something about this confusion and developed an international language that presents their product innovations.

Indeed, the GD&T (ASME Y14.5M-1994, 2009) is an international language that is used in engineering drawings to accurately describe the size, form, orientation, and location of part features. It is also a design-dimensioning philosophy that encourages designers to define a part based on how it functions in the final product or assembly. The intent of this new language is to enable engineers to say *what they mean* on a drawing and consequently improve their designs and at the same time lower cost. GD&T is a unique language that is used by manufacturing to determine the best manufacturing approach. On the other hand, it is also used by quality professionals

TABLE 20.1

Major Changes from ASME Y14.5-1994 to ASME Y14.5-2009

ASME 14.5-1994	ASME 14.5-2009
MMC/LMC applied to *both* geometric tolerances and datum references	MMC/LMC/applies *only* to geometric tolerances, while MMB/LMB applies only to the datums referenced, yet the symbols remain the same
RFS is the default requiring no symbol, and applies to *both* geometric tolerances and datum references	RMB is the default requiring no symbol, and *only* applies to the datum referenced
Flatness applies to the feature only and no material modifiers are allowed	Flatness can also be applied to noncylindical FoS, allowing MMC/LMC material modifiers to apply to the derived centerplane

Note: ASME is the acronym for American Society of Mechanical Engineers. Y14 is the standards group of the American National Standards Institute (ANSI) that handles technical drawings. Y14.5 is the standard associated with dimensioning and tolerancing. 2009 is the revision date for the standard.

including inspectors to determine proper setup and part verification. In other words, one may characterize the GD&T as a uniform language that is used throughout the organization and its suppliers to reduce guesswork, assumptions, and controversy in design and manufacturing. ASME Y14.5-2009 is a document that establishes uniform practices for stating and interpreting

- Dimensions
- Tolerances
- Related requirements on engineering drawings and related documents

To be sure, as a language there are many symbols and definitions included in the standard. It is beyond the scope of this book to include all the changes and requirements. However, the major changes from the 1994 version to the 2009 version are shown in Table 20.1. For all the changes and improvements, the reader should see appendix A of the actual ASME Y14.5-2009 standard.

Language of GD&T

In the GD&T language, three major categories define the characteristics. They are individual featured without datum reference, individual or related features, and related features with datum reference. Each one of these has a specific definition and a symbol. They are all shown in Table 20.2.

In addition to the classical symbols and concepts, the new 2009 standard includes the following new items of special importance:

TABLE 20.2

Summary of Categories Used in GD&T

For individual features	Form	Straightness	—
		Flatness	⏥
		Circularity	○
		Cylindricity	⌭
For individual or related features	Profile	Profile of a line	⌒
		Profile of a surface	⌓
For related features	Orientation	Parallelism	//
		Perpendicularity	⊥
		Angularity	∠
	Location	Position	⌖
		Concentricity	◎
		Symmetry	⌯
	Runout	Circular runout	↗
		Total runout	⌰

Source: Reprinted from ASME Y14.5 2009 by permission of the American Society of Mechanical Engineers. All rights reserved.

- Independency
- Continuous feature
- Spotface
- New for profile
  - All Over
  - All Over this Side of Parting Line symbols

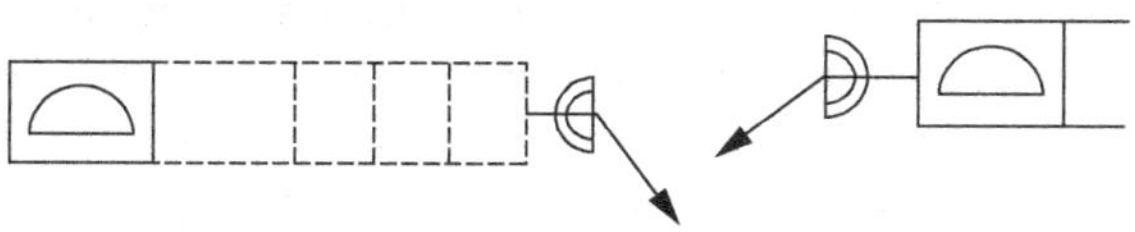

  - Circle U modifier
- Customized datum reference frames
  - The use of brackets in the DRF to identify the boundary of the datum feature simulator at a fixed size, for example,

⌖	Ø0.3	A	B	DⓂ[Ø7.5]

In order for the standard to be more flexible, simple, accurate, and precise in the interpretation, there are many more symbols that make it so. The list that follows is a partial one but very functional for most applications. If the reader is interested in all the symbols and their definitions, you should look at the actual standard appendix A and/or the specific sections of the standard.

The section is identified in parenthesis. All symbols and short definitions are reprinted from ASME Y 14.5-2009, by permission of The American Society of Mechanical Engineers. All rights reserved.

This is the symbol for All Around Symbol (3.3.19), indicating that a tolerance applies to surfaces all around the part.

This is the symbol for All Over Specification (8.3.1.6; 8.3.25). It is important to realize that this specification, whether in a general note or on the field of the drawing, applies *unless otherwise specified*.

This is the symbol for All Around This Side of Parting Line (3.14.1). To apply a requirement to all features all around one side of a parting line, the graphical symbol for all around this side of parting line is indicated on the leader line.

This is the symbol for All Over This Side of Parting Line (2.14.2). To apply a requirement to all features all over one side of a parting line, the graphical symbol for all over this side of parting line is indicated on the leader line.

∠ This is the symbol for Angularity (6.3.1). It is the condition of a surface, axis, or center plane, which is at a specified angle from a datum plane or axis.

⌒105 Arc Length (3.3.9) indicates that a dimension is an arc length measured on a curved outline. The symbol is placed above the dimension.

| 50 | Basic Dimension (1.3.23; 3.3.4) is used to describe the exact size, profile, orientation, or location of a feature. A basic dimension is always associated with a feature control frame or datum target.

↔ Between (3.3.11) is used to indicate that a profile tolerance applies to several contiguous features. Letters may designate where the profile tolerance begins and ends. These letters are referenced using the between symbol (since 1994) or the word *between* on drawings made to earlier versions of the standard.

◎ Concentricity (7.6.4) describes a condition in which two or more features, in any combination, have a common axis.

⌲ Conical Taper (2.13) is used to indicate taper for conical tapers. This symbol is always shown with the vertical leg to the left.

⟨CF⟩ Continuous Feature (2.7.5; 3.3.23). The note *continuous feature* or the continuous feature symbol is used to identify a group of two or more features of size where there is a requirement that they be treated geometrically as a single feature of size. Although the definition only mentions features of size, there is an example of CF being applied to a pair of planar features.

CR Controlled Radius (3.3.7) creates a tolerance zone defined by two arcs (the minimum and maximum radii) that are tangent to the adjacent surfaces. Where a controlled radius is specified, the part contour within the crescent-shaped tolerance zone must be a fair curve without flats or reversals. Additionally, radii taken at all points on the part contour shall be neither smaller than the specified minimum limit nor larger than the maximum limit.

⌴ Counterbore/Spotface (3.3.12) is used to indicate a counter bore or a spot face. The symbol precedes the dimension of the counter bore or spot face, with no space.

⌵ Countersink (3.3.14) is used to indicate a countersink. The symbol precedes the dimensions of the countersink with no space.

⌭ Cylindricity (5.4.4) describes a condition of a surface of revolution in which all points of a surface are equidistant from a common axis.

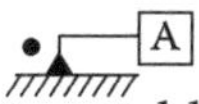

Datum Feature (1.3.16) is the actual component feature used to establish a datum.

Datum Target (4.24.1) is a specified point, line, or area on a part that is used to establish the datum reference plane for manufacturing and inspection operations.

↧ Depth/Deep (3.3.15) is used to indicate that a dimension applies to the depth of a feature. This symbol precedes the depth value with no space in between.

⌀ Diameter (3.3.7) indicates a circular feature when used on the field of a drawing or indicates that the tolerance is diametrical when used in a feature control frame.

⌱ Dimension Origin (3.3.17) signifies that the dimension originates from the plane established by the shorter surface and dimensional limits apply to the other surface.

\| ⌖ \| Ø0.5Ⓜ \| A \| B \| C \| Feature Control Frame (1.3.33; 3.4.1) is a rectangular box containing the geometric characteristics symbol, and the form, runout, or location tolerance. If necessary, datum references and modifiers applicable to the feature or the datums are also contained in the box.

⏥ Flatness (1.3.37; 5.42) is the condition of a surface having all elements in one plane.

Ⓕ Free State Variations (3.3.20) is a term used to describe distortion of a part after removal of forces applied during manufacturing.

Ⓛ Least Material Condition (LMC) (3.3.5) implies that condition of a part feature of size wherein it contains the least (minimum) amount of material; for example, largest hole size and smallest shaft size. It is opposite of maximum material condition.

Ⓘ Independency Symbol (2.7.3; 3.3.24) - is applied to the size dimension in order to invoke the principle of independency to regular features of size and override Rule #1.

Ⓜ Maximum Material Condition (MMC) (3.3.5) is that condition of a part feature wherein it contains the maximum amount of material within the stated limits of size. That is, minimum hole size and maximum shaft size.

Movable Datum Targets (4.24.6) symbol may be used to indicate movement of the datum target datum feature simulator.

5X Number of Places (1.9.6). The X is used along with a value to indicate the number of times a dimension or feature is repeated on the drawing.

// Parallelism (6.3.2) is the condition of a surface, line, or axis, which is equidistant at all points from a datum plane or axis.

Parting Lines (3.14) are depicted on casting/forging/molded part drawings as a phantom line extending beyond the part in applicable views, with the parting line symbol added.

⊥ Perpendicularity (6.3.3) is the condition of a surface, axis, or line that is 90° from a datum plane or a datum axis.

⌖ Position Tolerance (7.2) defines a zone within which the axis or center plane of a feature is permitted to vary from true (theoretically exact) position.

⌒ Profile of a Line (8.2.1.2) is the condition permitting a uniform amount of profile variation, ether unilaterally or bilaterally, along a line element of a feature.

⌓ Profile of a Surface (8.2.1.1) is the condition permitting a uniform amount of profile variation, ether unilaterally or bilaterally, on a surface.

Ⓟ Projected Tolerance Zone (7.4.1) applies to a hole in which a pin, stud, screw, and so on is to be inserted. It controls the perpendicularity of the hole to the extent of the projection from the hole and as it relates to the mating part clearance. The projected tolerance zone extends above the surface of the part to the functional length of the pin, stud, and screw relative to its assembly with the mating part.

R Radius (3.3.7) creates a zone defined by two arcs (the minimum and maximum radii). The part surface must lie within this zone.

(50) Reference Dimension (3.3.8) is a dimension usually without tolerance, used for information purposes only. It does not govern production or inspection operations. (It is equivalent to auxiliary dimension in ISO.)

RFS Regardless Of Feature Size (RFS) (1.3.48; 2.8; 2.8.1; 7.3.2) is the condition where the tolerance of form, runout, or location must be met irrespective of where the feature lies within its size tolerance.

○ Roundness (5.4.3) describes the condition on a surface of revolution (cylinder, cone, sphere) where all points of the surface intersected by any plane perpendicular to a common axis in case of cylinder and core (or passing through a common center in case of sphere) are equidistant from the axis (or center). Since the axis and center do not exist physically, measurement has to be made with reference to surface of the figures of revolution, which is the circular contour

↗ Runout (9.4.1) is the composite deviation from the desired form of a part surface of revolution through one full rotation (360 degrees) of the part on a datum axis.

⌳ Slope (3.3.18) is used to indicate slope for flat tapers. This symbol is always shown with the vertical leg to the left.

SØ Spherical Diameter (3.3.7) shall precede the tolerance value where the specified tolerance value represents spherical zone. In addition, a positional tolerance may be used to control the location of a spherical feature relative to other features of a part. The symbol for spherical diameter precedes the size dimension of the feature and the positional tolerance value to indicate a spherical tolerance zone.

SR Spherical Radius (3.3.7) precedes the value of a dimension or tolerance.

⌴SF⌴ Spot face (1.8.14). Counter bore and spot face previously used the same symbol. A spot face now looks like the counter bore symbol with the addition of the letters SF.

□ Square (3.3.16) is used to indicate that a single dimension applies to a square shape. The symbol precedes the dimension with no space between.

⟨ST⟩ Statistical Tolerance (3.3.10) is the assigning of tolerances to related components of an assembly on the basis of sound statistics (such as the assembly tolerance is equal to the square root of the sum of the squares of the individual tolerances). By applying statistical tolerancing, tolerances of individual components may be increased, or clearances between mating parts may be reduced. The increased tolerance or improved fit may reduce manufacturing cost or improve the product's performance, but shall only be employed where the appropriate statistical process control will be used. Therefore, consideration should be given to specifying the required C_p and/or C_{pk} or other process performance indices. Preference should be given to P_p and P_{pk}.

— Straightness (5.4.1) is a condition where an element of a surface or an axis is a straight line.

⌯ Symmetry (7.7.2) is a condition in which a feature (or features) is symmetrically disposed about the center plane of a datum feature.

Ⓣ Tangent Plane (3.3.21; 6.5) indicates a tangent plane is shown. The symbol is placed in the feature control frame following the stated tolerance.

X Target Point (4.24.2) indicates where the datum target point is dimensionally located on the direct view of the surface.

⌰ Total Runout (9.4.2) is the simultaneous composite control of all elements of a surface at all circular and profile measuring positions as the part is rotated through 360°.

▷ Datum Translation Symbol (3.3.26) indicates that a datum feature simulator is not fixed at its basic location and shall be free to translate.

Ⓤ Unilateral and Unequally Disposed Profile Tolerance (8.3.1.2) is used to indicate that a profile of a surface tolerance is not symmetrical about the true profile. The first value in the feature control frame is the total width of the profile tolerance. The value following the symbol is the amount of the tolerance that is in the direction that would allow additional material to be added to the true profile.

The GD&T standard provides many symbols as opposed to notes or verbiage because of the following advantages:

- Symbols have uniform meaning globally.
- Symbols are recognized internationally and will not have language barriers.
- Symbols can be quickly drawn and placed on the drawing where required.
- Symbols are compact.
- Geometric tolerancing symbols are part of the ANSI standard.

On the other hand, the disadvantages of defining a part heavily reliant upon notes are

- Require more space and time
- Require more explanations especially when notes originating in one country may need to be translated for use in another country
- Require explicit and very clear verbiage because notes could be misunderstood, depending on how the note is written

Therefore, symbols are used on drawings to help define a part. However, sometimes notes are necessary to clarify intent for unique situations or when no symbol exists to represent intent.

Types of Drawings

Generally, customers define the types of drawings. For example, in the automotive industry, you may have encountered the following three types, which are included in the Master Data Files, which are the official Documents of Record. They are

1. CtM (3D CAD data file). This is the master. There are two ways to establish this. (a) A CAD data file consisting of a 3D CAD geometry file containing complete product manufacturing information (PMI) annotation, and (b) a CAD data file consisting of a 3D CAD geometry file containing complete PMI annotation and a separate, yet associative, 3D CAD geometry file containing manufacturing information annotation. (Note that PMI is also referred to as Product Definition Data per ASME Y14.41-2003.) While there is no physical drawing for CtM, the CAD files will contain first frame information, such as (a) title block, (b) revision column, and (c) general notes.
2. DtMC (2D drawing and 3D CAD data file). This is drawing the master with CAD content. A combination of master data files consisting of a 3D CAD geometry file and a separate, yet associative, 2D drawing that contains a customer's title block and remaining PMI annotation not found in the 3D CAD geometry file. It may also have auxiliary views setup using descriptive geometry.
3. DtM (2D drawing). This is drawing the master. A 2D drawing contains a customer's title block and complete drawing definition including views and PMI annotation (dimensions and tolerances). It may also contain auxiliary views setup utilizing descriptive geometry. The 2D drawings utilize (a) 3rd angle projection and (b) orthographic projection. For 3rd angle and orthographic projection, the orthographic drawings have continuous points of view and the following characteristics: (a) separate views that are in line and projected with each other, (b) all views are at right angles from each other, and (c) all projections are straight lines.

The rationale for these three types is to make sure that special characteristics are identified properly and accurately. Special characteristics are identified through the design failure mode and effect analysis (DFMEA) and affect compliance to government regulations or the safety of the product. These are the only types of special characteristics to be identified on the drawing with the critical characteristic (CC) symbol, for example, inverted delta. However, in the CtM files, the verbiage *Control Item* can be used to replace the inverted delta symbol.

Title Block and Notes

A GD&T print has a title block and a note block. The title block contains the following information:

- Customer's trademark
- ISO hardcopy field protection
- Reference field optional
- Substance restriction field: substance restriction standard
- CAD field, which contains CAD file number
- CAD type field
- CAD location field
- Drawing scale
- Master field: Indicates which media is the master
- General information field: Contains orthographic view projection
- Customer's standard version field: Contains customer's version number
- Title field: Product drawing titles
- Drawing field: Contains the engineering part number
- Sheet number field: Sheet XXX of total number of sheets
- Left- and right-hand indicator
- Unit field: Manufacturing unit identification number
- Operation number field: Contains the manufacturing ID code for process
- Division field: Name of division to which the drawing is appointed (if applicable)
- Plant field: Name of plant or plant code
- Design field: ID code of the person who designed the object
- Detail field: ID code of the person who detailed the drawing
- Checked field: ID code of the person who checked the drawing
- Safety field: Contains ID code of safety personnel (manufacturing)
- Date field: Date that drawing was started
- Control item (inverted delta [or CC] symbol)

A print with GD&T guidelines also uses *notes* to supply necessary information that is not covered by dimensions. Notes are always located in the lower right-hand corner of the drawing and to the left of the title block. They contain

- One clear meaning and must be in English if English is the official language of doing business
- Supporting documentation (customer specific) notes such as
  - Performance specifications
  - Material specifications
  - Gauging requirements for restraint conditions
  - Diamond P/Diamond D requirements
  - Any exceptions

Notes do not contain

- Specific notes that require association to a specific feature on the part or assembly
- Supplier information
- Part numbers
- Part/assembly vehicle usage information; this information is not permitted anywhere on the drawing

Many interface dimensional requirements on drawings are verified for production part approval process (PPAP) at the component level. These are referred to as pass-through requirements and are often either left off the drawing to avoid having to verify twice or placed on the drawing as reference dimensions. It is important to remember here that the customer quality engineer can only enforce requirements placed on the customer's released drawings; the *Diamond P* symbol will link the customer drawing to the supplier internal drawing.

Diamond P and D Requirements

The Diamond P symbol shall be placed to the immediate right or left of each individual requirement, providing

- The requirement shall be unaffected by any subsequent manufacturing, installation, or assembly process that would require revalidation
- Generally, there are no other (customer's) drawings released showing a duplicate of the requirement such as a component drawing or a service part drawing
- The pass-through requirement note must be in the general notes column whenever the Diamond P symbol is used

On the other hand, the Diamond D designation will allow measurements to be deferred until the final assembly for noncritical elements.

- Prior to Diamond D, suppliers would often make this type of dimension a reference dimension, which the customer is not permitted to verify for quality assessment.
- The Diamond D designation is not allowed on any critical item designated as a high-impact characteristic (HIC), significant characteristic (SC), or critical characteristic (CC), nor would it be permitted on features that are used as datums at any stage of verification (component, subassembly, or final assembly).
- The Diamond D will allow verification where it is needed and would allow for full dimensioning on the drawing. If the feature ever came to be suspect in a quality investigation, it could be fully verified, which cannot happen with reference dimensions.

Tolerance Types

Per ASME Y14.5-2009, paragraph 1.3.60: Tolerance is the total amount a specific dimension is permitted to vary. The tolerance is the difference between the maximum and minimum limits.

Furthermore, paragraph 1.3.61–1.3.62 identifies three types. They are

1. Bilateral: A tolerance in which variation is permitted in both directions from the specified dimension. (Note: The tolerance is not required to be equal in both directions.)
2. Geometric: The general term applied to the category of tolerances used to control size, form, profile, orientation, location, and runout.
3. Unilateral: A tolerance in which variation is permitted in one direction from the specified dimension.

Everyone recognizes that when one uses tolerances there are issues of limits as well as plus/minus tolerance dimensions. Therefore, some recommendations are provided as to when good or poor applications are applicable. First, *good* applications are in the areas where

- Size dimensions (features of size [FoS]) are applicable
- Noncritical radii sizes are identified
- Noncritical chamfers are available

Poor applications are in any of the three categories:

- Form
- Orientation
- Location

When geometric tolerances are directly applied to features, they are shown as

In the case of form:

In the case of orientation:

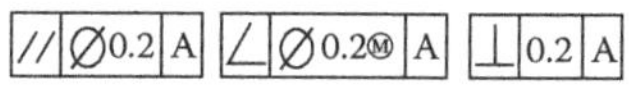

In the case of location:

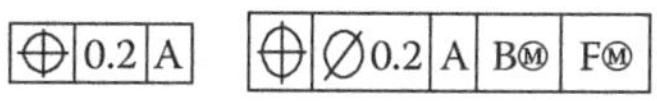

In the case of profile: 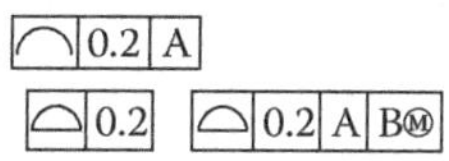

In the case of runout: 

Basic Dimensions

Proper application of GD&T will ensure that the allowable part and assembly geometry defined on the drawing leads to parts that have the desired form, fit (within limits), and function as intended. To do that, there are some fundamental rules that need to be applied (these can be found on pages 7–8 of the 2009 edition of the standard):

- All dimensions must have a tolerance. Every feature on every manufactured part is subject to variation; therefore, the limits of allowable variation must be specified. Plus and minus tolerances may be applied directly to dimensions or applied from a general tolerance block or general note. For basic dimensions, geometric tolerances are indirectly applied in a related feature control frame. The only exceptions are for dimensions marked as minimum, maximum, stock, or reference.
- Dimensioning and tolerancing shall completely define the nominal geometry and allowable variation. Measurement and scaling of the drawing is not allowed except in certain cases.

- Engineering drawings define the requirements of finished (complete) parts. Every dimension and tolerance required to define the finished part shall be shown on the drawing. If additional dimensions would be helpful, but are not required, they may be marked as reference.
- Dimensions should be applied to features and arranged in such a way as to represent the function of the features. Additionally, dimensions should not be subject to more than one interpretation.
- Descriptions of manufacturing methods should be avoided. The geometry should be described without explicitly defining the method of manufacture.
- If certain sizes are required during manufacturing but are not required in the final geometry (due to shrinkage or other causes), they should be marked as nonmandatory.
- All dimensioning and tolerancing should be arranged for maximum readability and should be applied to visible lines in true profiles.
- When geometry is normally controlled by gauge sizes or by code (e.g., stock materials), the dimension shall be included with the gauge or code number in parentheses following or below the dimension.
- Angles of 90° are assumed when lines (including center lines) are shown at right angles, but no angular dimension is explicitly shown. (This also applies to other orthogonal angles of 0°, 180°, 270°, etc.)
- Dimensions and tolerances are valid at 20°C/101.3 kPa unless stated otherwise.
- Unless explicitly stated, all dimensions and tolerances are only valid when the item is in a free state.
- Dimensions and tolerances apply to the full length, width, and depth of a feature including form variation.
- Dimensions and tolerances only apply at the level of the drawing where they are specified. It is not mandatory that they apply at other drawing levels, unless the specifications are repeated on the higher-level drawings.

Specifically, basic dimensions are considered:

- When a dimension is specified on a drawing as BASIC, it is a theoretical value used to describe the exact size, shape, or location of a feature.
- Basic dimensions on customer's drawings are shown in rectangular boxes.
- A BASIC dimension only provides half of the requirement. The tolerance portion of the requirement comes from the geometric controls in the feature control frame.

In some cases, reference dimensions are considered.

- When a dimension is duplicated on a print to add clarity, the nominal value is specified between parentheses without tolerance.
- This type of dimension is for *reference* only and is not required to be measured or gauged.
- For DtMC drawings, sometimes it is necessary to put dimensions on the drawing that were queried from the CAD model. These dimensions are referenced basic dimensions and consist of a basic dimension between parentheses. Also, these dimensions are for *reference* only and are not required to be measured or gauged.
- If a discrepancy arises between the referenced basic dimension and the CAD model, the CAD model is Master.

Line Conventions on Production Drawings

In the old blueprint, the construction of lines was very important. The significance continues in the GD&T protocol. In the GD&T approach, there are eight different types of lines used on manufacturing prints. Each type of line has a specific purpose. The eight line types are

1. Visible lines graphically define the part.
2. Hidden lines show features not visible in a view.
3. Center lines define the central axis or plane.
4. Dimension lines direct attention from feature to dimension.
5. Projection lines direct the eye to related views.
6. Cutting plane lines show location of a cut section.
7. Phantom lines show datum target areas and projected tolerance zones.
8. Leader lines direct the eye from features to notes or dimensions.

Feature of Size (FoS)

FoS has two categories: regular and irregular. The regular feature of size is defined as a circular element or one cylindrical or spherical surface, contains a set of two opposed parallel elements or opposed parallel surfaces, all of the above having a directly tolerance dimension (size limits), and a repeatable center plane or center point can be derived. The definition can be simplified as a way of describing things like a hole, pin, a part thickness, or other feature that can be measured directly for size. On the other hand, the irregular feature of size is new. It is defined as "a directly toleranced feature or collection of features that may contain or be contained by an actual mating envelope."

In addition to the above definitions, there are two rules of which the individual using GD&T must be aware. Rule #1 deals with perfect form at maximum material condition (MMC). When a regular FoS is at its MMC, it cannot extend beyond its boundary of perfect form at MMC. Rule #1 does not apply when

- FoS is not involved or when the FoS is irregular or if listed in "Other exceptions to Rule #1"
- Stock sizes: materials produced to industry or government standards (e.g., 2×4s, stock steel, tubing, etc.)
- Parts with a straightness or flatness tolerance added to the size dimension
- Parts that use the independency symbol (which will be discussed later)
- Nonrigid parts, such as o-rings, molded or body panels (these parts may need to be restrained when inspected)

Rule #2 RFS/RMB is the default. Here there are two possibilities: (a) Regardless of feature size (RFS) and (b) regardless of material boundary (RMB). In the case of RFS, For a FoS, if MMC or least material condition (LMC) is not specified, RFS applies. The geometric tolerance will not vary from the stated value.

In the case of RMB, For a referenced datum feature, when maximum material boundary (MMB) or least material boundary (LMB) is not specified, RMB applies. It establishes a variable boundary for the datum feature simulator. Unless otherwise specified, the default conditions for all dimensions on the drawing are

1. Measured without restraint in the free-state condition. Remember that the base assumption is that all dimensions and tolerances are in a free-state condition. When parts are measured in free-state, additional forces (the only exception is gravity) cannot be used to change and/or deflect the part.
2. After heat-treating.
3. After all specified coatings have been applied.

ASME Y14.5-2009 Section 1.4 (m) Fundamental Rules state that free-state is the default condition as: "Unless otherwise specified, all dimensions and tolerances apply in a free-state condition." This rule can be overwritten by placing a restraint note on the drawing describing the forces needed to ensure that a component will meet tolerance requirements when in the assembled condition. Other fundamental rules to be aware of in the same section are paragraphs (d), (l), and (n). For the entire list, see the standard. Specifically, the mentioned paragraphs are

- (d) Dimensions shall be selected and arranged to suit the function and mating relationship of a part and shall not be subject to more than one interpretation.
- (l) Unless otherwise specified, all dimensions and tolerances are applicable at 20°C (68°F) in accordance with ANSI/ASME B89.6.2. Compensation may be made for measurements made at other temperatures.
- (n) Unless otherwise specified, all dimensions and tolerances apply for the full depth, length, and width of the feature.

Maximum Material Condition (MMC)

MMC is the point at which a feature contains the greatest amount of material within its acceptable size limit. The smallest acceptable hole and the largest acceptable shaft are examples of MMC.

- MMC can be applied to both internal and external features.
- When MMC is applied to an external feature like a shaft, this will be the largest permissible diameter.
- When MMC is applied to an internal feature like a hole, this will be the smallest permissible diameter.
- MMC means that the part weighs the most.

If MMB modifier on referenced datum features is used, the MMB modifier (the letter M in a circle) on the datum referenced allows the part to *shift* with respect to the fixed size datum simulators (gauge) as the datum features depart from the MMB limits. The referenced datum at MMB establishes the datum feature simulator boundary at a fixed size (virtual condition), as shown:

⌖	∅ 0.3	A	B	DⓂ

On the other hand, if the boundary size is unclear or another fixed size boundary is desired, the boundary may be specified in the FCF as shown:

⌖	∅0.3	A	B	DⓂ [∅7.5]

Least Material Condition (LMC)

LMC is the point at which a feature contains the least amount of material within its acceptable size limit. The largest acceptable hole and the smallest acceptable shaft are examples of LMC.

- LMC can be applied to both internal and external features.
- When LMC is applied to an external feature like a shaft, this will be the smallest permissible diameter.
- When LMC is applied to an internal feature like a hole, this will be the largest permissible diameter.
- LMC means the lightest part.

If LMB modifier on referenced datum features is used, the LMB modifier (the letter L in a circle) is used on datums and allows the part to *shift*. The referenced datum at LMB establishes the datum feature simulator boundary (virtual condition). In this case, functional gauging cannot be used. This is typically a coordinate measuring machine (CMM) check. For example,

⌖	∅0.3	A	B	DⓁ

On the other hand, regardless of material boundary (RMB) or referenced datum features, the RMB modifier is the default modifier with respect to feature control frames (FCF) referenced datums. There is no symbol for RMB. This shown as

⌖	∅0.3	A	B	D

If FoS datums referenced at RMB are mated to their variable size datum feature simulator without regard for their size variation, no datum feature shift is necessary.

Virtual Condition

Virtual condition (VC) is the boundary created by a feature's size and related geometric tolerance. It also represents the worst condition the mating part will experience. VC is calculated by either of two methods. They are

1. Internal features (hole): MMC Hole − Geometric Tolerance @ MMC = VC
2. External features (shaft): MMC Shaft + Geometric Tolerance @ MMC = VC

Datum and Target Datum Symbols

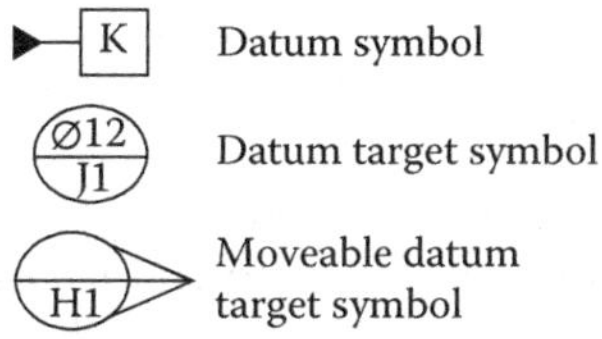

The triangle in the datum symbol and the arrowheads in the runout symbols may be left open or filled; filled is preferred with most customers. It is of paramount importance here to make sure that the letters I, O, and Q are not used.

Tolerance Modifying Symbols

The tolerance can have modifiers, which affect the shape of the tolerance zone or how it is applied. The most common symbols are the following six. For the complete list, the reader is encouraged to see the actual standard.

□ Ø Ⓕ
SØ Ⓤ Ⓟ

Independency

Independency allows the boundary of perfect form at MMC to be violated:

- When FoS is dimensioned with the symbol Ⓘ, only a two-point measurement is needed to verify size.
- Form (flatness, straightness, circularity, and cylindricity) is not controlled and an additional control may be required.

Feature Control Frames (FCF)

FCF is a series of compartments containing symbols and values that describe the tolerance of a feature. The order and purpose of these compartments follow a consistent standard. In essence, FCF provide information on geometric control. A typical FCF is shown here, followed by some general comments.

Characteristic	Tolerance	Datum reference	Datum reference	Datum reference

Form characteristics define straightness, flatness, circularity, and cylindricity and do not reference datums. Next to the tolerance, we usually find the tolerance modifier (MMC or LMC). However, when there is no modifier, RFS is default—no symbol. The datum reference establishes the boundary for the datum simulator (MMB or LMB). Note: RMB is default—no symbol. Generally, there are no more than three datum references in a FCF. Not every FCF will have datums or modifiers. Not all geometric controls can use modifiers.

Datum

Perhaps one of the most important concepts in the GD&T standard is by far the concept of datum. The reason for its importance is the fact that the datum is used as the reference point for a specific measurement. Without that reference, the measurement may be miscalculated. According to the standard, we have the following official definitions:

Datum: ASME Y14.5-2009 [1.3.13]. It is a theoretically exact point, axis, line, plane, or combination thereof, derived from the theoretical datum feature simulator. A datum is required to aid gauge studies (repeatability, reproducibility, and stability. In essence, datums are required to establish foundational dimensional controls for a part or assembly.

Datum features: ASME Y14.5-2009 [1.3.16]. It is a physical portion of the part that directly relates to the theoretical datum. It is an actual feature of a part that establishes a datum and is identified with either a datum feature symbol or datum target symbol.

Datum feature simulators: There are two types.

1. Theoretical: The theoretical perfect boundary used to establish a datum from a specified datum feature. A theoretical or mathematical representation of the datum feature in an electronic format (CAD model, CMM program, etc.)
2. Physical: The equipment contacting the datum feature that establishes the simulated datum. A physical representation of the mate for the datum feature. (Note: Keep in mind that V-Blocks, Diamond Pins and 3-Jaw Chucks do not reliably and repeatedly establish a datum axis. These devices add unintended and unmeasured variability that can lead to manufacturing issues.)

At this point, beyond the official definition of a datum, the user of GD&T must be aware of two more concepts. The first is the relationship of datum features and the second is what goes in the datum reference frame (DRF) box. A datum feature is a physical feature that acts as an acceptable substitute for a datum. Datum features relate the various features of the part to each other. Three relationships of the datum feature must be considered. They are

1. Primary datum features establish basic stability for the part with respect to a functional mating feature and typically control three degrees of freedom.
2. Secondary datum features establish orientation to the primary datum features and typically control two degrees of freedom.
3. Tertiary datum features establish orientation and location to both the primary and secondary datum features and typically control one degree of freedom.

On the other hand, a DRF is three imaginary planes perpendicular to one another that are mapped onto the part to relate features to each other. Specific requirements that should be kept in mind are

- A maximum of three datums can be referenced.
- Specify enough datums to sufficiently constrain the part.
- Use as many letters as needed. The letters do not have to follow alphabetical order.
- Letters I, O, and Q cannot be used.

In any discussion of datums, one must also consider the effects of datum feature shift. This happens when

- FoS datums are referenced at MMB or LMB, datum feature shift occurs where the part is allowed to move relative to the datum feature simulators
- Datum feature shift is an additional tolerance that must be considered when establishing gauging or when calculating tolerance stackups of component alignment or minimum sealing surfaces of mating parts
- Datum feature shift can be easily evaluated on functional gauging; it is more difficult to assess on a CMM

Datum Targets

ASME Y14.5-2009 4.24 defines datum targets as designated points, lines, or areas that are used in establishing datum controls. A datum target represents the datum feature simulator (otherwise known as the gauge) contacting the part. It is not on the part itself. The part, however, must contact all datum targets. Datum targets can be fixed or moveable. They should be used when (1) only a portion of a feature is functional as a datum feature, and (2) inherent irregularities cannot be effectively used to establish a datum. The datum target is identified with the symbol ⊖. The top half specifies a point, line, or area. The circle contains size and shape. The bottom half of the circle usually has a letter and a number. The letter designates the datum, followed by the number of the target in sequence beginning with 1 (e.g., A1).

Sometimes when dealing with datum targets, one will come across customer-specific requirements that are associated with datum target dimensions and moveable datum targets. For example, a customer may state the location and size of datum targets to be defined with basic dimensions, except when (1) the datum target is moveable, (2) expandable, or (3) collapsible; (4) the customer may clarify that for basic dimensions, established tooling or gauging tolerances apply; or (5) in the exception cases, the motion is limited by the part/assembly tolerance. Note: For information on datum feature simulator tolerances and tolerance relationships between the simulators, see ASME Y14.43.

In the case of moveable datum targets, the ASME Y14.5-2009, paragraph 4.24.6 tells us that when the direction of movement is not normal to the true profile or when the direction is unclear, dimensions and/or notes indicating direction of movement shall be added to the drawing. This symbol (H1), also known as the *birdbeak*, indicates the movement of the datum target feature simulator. The direction of movement is typically normal to the true profile.

Things to Remember: Datums

- Always verify that the selected datum is a real feature that can be measured.
- Verify that the selected features have form and size controls applied.
- Ensure that the primary, secondary and tertiary datum features been related with location and orientation controls.
- Verify that the features selected are permanent. If the features are not permanent, determine how the relationship will be measured when they are gone.
- Datum targets may require additional datum references to ensure repeatability.

Things to Remember: Form

- No datum reference allowed
- Only one feature may be controlled with form. A seal groove may create two surfaces and profile will need to be used instead of flatness

- Straightness tolerance of an axis RFS and at MMC can override Rule 1
- Of all the form controls, only straightness tolerance of an axis can use the MMC modifier. However, only attribute data will be obtained from the gauge (fits or does not fit)
- Do not use form tolerances greater than the limits of size when specifying flatness at RFS

Orientation

With orientation controls, all tolerance zones are referenced to a datum or datums with a basic angle of

0°	//
90°	⊥
ALL°	∠

Rules Applied to Orientation

- Lines drawn parallel have a basic angle of 0°.
- Lines drawn perpendicular have a basic angle of 90°.
- In 3D space, angles other than parallel can appear to change depending on the models rotation and slight angles may be difficult to detect.
- Angles that are not parallel or perpendicular must be specified on the drawing or queried in the CAD model to avoid confusion.

Very often customers may have their own special orientation requirements. Therefore, it is important that the supplier is aware of them.

Things to Remember: Orientation

- Orientation controls do not provide location for features.
- Location and profile controls or toleranced dimensions are the only ways to locate surfaces and features.
- Orientation tolerances should be significantly smaller than the position tolerance to make it a cost-effective control.
- Make sure sufficient datums are referenced for full orientation control.

Projected Tolerance Zone (Ⓟ)

- Since the positional tolerance zone controls location and orientation, the projected tolerance zone is required to transfer the tolerance zone to the area used by the mating part.
- Section [7.4.1] of the standard specifies where a composite or multiple segment feature control frame is used. The projected tolerance zone symbol shall be shown in all applicable segments.
- If a projected tolerance zone is not used, the minimum clearance hole must be calculated using the formula shown in the ASME Y14.5-2009 appendix B5, or similar method.

The AMSE standard paragraph 7.4.1 also identifies an interference, which can occur where a tolerance is specified for the location of a threaded or press-fit hole, and the hole is inclined within the positional limits. Unlike the floating fastener application involving clearance holes only, the orientation of a fixed fastener is governed by the inclination of the produced hole into which it assembles. Therefore, the projected tolerance zone value is defined as a minimum height, which is equal to the maximum material thickness of the mating clearance hole. On the other hand, Section [7.4.1.2] (p. 120) states that the specified position value for the projected tolerance zone is a minimum and represents the maximum permissible mating part thickness, or the maximum installed length or height of the components, such as screws, studs, or dowel pins. Furthermore, [7.4.1.2] (p. 121) states that the minimum extent and direction of the projected tolerance zone are shown in a drawing view as a dimensioned value with a heavy chain line drawn closely adjacent to an extension of the center line of the hole.

As for position surface (boundary) interpretation, the Standard states that (1) whenever a discrepancy occurs between surface (boundary) or axis interpretation, the surface interpretation is the default and (2) while the *boundary* notation is not required, it is recommended whenever the axis interpretation of the FoS is not clear. This concept is especially helpful for irregular FoS, such as slots and tubes/hoses, and so on.

Things to Remember: Position

- Position is the measurement of permissible variation from theoretical design intent (exact location).
- Location of one or multiple FoS are relative to one another or to specified datums.
- Basic dimensions establish the theoretical design intent (exact location).
- Positional tolerance applies only to the FoS location, not to the basic dimensions.

- Positional tolerancing is applied at RFS, MMC, and LMC.
- Axis and surface interpretations:
  - Axis: Zone defining variation allowance of center, axis, or center plane of a FoS from an exact location
  - Surface: A VC boundary, located at exact location which may not violate FoS boundary

Profile Controls

Profile controls are a group of powerful geometric tolerances that control the size, location, orientation, and form of a feature. Profile tolerances can be either independent or related. They use basic dimensions or a CAD model defined as CAD Basic. Profile was originally intended to control irregular surfaces; however, it has expanded to include many other applications.

Circle U Modifier (Ⓤ)

Whether it is applied to a drawing or CAD model, the Circle U symbol eliminates the need for support geometry and simplifies the profile callout. Both methods are shown in Figure 20.1.

Things to Remember: Profile

- Profile can be used with or without referencing datums.
- The profile tolerance is applied RFS. Currently, no modifiers are allowed in the tolerance zone with the exception of Circle U.
- Unless otherwise specified, the profile tolerance is equal bilateral.
- The profile tolerance can also be unilateral or unequally disposed by adding support geometry showing the amount of tolerance that is applied outside of the part or by using the Circle U modifier.
- Referenced datums are allowed to have material modifiers, yet the added tolerance needs to be considered.

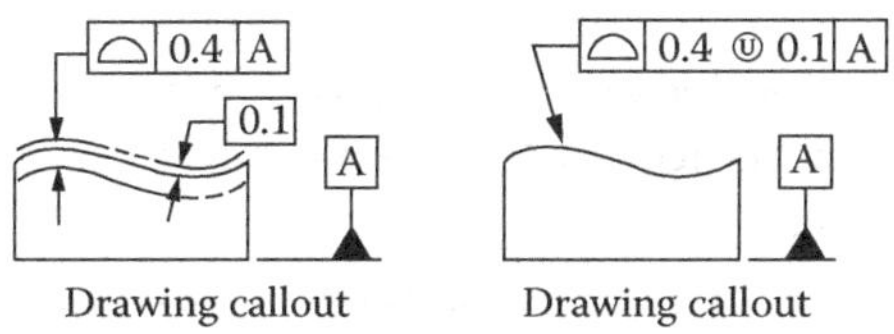

FIGURE 20.1
Options of drawing callout the circle U modifier.

Always remember, "Shift Happens!" Therefore, to minimize the effects of datum shift, particularly when the datum features have a large size tolerance, reference datums at RMB or decrease the range of the datum feature size tolerance.

Coaxial Controls Overview

- Circular runout and total runout are two low cost checks that quickly provide an indication of the coaxiality of the considered feature on an RFS basis. This control may be ideal for rotating parts.
- Position with the tolerance evaluated at MMC is ideal for parts that must fit together and typically are not rotating. If rotation or balance is a consideration, the extreme conditions resulting from the bonus of MMC should be evaluated.

Key Elements to Review GD&T by Asking the Following Questions

- Do the tolerances reflect the supplier's process capability?
- Does the drawing contain all the key dimensions and feature control frames critical to the mating components to ensure stackups can be completed?
  - Often, these are incomplete or missing.
  - There are many examples where tolerances are missing or are implied in the title block.
  - When these controls are missing, the customer cannot predict the worst case conditions of the system or develop a variation simulation analysis (VSA) model.
- Are the datum targets defined and located?
  - If they are not located with basic dimensions or their sizes called out, one is forced to make assumptions for inspection or manufacturing.
  - Repeatability is lost when the targets lack basic dimensions.
  - If only the datum target point symbol is shown, point contact will be assumed where no area is specified.

Even though this international language is unique in the sense that it is very precise and avoids many problems from design to manufacturing, it involves many symbols and terms. In fact, some have called it very confusing and overwhelming. Nevertheless, understanding how to apply and interpret GD&T correctly will help anyone who is interested in improvement of both process and cost. Some of the benefits are

- Create clear, concise drawings
- Improve product design
- Create drawings that reduce controversy, guesswork, and assumptions throughout the manufacturing process
- Effectively communicate or interpret design requirements for suppliers and manufacturing

Datums are features (points, axis, and planes) on the object that are used as reference surfaces from which other measurements are made. Datums are used in designing, tooling, manufacturing, inspecting, and assembling components and subassemblies.

- Features are identified with respect to a datum.
- Always start with the letter A.
- Do not use letters I, O, or Q.
- May use double letters AA, BB, and so on.
- This information is located in the feature control frame, which looks like the following:

⌖	∅ .003 Ⓜ	A

Datums on a drawing of a part are represented using the symbol shown as B. A primary datum is selected to provide functional relationships, accessibility, and repeatability.

- Functional relationships
    - A standardization of size is desired in the manufacturing of a part.
    - Consideration of how parts are orientated to each other is very important.
- For example, Legos are made in a standard size in order to lock into place. A primary datum is chosen to reference the location of the mating features.
- Accessibility
    - Does anything, such as shafts, get in the way?

- Repeatability
  - For example, castings, sheet metal, and so on (The primary datum chosen must ensure precise measurements. The surface established must produce consistent measurements when producing many identical parts to meet requirements specified).
  - Secondary and tertiary datum (all dimensions may not be capable to reference from the primary datum to ensure functional relationships, accessibility, and repeatability).
  - Secondary datum: Secondary datums are produced perpendicular to the primary datum so measurements can be referenced from them.
  - Tertiary datum: This datum is always perpendicular to both the primary and secondary datums, ensuring a fixed position from three related parts.

MMC (M) and LMC (L)

MMC (M)

This is when a part will weigh the most. Permits greater possible tolerance as the part feature sizes vary from their calculated MMC.

- MMC for a shaft is the largest allowable size.
  - MMC of Ø0.240±.005?
- MMC for a hole is the smallest allowable size.
  - MMC of Ø0.250±.005?
- Ensures interchangeability
- Used
  - With interrelated features with respect to location
  - Size, such as, hole, slot, pin, and so on.

LMC (L)

This is when a part will weigh the least.

- LMC for a shaft is the smallest allowable size.
  - LMC of Ø0.240±.005?
- LMC for a hole is the largest allowable size.
  - LMC of Ø0.250±.005?

Regardless of Feature Size (RFS)

RFS requires that the condition of the material *not* be considered. Therefore, RFS is used when the size feature does not affect the specified tolerance and is valid only when applied to features of size, such as holes, slots, pins, and so on, with an axis or center plane. Additional items of concern are

- Position tolerances. A position tolerance is the total permissible variation in the location of a feature about its exact true position. The exact position of the feature is located with basic dimensions and the position tolerance is typically associated with the size tolerance of the feature.
- For cylindrical features, the position tolerance zone is typically a cylinder within which the axis of the feature must lie.
- For other features, the center plane of the feature must fit in the space between two parallel planes.
- Datums are required.
- MMC: The condition where a size feature contains the maximum amount of material within the stated limits of size, that is, largest shaft and smallest hole.
- LMC: The condition where a size feature contains the least amount of material within the stated limits of size, that is, smallest shaft and largest hole.
- Tolerance: Difference between MMC and LMC limits of a single dimension.
- Allowance: Difference between the MMC of two mating parts (minimum clearance and maximum interference).
- Basic Dimension: Nominal dimension from which tolerances are derived.

Features That Require Datum

- Orientation
  - Perpendicularity
  - Angularity
  - Parallelism
- Runout
  - Circular runout
  - Total runout
- Location
  - Position
  - Concentricity
- Symmetry
  - Perpendicularity

- Angularity
- Parallelism

References

ASME Y14.5-1994. *Dimensioning and Tolerancing*. New York: The American Society of Mechanical Engineers.

ASME Y14.5-2009. *Dimensioning and Tolerancing*. New York: The American Society of Mechanical Engineers.

ASME Y14.43-1011. *Dimensioning and Tolerancing Principles for Gages and Fixtures*. New York: The American Society of Mechanical Engineers.

Selected Bibliography

http://www.tec-ease.com/gdt-terms.php.

Olson, D. (2012). The GD&T model. *Quality*. October, 35–38.

Tandler, B. (2012). Datum feature: Selection and use. *Quality*. October, 16–17.

Tandler, W. (2014). Corporate GD&T implementation. *Quality*. August, 22–23.

Wion, D. (2014). Don't forget about fixturing. *Quality*. July, 25–27.

21

8D

Overview

In Chapter 13, we discussed the issue of problem solving, and 8D was mentioned as one of the approaches to problem solving. In this chapter, we are going to discuss this particular methodology in more detail, because it is used very frequently and by many suppliers in many different industries. Fundamentally, the methodology actually is a nine-step process even though it is identified as an eight-step approach. It was developed by Ford Motor Company in the mid-1980s and has been revised several times since then. The final revision added one more step and that made it a nine-step approach. However, since everyone was accustomed to 8D, Ford decided to rename it global 8D (G8D) to reflect the new approach. The new approach has very few changes other than the addition of the D0 step. We have chosen to use the old designation of 8D only because most suppliers are accustomed to that name, even though in reality the G8D methodology is used. Therefore, the goal of 8D as we will explain it is equal to G8D, and its purpose is to eliminate defects so that they do not recur. A defect by definition here is a problem; a concern; an opportunity. The steps are identified as D0–D8; D stands for discipline and the numerical numbers 0–8 stand for the actual steps. In summary, the steps are

D0: Prepare for the 8D process: In response to a symptom, evaluate the need for the 8D process. If necessary, provide emergency response action (ERA) to protect the customer and initiate the 8D process.

D1: Establish team: Establish a small group of people with the process and/or product knowledge, allocated time, authority, and skill in the required technical disciplines to solve the problem and implement corrective actions. The group must have a designated champion and team leader. The group begins the team building process.

D2: Describe the problem: Describe the internal/external customer problem by identifying "what is wrong with what" and detail the problem in quantifiable terms. The more precise the operational definition of the problem is, the higher the probability of solving the problem.

D3: Develop the interim containment action (ICA): Define, verify, and implement the interim containment action (ICA) to isolate effects of the problem from any internal/external customer until permanent corrective actions (PCAs) are implemented. Validate the effectiveness of the containment actions.

D4: Diagnose problem. Define and verify root cause and escape point: Isolate and verify the root cause by testing each possible cause against the problem description and test data. Also isolate and verify the place in the process where the effect of the root cause should have been detected and contained (escape point), but was not.

D5: Choose and verify PCAs for root cause and escape point: Select the best permanent corrective action to remove the root cause. Also, select the best permanent corrective action to contain the effect of the root cause. Verify that both decisions will be successful when implemented without causing undesirable effects.

D6: Implement and validate PCAs: Plan and implement selected permanent corrective actions. Remove the ICA. Validate the actions and monitor the long-term results.

D7: Prevent recurrence: Modify the necessary systems including policies, practices, and procedures to prevent recurrence of this problem and similar ones. Make recommendations for systemic improvements, as necessary.

D8: Recognize team and individual contributions: Complete the team experience, sincerely recognize both team and individual contributions, and celebrate.

8D Application Criteria

Contrary to the opinion of many practitioners, 8D is not necessary in all situations. There are six definite guidelines when the 8D methodology is appropriate and applicable. They are

1. There is a definition of the symptom. The symptom has been quantified.
2. The 8D customer who experienced the symptom and, when appropriate, the affected parties have been identified.

3. Measurements taken to quantify the symptom demonstrate that a performance gap exists and/or priority (severity, urgency, growth) of the symptom warrants initiation of the process.
4. The cause is unknown.
5. Management is committed to dedicate the necessary resources to fix the problem at the root-cause level and to prevent recurrence.
6. Symptom complexity exceeds the ability of one person to resolve.

If any one of these items is present, then an 8D is appropriate. All 8Ds require some common tasks that optimize the resolution of the problem. These common tasks need continual consideration throughout the problem-solving process using the 8D approach. Some of these considerations have to do with making sure that the team is responsible to

1. Document the changes
2. Review team composition
3. Review measurables
4. Determine if service action is required
5. Review assessing questions
6. Update the 8D report itself
7. Recognize that a change has occurred. There are three possibilities as shown in Figure 21.1. The solid line is the expectation performance or desired performance and the dotted line is the actual performance. The three possible changes are
   a. Something has changed gradually.
   b. Something has changed abruptly.
   c. Something new is expected that has never been there before.

D0: Prepare for the Problem-Solving Process

In response to a symptom, evaluate the need for the 8D process. If necessary, provide ERA to protect the customer and initiate the 8D process.

The *application criteria* identify conditions that justify the use of the 8D process. This justification is necessary to allocate resources because the demand for resources in terms of people, time, money, experience, and so on is usually greater than what is available.

Many problem-solving techniques and processes are available. Matching the right problem with the right process is one way to avoid wasting limited resources. Using those resources effectively is also important. 8D requires the use of many resources, but it offers many benefits. Of special interest

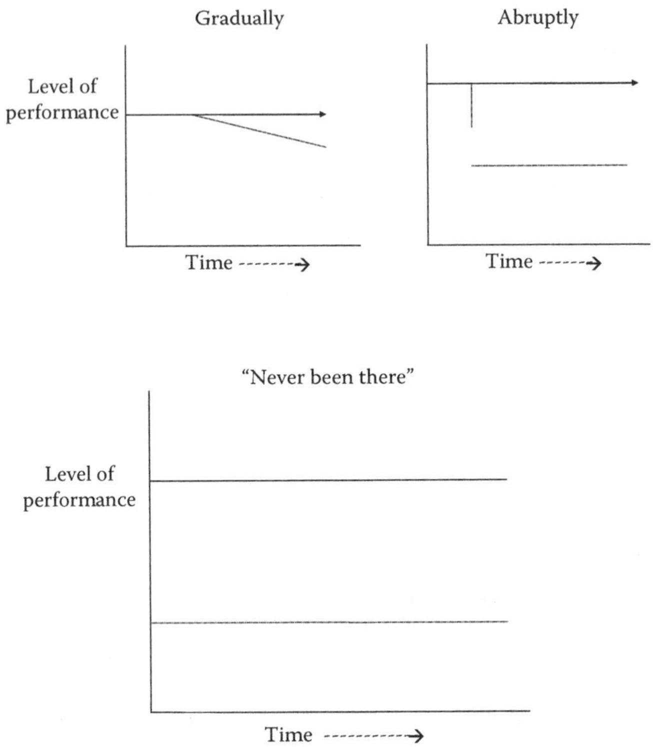

FIGURE 21.1
Pictorial view of the three possibilities of something changed.

here is that it is the choice of the supplier to use whatever method it wants to use to identify and correct the defect. However, some customers demand (require) that the results of the supplier's investigation must be in the customer's 8D format.

In this stage, two key questions have to be addressed. The first one deals with ERAs. Typical questions may be

- Are ERAs necessary?
- Is a service action required as part of the emergency response?
- How was the ERA verified?
- How was the ERA validated?

The second question deals with the application criteria. Typical questions are

- Will the new 8D duplicate an existing 8D?
- How well does the proposed 8D meet the application criteria?

- Is there a definition of the symptom? Has the symptom been quantified?
- Has the 8D customer who experienced the symptom and, when appropriate, the affected parties, been identified?
- Have measurements been taken to quantify the symptom demonstrated that a performance gap exists and/or has the priority (severity, urgency, growth) of the symptom warranted initiation of the process?
- Is the cause unknown?
- Is management committed to dedicating the necessary resources to fix the problem at the root-cause level and to prevent recurrence?
- Does the symptom complexity exceed the ability of one person to resolve the problem?

In some cases, additional basic questions may have to be asked such as

- Have all changes been documented (e.g., FMEA, control plan, process flow)?
- Do we have the right team composition to proceed to the next step?
- Have we reviewed the measurable?
- Have we determined if a service action is required?

D1: Establish the Team: Process Flow

The purpose of this step is to establish a small group of people with the process and/or product knowledge, allocated time, authority, and skill in the required technical disciplines to solve the problem and implement corrective actions. The group must have a designated champion and team leader. The group begins the team building process.

It is imperative that everyone understands that the original *group* has to develop into a *team* for an effective problem resolution. This means that the team has to go through the traditional development of forming, storming, norming, performing, and adjourning (Tuckman 1965, Blanchard and Parisi-Carew 2009, Abudi 2010).

The reason for team development is that problem solving implies changes will be made to resolve an undesired condition, which implies performance based on synergy. Comprehensive or synergistic problem-solving approaches acknowledge the need to make provisions for the entire problem-solving effort. Such provisions may include blending the organization's needs with the problem solvers' skills, behaviors, and personality styles.

Step D1 addresses part of the requirements for successful problem-solving efforts. Of course, some problems can be tackled by one person, but, generally speaking, resolving a concern at the systemic level requires a cross-functional group—a team.

A problem-solving team may be created with specific tasks or objectives in mind. However, when people with different experiences, skills, knowledge, and priorities are assembled, human relations issues tend to arise and interfere with completion of assigned tasks.

D1's primary function is to prevent these human relations issues from developing. During D1, roles, goals, and responsibilities are clarified. In group problem-solving situations, the team is well advised to establish guidelines for group and individual behaviors.

At a minimum, the owner of the concern (called the champion), the one with decision making authority, must be strongly committed to the problem-solving effort. Fixing the problem at the root-cause level and subsequent prevention efforts are impossible without this commitment. The champion's critical role is to empower the problem solver to complete the prescribed problem-solving steps. The champion also helps remove barriers and procures resources needed to resolve the concern.

Team Composition

For the team to function properly and effectively, there are specific roles and responsibilities that the participants must have. Some of the key ones are as follows.

Champion

The champion is the person who has the authority to make (and implement) decisions regarding containment actions, permanent corrective actions, and prevention of recurrence. He or she also supports the problem-solving efforts of the team throughout the process. The champion also sets priorities of specific tasks as well as being the individual with power and influence to diffuse any bottlenecks that may occur during the problem-solving process. Some of the key duties are

- Tasks the team with the 8D method
- Helps remove organizational barriers to the 8D method
- Helps procure resources the team requires to complete the process
- Reviews assessing questions with the team
- Exercises authority to implement team recommendations
- Supports the organization as a point of contact regarding the 8D process
- Forwards the 8D report to concerned departments

Team Members

Team members are those who are selected to participate on the 8D team due to their *expertise* and *knowledge.* They are the subject matter experts who do the work of each objective. The members might change during the 8D process. They are responsible for

- Doing the investigative work
- Developing plans
- Using their judgment, skills, experience, and knowledge
- Finding the answers
- Developing recommendations at each 8D step
- Executing the implementation actions

Team Leader

The team leader is the individual responsible for leading team members through the 8D process. This person has leadership and interpersonal skills. The team leader *is not* a chairperson of the team. It is a person who makes sure that synergy is cultivated for the optimum solution. Typically, the team leader

- Leads the teams to complete each 8D objective
- Develops agendas for meetings and team activities
- Conducts or leads the meetings
- Asks (not generally answers) the 8D process questions
- Manages the team in accordance with established team effectiveness guidelines
- Supports the champion as the primary point of contact

Recorder

The recorder is a team member who generates, holds, and publishes team reports (such as 8D reports, meeting minutes, agendas, action plans, etc.). Typically, the recorder

- Controls documents for the team
- Tends to be an administrative support person for the team
- Takes responsibility for creating and distributing meeting minutes and reports in a timely manner

Facilitator

The facilitator's primary task is to assist the team with interpersonal relationship issues such as resolving conflict, arriving at consensus (when appropriate), confronting counterproductive team behaviors, and so on. The facilitator's role is optional and depends on the style of the leader. If the leader is not open minded, if the leader is biased toward a particular approach, and if the leader does not allow for 100% participation by everyone, then a facilitator is necessary. Typically, the facilitator

- Works with the team to resolve conflicts
- Shows team members ways to work together while maintaining individual self-esteem
- Coaches individual team members, leader, and champion on how to be more effective in team situations
- Helps the team develop team effectiveness guidelines
- Provides feedback to the team on team performance behaviors
- Helps the team apply techniques such as the decision making process, creative processes, and so on, but is neutral in decision making situations

D1: Process Guidelines

- The purpose of D1 is to form and develop the team with the required experience for the 8D process. During D1, goals, roles, and responsibilities are developed and defined.
- The type of problem usually defines what experience and skills are needed and, therefore, who needs to be on the team.
- All members must contribute, use their experience, and help gather facts.
- Team members may change, as the team's required experience changes. An ideal team size is about seven and generally, but not necessarily, ranges from five to nine members to avoid ties.
- Team positions include leader, recorder, members, and facilitator.
- The roles on an 8D team are interdependent. No one can do it alone.
- The champion is the person who owns the problem, supports the team's efforts, uses authority and influence to help the team with organizational barriers, and helps get resources a team needs to complete each 8D step.
- The entire 8D team works with the champion's help, involvement, and agreement.
- The champion should conduct the first meeting to describe what is needed and why the problem needs to be resolved, and outline what is expected of the team.
- The leader is expected to lead the team, not have all the answers. Critical competencies for the leader are people-handling and leadership skills.

- At D1, there are more questions than definitive answers about the problem. This is normal. Better data regarding the problem will be created as the team progresses through the 8D process. At D1, the team members, champion, and others are selected using the best currently available information.

For an optimum selection of team members, several key questions should be addressed before closure of this step. They deal with logistical issues (warming up) as well as membership questions. Typical ones are as follows.

Warming Up

- What have you done to make the room user-friendly?
- When and where will the team meet?
- What has been done to help team members build their relationships with each other?
- What has been done to help team members focus on the team's activity?
- Has the purpose of the meeting been stated?
- Has the team been informed of the agenda for the meeting?

Membership

- Are the people affected by the problem represented?
- How is the 8D customer's viewpoint represented?
- Does each person have a reason for being on the team?
- Is the team large enough to include all necessary input, but small enough to act effectively?
- Does the team membership reflect the problem's current status?
- Do the team members agree on membership?
- What special skills or experience will the team require in order to function effectively?
- Have the team's goals and membership roles been clarified?
- Does the team have sufficient decision making authority to accomplish its goals?
- How will the team's information be communicated internally and externally?
- Do all members agree with and understand the team's goals?
- Has the designated champion of the team been identified?
- Has the team leader been identified?

- Are team members' roles and responsibilities clear?
- Is a facilitator needed to coach the process and manage team consensus?

D2: Describe the Problem

D2's purpose is to describe the internal/external customer problem by identifying what is wrong with what and detail the problem in quantifiable terms. The word *problem* may refer to cause, concern, defect, or consequences of the defect's occurrence. Therefore, the word *problem* is too general a term. That is why D2 makes a clear distinction between the cause and the effect. The function of D2 is to factually describe the problem with a correct *operational definition*. In 8D terminology, the problem is a deviation from expectation, a special cause, a distribution within control limits that are too wide, or any unwanted effect where cause is unknown.

In order to factually describe the problem, we need data. The 8D process is a data-driven process. The focus of D2 is to obtain an accurate and unbiased description of the object and defect. The information collected during D2 is critical to successful problem resolution. Historically, problem-solving proponents have advocated that defects be described using either rationally based or experientially based techniques. Each approach has advantages and limitations. The 8D process advocates a *synergistic* approach to problem description. The result is a factually based, team-generated problem description.

To generate a fact-based approach, it is necessary to allocate the appropriate time for understanding and gathering the appropriate and applicable data. The time spent in D2 is always recovered (many fold) by the time saved in subsequent 8D steps. Unfortunately, this feature of D2 is not obvious until later in the process. Teams that quickly pass through or skip D2 (and so have not adequately described the problem) have solved the wrong problem.

The initial D2 step develops a problem statement, which is then expanded by the factual listing of the problem's profile. Critical features of the problem are made visible using an accurate process flow diagram, which depicts the entire process (including any existing informal fixes). This body of information is further enhanced by developing a complete cause and effect (fishbone) diagram. Supplemental, effect-specific techniques (often experienced based) are useful.

D2 includes various types of reports and description approaches. Completing all these fact-gathering techniques may seem excessive. In reality, the synergism created by combining these techniques is of enormous benefit from D3 through D7. These remaining objectives tend to be realized more quickly, more effectively, and with far less frustration.

A significant byproduct of this synergism is the positive effect on the team members' behaviors. Team members appreciate each member's experience, resist the tendency to jump to conclusions, and realize the benefits of teamwork. The need for a leader remains, but the team is prepared to proceed from a complete and factual (not opinionated) foundation.

In this step, the team may want to use the 3×5 Why method (Figure 21.2). In Chapter 13, we mentioned it and gave a pictorial view of it. Here we show the reader a simpler form with some of the characteristics.

In the process of determining the underlying root cause of the process (specific), escape (detection), and systemic issues, the Three-Legged 5 Whys is a much more focused approach to problem solving. It focuses on key generic questions such as

- How was it created?
- How did we miss it?
- How did our system fail?
- Ask, "Why?" as many times as necessary (usually five times).
- Check your analysis with the "therefore" test in a backward flow.

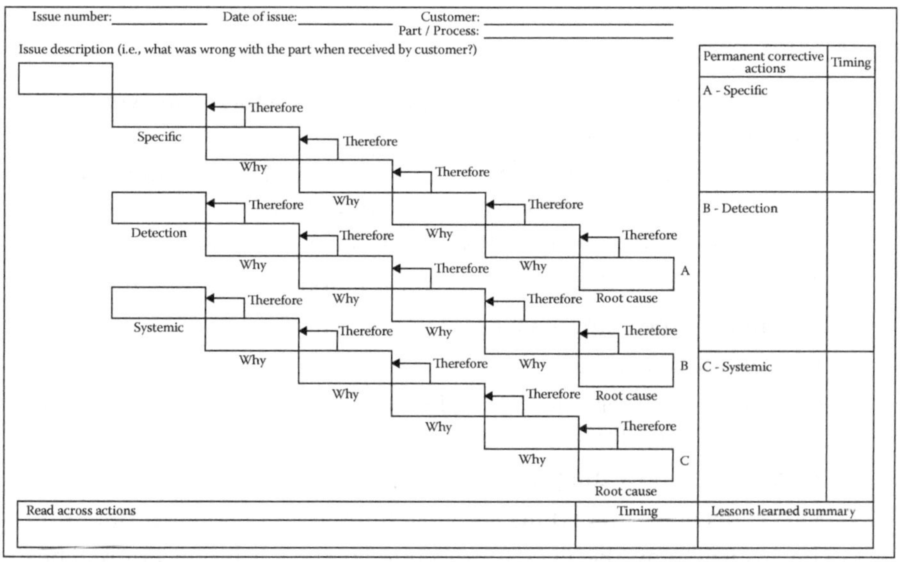

FIGURE 21.2
Three-Legged 5 Whys analysis format.

Depending on the answers, the analysis goes further dealing with the following:

- Direct root cause: How did the problem occur?
  - Start with the actual problem.
  - Keep the focus simple.
- Detection root cause: How did the problem escape?
  - Focus on inspection system.
  - How did it fail?
  - Did we account for a problem in our detection processes?
- Systemic root cause: Why didn't our systems protect the customer?
  - Start with the specific root cause.
  - Look at the problem from a bigger perspective.
  - Find organizational flaws.

Obviously, these questions are only a guideline. In the real process, there will be more in both volume and depth. Here we are emphasizing that this specific method allows the team to (1) drive corrective actions for all 3 Legs, (2) separate *noise* from real issues, (3) force people to *dig deep* into processes, procedures, and data, (4) make management issues visible, and (5) provide focus for the efforts of the cross-functional team action and systemic issues.

D2: Process Guidelines

- The purpose of D2 is to define the problem for which you eventually want to determine the root cause.
- D2 is not intended to determine the root cause. The information will be used later (at D4) to determine the root cause.
  - D2 is the observations collection step.
- D2 establishes the data that is the foundation for all other 8D steps.
- D2 demands the team develop
  - A problem statement
  - A problem description using IS and IS NOTs (what, where, when, how big); see Table 21.1
  - A process flow diagram
  - A cause and effect diagram (also known as a fishbone diagram)
- The problem may also need to be defined using additional approaches (for example, SPC data, customer satisfaction, and field reports).
- To complete this step, the team may need to obtain missing information.

TABLE 21.1

Typical Format for the IS/IS NOT

IS	IS NOT
What	*What*
Name the object that is having the trouble.	Name objects that are similar in shape, composition, form, or function that could have the same trouble but are not.
Name the trouble or problem that the object is having.	Name other kinds of troubles or problems this object could be experiencing, but is not experiencing.
Where	*Where*
Name the place where this object with the trouble can be found.	Name other places where this object can be found without the trouble.
Name the place where this problem first showed up.	Name places where the problem could first have shown up but did not.
Name all other places where this problem occurs (or has occurred).	Name similar places where this problem has never occurred.
Where on the object is the problem occurring (inside, outside, top, bottom, etc.)?	Where on the object could the problem occur, but is not?
Where in the process flow did the problem first develop?	Where in the process could the problem have first developed, but did not?
Describe all other places where the object and the trouble can be found.	Use similar units of measure to describe where the object and the trouble are absent.
When	*When*
When with respect to time did the trouble/problem first occur?	Using the same unit of time, when could the trouble and/or the problem first occurred, but did not?
Describe the above information concerning day, month, year, time of day, and so on.	Describe the above information concerning matching day, month, year, time of day, and so on when the trouble did not occur.
Describe any patterns concerning time.	Describe corresponding units of measure to describe when the trouble did not occur.
Describe when in the process the trouble/problem first occurred.	Describe when the trouble/problem could have first occurred, but did not.
Describe when in the life cycle the problem first occurred.	Describe when in the life cycle the problem could have first occurred, but did not.
Describe other places in the process and life cycle where the trouble or problem was observed.	Describe when it was not.

Note: Consider all units of time such as hours, days, minutes, seconds, shifts, periods, quarters, years, and so on. Also consider sequences in operations as would occur in a process flow diagram.

(*Continued*)

TABLE 21.1 (Continued)

Typical Format for the IS/IS NOT

IS	IS NOT
How Big	*How Big*
Describe the size of the problem/trouble/effect.	Using a similar unit of measure describe the limits of the problem/trouble/effect.
Describe the number of objects that have or have had the trouble/problem/effect.	Describe the number that could have had the trouble/problem/effect, but did not.
Describe the magnitude of the trouble in terms of percentages, rates, patterns, trends, yield, dimensions, and so on.	Describe what it could have been, but is not.
Describe the number of defects per object.	Describe what it could have been, but is not.
Describe the physical dimensions of the defect or problem.	Describe in physical dimensions what the defect could demonstrate, but has not.
Ask: 1. Are there any other facts that do not appear on the Is/Is Not tool? 2. Who can we talk to about the Is/Is Not facts?	

- Recommended attachments to the 8D Report at D2 include IS/IS NOTs, 3×5 Why and the cause and effect diagram.
- The team and champion use their judgment to determine how much information is sufficient.
- The process flow diagram should include the initial actions taken to compensate for the problem when it first occurred.
    - Note all temporary or informal fixes on your process flow. Do not remove them, just note them.
    - Start where the defect can be seen first, and then work your way back from there to the beginning of the process.
- Use only one 8D per problem. Do not use a single 8D on many different problems. Each problem probably has a different root cause.

How to Develop a Problem Description

To develop the correct problem statement, there are at least two fundamental issues that have to be answered. The first is, "What is wrong with what?" If the answer is "No," the answer to, "What is wrong with what?" is your problem statement. If the answer is "Yes," repeat the question, "What is wrong with what?" until reaching the statement where the cause is unknown. To develop the problem description, apply the IS/IS NOT method using the format and questions in Table 21.1. The second is, "Do we know why this is happening?"

Once you know the answers to these two questions, the process for developing the problem description is as follows:

Step 1. Develop the problem statement.

- Develop a simple, concise statement that describes who or what object shows the specific defect of which you want to determine the root cause.
- Intended as a starting point for further description, the statement must be very concise and vivid.
- Use the repeated whys to produce the lowest rung in what might be a series of events, a chain reaction.

Step 2. Develop the problem description.

- Describe in quantifiable, factual terms the extent of the defect. List all facts with regard to who, what, where, when, and how big (magnitude).
- Describe in quantifiable, contrasting dimensions where the symptom is not manifested or visible.

Step 3. Identify information that needs to be confirmed or collected.

- Obtain the information from knowledgeable people by asking the same questions used in Step 2.

Sometimes in conducting the IS/IS NOT analysis, it is necessary to do a comparison analysis for identifying differences in

- Facts
- Unique items to the IS
- Facts that have not already been stated in the IS column
    - To uncover any differences the key question is, What is different, unique, peculiar, special, or true only of the IS when compared to the companion IS NOT? A good guideline to develop differences is to:
        - List all facts without prejudice to cause or potential cause
        - Consider the categories of people, methods, materials, machines, measurement, and Mother Nature (environment)

A second analysis that is often necessary is the comparison of changes in

- Facts
- Specific items related to differences

- To uncover the changes, the key question is, What has changed in, on, around, or about the difference and when did the change occur? A good guideline to develop changes is to:
  - List all changes that occurred related to difference regardless of date or their potential for cause
  - Consider the categories of people, methods, materials, machines, measurement, and Mother Nature (environment)

Once the IS/IS NOT analysis is finished, the team is ready to begin the identification and/or the development of possible cause (change-how theories). At this stage, theories are

- Statements of ways that the changes may have created the trouble
- A limited form of brainstorming
- Simply a listing of possible causes not probable causes
  - To develop theories, the key question is, How could this change have caused the effect on the object? (Alternatively, in what ways might this change have caused the effect on the object?) A good guideline to develop change-how theories is to
    - List each theory individually.
    - Do not reject or qualify the theory based on its practicality or probability.
    - List single change/single variability theories first.
    - Develop more complex variability theories second.
    - Continue to prompt the problem-solving team with the above question until all possible theories are developed.
    - Defer critical thinking until the next step.

At this stage, the team is ready to check the theories. This is done with trial runs such as

- Critical evaluations of theories against the sets of IS/IS NOT data
- Test of the plausibility, not remote possibility, of each theory
- A test of the likelihood of the theory
- A process of elimination
  - To test the theory, the key question is, Does the change–how theory explain both the IS and the IS NOT? A good guideline to develop change–how theories is to
    - Test the theory against each individual set of IS and IS NOT.

- If the theory accounts for both the IS and the IS NOT phenomena, record with a + (plus) symbol.
- If the theory cannot plausibly explain either the IS or the IS NOT phenomenon, record the result with a – (minus) symbol and note the reason for the rejection.
- Test the theory against all sets of IS and IS NOTs. Do not stop testing a theory prematurely.
- Test all the theories in and of themselves. Consider each "change-how" theory a separate theory.
- Avoid testing interactive changes and highly complex theories to the end of the trial runs. Test the simplest theories first.
- If the plausibility of the theory is uncertain (+ or –) use a question mark (?) to record the uncertainty. Note why the theory is uncertain.

This simple approach does not suggest that the team may not use any advanced statistical techniques. The team is free to use any methodology or any tool that may provide the problem.

To finish the requirements of the D2, the last item should be to verify the root cause by

- Describing the optimum method to passively, and then actively, verify the root cause
- Conducting the verification in the appropriate setting
- Recording the results

D2: Evaluating Questions

When the team feels comfortable with having completed the D2 step, it is a good practice to review the discussion with at least the following basic questions:

- Can the symptom be subdivided?
- Has a specific problem statement been defined (object and defect)?
- Have repeated whys been used?
- What is wrong with what?
- Do we know for certain why that is occurring?
- Has IS/IS NOT analysis been performed (what, where, when, how big)?
- When has this problem appeared before?
- Where in this process does this problem first appear?

- What, if any, patterns are there to this problem?
- Are similar components and/or parts showing the same problem?
- Has the current process flow been identified? Does this process flow represent a change?
- Have all required data been collected and analyzed?
- How does the ERA affect the data?
- Is there enough information to evaluate for potential root cause?
- Do we have physical evidence of the problem?
- Has a cause and effect diagram been completed?
- Does this problem describe a something changed or a never been there situation?
- Has the problem description been reviewed for completeness with the 8D customer and affected parties? (Has concurrence been obtained from the 8D customer and the affected parties?)
- Should this problem be reviewed with executive management?
- Should financial reserves be set aside?
- Should any moral, social, or legal obligations related to this problem be considered?

D3: Develop the Interim Containment Action (ICA)

The D3's purpose is to define, verify, and implement the ICA to isolate effects of the problem from any internal/external customer until PCAs in D6 are implemented. Validate the effectiveness of the containment actions. In other words, the D3's predominant function is to protect the customer from the problem until it is resolved at D6.

The customer is whoever gets your output as their input. The customer could be internal to the company as well as external. If the customer cannot accept the defect, then an interim containment action (CA) is required. An ICA is intended to protect the customer from getting the problem because a root cause is not yet known. It must be understood that an ICA is *always* a temporary measure. It is a non-value-added operation. Commitment to completing the 8D remains in force. An ICA is effective when, from the customer's point of view, the defect is no longer evident. ICAs are often compared to quick fixes or hidden factories, but containment actions are actually

- Verified to work before implementation
- Validated that the ICA is working after implementation

- Validated independently, in addition to the customer's confirmation of effectiveness
- Documented to exist (i.e., made visible) by amendments to the organization's process flow diagrams, procedures, process instructions, and so on
- Replaced by a PCA at D6
- Implemented without creating other troubles downstream
- Formal temporary fixes; care is exercised to ensure that all supporting functional areas/departments are involved in planning and implementation
- Implemented only with the clear approval of the champion and customer
- Managed through comprehensive planning and follow up

Another important feature of D3 is that prudence may require containment action (an ERA) before Dl and D2 are completed. However, after D2 is completed, the ERA should be reviewed. The D2 data may suggest a more effective interim containment action (DS3).

An ICA should not be implemented by a team without the champion's knowledge or involvement. A subtle but significant feature of the 8D process is that the authority to implement an ICA is retained by the champion. It is not delegated to the team. Likewise, the champion is responsible for cross-functional communications. Some team problem-solving guidelines imply the team itself would assume this responsibility. The 8D process promotes action within the organization's chain of command. It does not promote creation of an artificial command structure, which bypasses the normal authority structure.

D3: Process Guidelines

As mentioned earlier, the fundamental purpose of D3 is to protect the customer from receiving a defective part while the supplier is determining the source of the cause. Steps and questions during D3 include

- Is the D3 step necessary? Sometimes it may not be, especially if the ERA is sufficient to control the product.
- If implemented, can the D0 ERA be improved?
- Which action is best?
- Choose the best containment action.
- Will the containment action work (i.e., eliminate 100% of the problem from reaching the customer)? (Is verification valid and accurate?)
- Follow the management cycle–PDCA cycle

- Plan (in detail) the containment action and verify.
- Implement the containment action.
- Record the results.
- Evaluate the ICA; did it work? (Validation)
- Document the actions and communicate how to do it to all who need to know about it.
- Continue to monitor its effectiveness throughout the time it is in place.
- The champion must be involved before the ICA is implemented and during implementation.
- This is the only optional step in the entire 8D process.
- It is temporary, not permanent. It is costly, not cost-effective. It is containment, not mistake proof.
- D3 works against the problem, it is not the root cause.
- It increases costs while the ICA is used.
- D3 may be done while other members of the team are working on D4 (Determine Root Cause).
- D3 may require additional members be added to the team in order to complete all of D3.
- It looks like a quick fix, but its actions are documented, later removed, and checked for effectiveness. The team must ensure that the actions will not create other problems for departments and customers downstream of the D3 action.
- More than one D3 action may be required to fully protect the customer (internal or external).
- After D3, the team's membership might change. This is true at all steps.
- D3 may occur before Dl or D2. If done before D2, then the D3 ICA should be reviewed after the D2 data is compiled. Better D2 data may prescribe the need for an improved D3 ICA.

D3: Evaluating Questions

When the team feels comfortable with having completed the D3 step, it is a good practice to review the discussion with at least the following basic questions:

- Are ICAs required?
- Is a service action required as part of the ICA?
- What can we learn from the ERA that will help in the selection of the best ICA?
- Based on consultation with the 8D customer and champion, have criteria been established for ICA selection?

- Based on the criteria established, does the ICA provide the best balance of benefits and risks?
- How does this choice satisfy the following conditions?
  - The ICA protects the customer 100% from the effect.
  - The ICA is verified.
  - The ICA is cost-effective and easy to implement.
- Have the appropriate departments been involved in the planning of this decision?
- Have appropriate advanced product quality planning (APQP) tools (e.g., FMEA, control plans, instructions) been considered?
- Have plans, including action steps, been identified? (Who needs to do what by when?)
- Has a validation method been determined?
- Does the customer have a concern with this ICA (is customer approval required)?
- Have we identified what could go wrong with our plan and have preventive and contingency actions been considered?
- Are implementation resources adequate? This is a very important consideration especially when third-party auditors (inspection houses) do the ICA.
- Does the validation data indicate that the 8D customer is being protected?
- Can the ICA effectiveness be improved?

D4: Define and Verify Root Cause and Escape Point

D4's purpose is twofold. The first is to isolate and verify the root cause by testing each possible cause against the problem description and test data. The second is to isolate and verify the place in the process where the effect of the root cause should have been detected and contained (escape point) but was not.

To appreciate the rationale of the D4, we must understand that a situation was once problem free, but now is not. When this happens, the situation is considered a *change-induced condition*. The process of identifying this change is called problem solving. Various problem-solving techniques exist; see Chapter 13. However, they are all based on three basic strategies:

1. Fact-based, deductive approaches
2. Experientially based approaches
3. Creative approaches

A working knowledge of various problem-solving techniques is desirable. The single verified reason that accounts for the problem is the *root cause*. The cause explains all facts about the problem. It is verified by the ability to make the problem come and go on demand.

Determining Root Cause

Information compiled at D2 is used to identify a set of possible causes. At D4, the root cause is discovered. Problem-solving techniques are used to reduce the time and confusion to systematically deduce the cause. All subsequent 8D objectives depend on the accurate diagnosis of the root cause. Therefore, verification of the root cause is critical to the success of the 8D process.

Few things are more damaging to the problem-solving process than assigning blame. Blaming leads to defensiveness and facts are then obscured or kept hidden. Misinformation is often generated as a defensive measure. Therefore, avoid *all blame* and concentrate on the process!

D4: Guidelines

As mentioned earlier, the fundamental purpose of D4 is to determine the root cause and escape point of the problem. Therefore, the focus is to identify both and verify them as the correct ones. Once identified, they are fixed at D6. This step identifies and evaluates the correct process of problem solving using the information of D2. The basic guidelines begin by the appropriate definition of the root cause, which is the single verified reason that accounts for the problem and is followed by the escape point, which is the earliest location in the process, closest to the root cause, where the problem should have been detected but was not. Other guidelines may deal with

- Verification, which helps to make certain that permanent corrective actions are directed at the root cause and escape point. Time, money, effort, and resources are not wasted on false causes.
- Band-aids, which can mask information needed to find the root cause. Look out for band-aid fixes.
- The *true* root cause. Usually, the root cause is one change that caused the problem. If previous problems were never fixed at the root cause level, then the problem may be the result of more than one change.
- Deductive reasoning (first) to identify the possible causes. If this method does not work (due to missing information), then use the cause and effect diagrams and team members' experience to pursue other possible causes.
- Appropriate and applicable facts. You cannot identify root cause without facts. The IS/IS NOT data must be correct and the process flow diagram must be correct and up to date.

- All pertinent facts that are directly related to the root cause. The root cause should explain all known facts about the problem. Unexplained facts often indicate the presence of another root cause that creates a similar problem. Ideally, you should be able to make the problem come and go to prove you have identified the root cause.
- New surprises. When the root cause is uncovered, other troubles (which went unnoticed) are sometimes made visible. These other troubles create the need for an improved ICA in D3 and/or a return to D2.

D4: Evaluating Questions

When the team feels comfortable with having completed the D4 step, it is a good practice to review the discussion with at least the following basic questions:

- Has the factual information in the problem description been updated?
- What sources of information have been used to develop the potential root-cause list?
- Is there a root cause (a single verified reason that accounts for the problem)?
- What factors changed to create this problem? What data are available that indicate any problem in the manufacturing or design process?
- How did we verify this root cause?
- Does this root cause explain all the facts compiled at D2?
- Has the root-cause analysis gone far enough? (Do we need to know why this root cause happened?)
- Is there more than one potential root cause?
- Does each item on the potential root cause list account for all known data? Has each item been verified (used to make the defect come and go)?
- How did you determine assignment of percentage contribution?
- Combined, do the items on the potential root cause list account for 100% of the problem? (Is the desired performance level achievable?)
- If the level is achievable, has the team considered and reviewed with the champion the benefit of developing a separate problem description (and, by definition, separate 8D) for each potential root cause?

- If the level is not achievable, has the team considered and reviewed with the champion the benefit of alternate problem-solving approaches?
  - Approaches independent of an 8D
  - Approaches as a compliment to an 8D
- Does a control system exist to detect the problem?
- Has the current control system been identified? Does this control system represent a change from the original design?
- Has it been verified that the control system is capable of detecting the problem?
- Is the identified control point closest to the root cause/potential root cause?
- Is there a need to improve the control system?

Generic Simple Root-Cause Analysis

Root-cause analysis is based on a questioning process focusing on the what, where, when, and how big, and answering the following concerns about the cause using the IS/IS NOT method, identify differences, identify changes, and testing the outcome. Figure 21.3 shows the flow of this analysis.

Determination of the Root Cause

In cases where *should* and *actual* were once the same but now are different, the root cause will be a change of some type. The definition of root cause is the single verified reason that accounts for the problem.

Two deductive approaches are available to lead the problem solver to the root cause. Both approaches are based on a thorough defect profile built around the problem statement. Both of these deductive approaches are a series of questions that yield answers to which another question is applied. The result is a steady reduction of the number of possible causes to be investigated.

For comparison, the steps of the two approaches are shown side-by-side in Table 21.2, having been anchored by the scope (problem statement and problem description). The two approaches are identical with one exception. Method B has an addition inserted in the third step.

Method A is best used when all changes are known. On the other hand, Method B with the additional step eliminates consideration of changes common to both the IS and IS NOT listed in the problem description. Furthermore, Method B provides a hint where to investigate for hidden changes.

During either Method A or B, the problem solver may have to collect missing information. The problem solver can use probing questions to gather

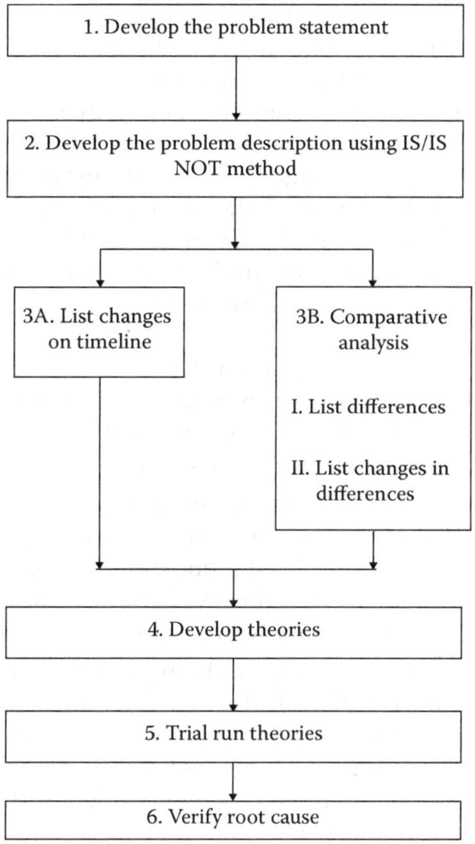

FIGURE 21.3
Determine root cause of the problem.

TABLE 21.2
Typical Comparison of Determining Root Cause Using Two Approaches

Approach A	Approach B
1. Establish a problem statement.	1. Establish a problem statement.
2. Develop a problem description.	2. Develop a problem description.
3. List all known changes on timeline.	3a. List differences.
	3b. List changes in differences (only).
4. Develop theories based upon the changes.	4. Develop theories based upon the changes.
5. Trial run the theories.	5. Trial run the theories.
6. Verify the root cause/potential root cause.	6. Verify the root cause/potential root cause.

information from people who do the work (operators) and are more familiar with details of the problem. Frequently, these people have essential information that they may not recognize as significant.

Test each theory against each discrete set of IS/IS NOTs listed in the problem description (Step 2) for every element of what, where, when, and how big. Specifically, ask, "Does this 'change-how' theory completely explain both the IS and IS NOT?" With this question, the team is asking, "If this theory is the cause of the effect, do the factual data in the problem description explain fully why the effect manifests itself in the IS dimension but never manifests itself in the IS NOT dimension?" There are three possible outcomes:

1. Explain fully: If the data explain fully why the effect manifests itself in the IS dimension but never manifests itself in the IS NOT dimension, then a plus (+) symbol should be entered in the test matrix for that element. This theory fully explains this element.
2. Cannot explain: If the data cannot explain why the effect manifests itself in the IS dimension and/or never manifests itself in the IS NOT dimension, then a minus (–) symbol should be entered in the test matrix for that element. This theory definitely does not explain this element and therefore cannot be the cause.
3. Insufficient data: If the theory could explain the effect but there are insufficient data to explain fully why the effect manifests itself in the IS dimension and/or never manifests itself in the IS NOT dimension, then a question mark (?) symbol should be entered in the test matrix for that element and a comment added at that point to indicate that further data collection or analysis is required. This theory could explain this element, but more data or analysis is required. If, in fact, this is the situation, the team needs to
   a. Repeat the test question against each pair of IS/IS NOTs until a minus symbol is encountered for a theory, when it should be discounted and a further theory examined. The use of the minus symbol is to exclude unfeasible theories from further consideration.
   b. If only one theory passes through the trial run (with all pluses or a combination of pluses and questions marks), then proceed to Step 6 to verify the roof cause included in this theory.
   c. In practice, particularly where multiple potential root causes are considered, more than one theory may pass the trial run with a combination of pluses and question marks. In those cases, and where it is feasible and practical to do so, collect and analyze the missing data, and re-examine the theory to resolve the question marks to either pluses or minuses and proceed as above. If it is not feasible or practical to collect and analyze the

missing data, proceed to Step 6, starting with the theory with the most pluses.

Verify Root Cause

Verification of the possible causes can only be done in the real world, where the defect is occurring. Root-cause verification cannot be done on a problem-solving form. Verification can be done in two ways: passive or active.

1. Passive verification is done by observation. With passive verification, you look for the presence of the root cause/potential root cause without changing anything. If you cannot prove the presence of the root cause/potential root cause variable, then chances are great that this possible cause is not the root cause.
2. Active verification is a process where the problem solver seeks to make the defect come and go using the variable thought to be the root cause/potential root cause. Both coming and going are important tests to confirm root cause/potential root cause.

Helpful Supplemental Tools

The two deductive approaches to determine root cause are based on the availability of all relevant facts. They also presume to contain *myths* or information that is not fact. Sometimes, despite the best efforts of asking the right questions of the right people, the needed facts cannot be found. Supplements to the two deductive approaches include, but are not limited to

- Statistical process control data
- Process flow diagrams
- Process flow cause and effect diagram
- Possible causes listed on a cause and effect diagram
- A subject matter expert's experience
- Part analysis of the defect
- Distinction analysis
- Group brainstorming
- Force field analysis
- 5 Whys analysis
- 3×5 Whys analysis

Each of the above exhibit both benefits and liabilities when used with a purely deductive method. For more tools, see Appendix V.

D5: Choose and Verify PCAs for Root Cause and Escape Point

The purpose of D5 is to select the best permanent corrective action to remove the root cause. Also, select the best permanent corrective action to eliminate the escape point. Finally, another salient purpose of the D5 is to verify that both decisions will be successful when implemented without causing undesirable effects.

The 8D process is applied to fix a problem at the root-cause level and prevent its recurrence. In change-induced conditions, the root cause is introduced by a change. In an overwhelming number of cases, the change is introduced by someone's decision. At the time of the decision, the problem is not anticipated. The D5 objective is designed to avoid a similar sequence of events. The historical function of D5 is to make the best decision on how to remove the root-cause variable that created the defect. Because decisions potentially lead to other effects, the decision needs to be thoroughly examined before its implementation at D6. Customers and managers must be involved.

The decision is first examined for both its benefits and its risks. Then, the decision is tested (verified) to ensure it will be successful when implemented at D6. The D5 objective ends with a commitment to full-scale implementation of this decision at D6. The decision making process depends on the experience of the decision makers and the criteria that are applied. This is especially true during the risk analysis phase of the decision making process. *Experience* is also critical to the verification segment of D5. Experience guides how verification should be conducted and how much data is appropriate. (Of special note here is that the champion's commitment is of paramount importance and is essential. The criteria on which the decision is based are typically a synthesis of the champion's criteria and the detailed experience of the team members.)

D5: Process Guidelines

As mentioned earlier, the fundamental purpose of D5 is to select the best corrective action to eliminate the root cause and escape point of the problem. Therefore, more than one permanent corrective action may be required to resolve 100% of the problem. The team should remember that the D5 decision is not implemented until D6. In addition, the team should consider having risk analysis conducted by those who must implement the corrective action. Some implied actions of D5 are

- Check the composition of the team. Does the team have the right experience to make decisions?
- Decide what is the best permanent corrective action.

- Evaluate each choice from both its benefits and liabilities.
- Test, if necessary, the decision to determine that the choice (action) will, in fact, work.
- If this PCA cannot be implemented quickly, will the D3 ICA last long enough?
- Get approval (through the champion) to proceed with the choice. Implement the decision at D6.

In addition to these implied actions, it is necessary to recognize that

- Sometimes the best permanent corrective actions cannot be implemented due to cost, limited resources, and so on Sometimes the root cause cannot be eliminated. In these cases, the best decision is to continue using the D3 ICA. This measure is a compromise, but is sometimes the only resolution. This should be considered an exception for D5, not the norm. Even if this action is appropriate, D6, D7, and D8 should still be completed.
- Sometimes the best PCA will require time before it can be implemented. In this case, the D3 ICA will continue until replaced by the D6 PCA. The 8D process permits a redesign of the ICA. This redesigned ICA might be more effective and less costly than the initial D3 ICA.
- Sometimes during D5 verification, the presence/evidence of another root cause, which was not discovered during D4, may be uncovered. This is indicated when the verified PCA does not remove 100% of the unwanted effect.

D5: Evaluating Questions

When the team feels comfortable with having completed the D5 step, it is a good practice to review the discussion with at least the following basic questions:

- What criteria have been established for choosing a PCA for the root cause and escape point? Does the champion agree with these criteria?
- Is a service action required as part of the PCA?
- What choices have been considered for the PCAs?
- Have we overlooked better choices?
- What features and benefits would the perfect choice offer? How can we preserve these benefits?
- Do we have the right experience on this team to make this decision?
- What risks are associated with this decision and how should they be managed?

- Does the champion concur with the PCA selections?
- What evidence (proof) do we have that this will resolve the concern at the root-cause level?
- Did our verification approaches make allowances for variations in the frequency (or patterns) created by the cause?
- Which variables did we measure during the verification step? Do these indicators constitute sound verification?
- What are the possibilities that this choice, once implemented, will create other troubles?
- Can our customer live with this resolution?
- Will our containment continue to be effective until our choice can be implemented?
- What resources will be required for PCA implementation? Do we have these resources?
- What departments will need to be involved in the planning and implementation of this decision?
- Have actions been considered that will improve the ICA prior to PCA implementation?

D6: Implement and Validate PCAs

The purpose of D6 is to plan and implement selected permanent corrective actions, remove the ICA, and monitor the long-term results. In D5, the PCA was chosen and verified. In D6, the PCA is implemented and the problem is corrected. In D6, the team is responsible for detailed planning, implementation, and evaluation.

One of the major responsibilities of the champion's authority is the requirement to authorize making changes to resolve the problem. Generally, even relatively minor changes will require cross-functional participation. For example, changing the length of a purchased bolt impacts the documentation of purchasing, quality assurance, material control, inventory control, engineering, and production departments. In this simple example, the champion might be the production department manager, but the production department still requires communication and close coordination.

Cross-functional involvement at D6 also means cross-functional planning to implement the D6 correction. During D6, team membership may expand temporarily to facilitate complete planning. This planning may or may not require a group meeting on the subject, but some form of contact with other departments is essential for successful implementation. At a minimum, the existing team must identify who (in which department) needs to be involved in planning the implementation of D6.

Removing the ICA

Embedded in D6 is removal of the D3 ICA. The ICA is no longer needed because the PCA removes the root cause variable. Continuing the ICA wastes important resources. Further, the ICA has masked the problem, so validation of the D6 PCA would be questionable.

D6 planning must account for when to terminate the D3 ICA and how to validate results of the D6 PCA. Technically, this validation is continuous and ongoing, but validation is likely to be more frequent at first, tapering off over time. In simple terms, the events of D6 could be expressed by the model in Figure 21.4.

The team members' and champion's active involvement occur at each event, although the activities of each tend to be different.

D6: Process Guidelines

The critical task of updating all documents, procedures, process sheets, and so on is too frequently overlooked. The D6 PCA fixes a problem by making changes. If the PCA change *is not* documented in the normal systems, practices, and procedures, then additional corrections will likely be necessary. After all, often a problem recurs when someone reads an obsolete instruction. Related problems also occur, such as ordering obsolete parts and being unable to order needed parts that are not recorded on the bill of materials in the material control system.

As mentioned previously, the purpose of D6 is to implement the decision made at D5. However, this implementation is done via the management systems. They are not bypassed. The implied actions of D6 include:

- Check the composition of the team. Is the needed experience on the team?
- Verify that the decision needs to be planned, implemented, monitored, and evaluated for effectiveness.

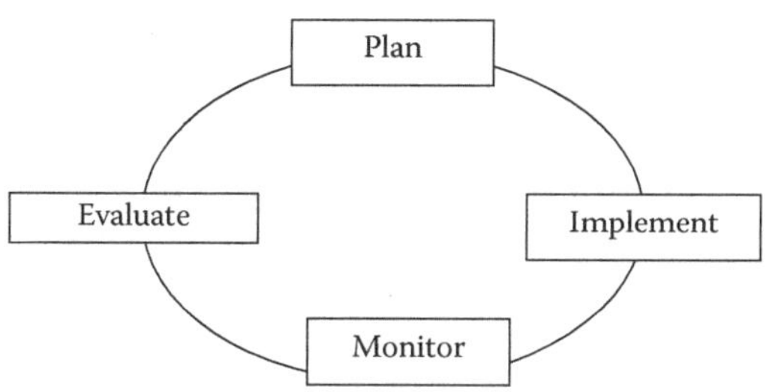

FIGURE 21.4
Validation model of removing the ICA.

- Update any changes implemented during D6. All updates must be documented (formally) in process flow sheets, procedures, practices, instructions, standards, spec sheets, FMEAs, control plans and so on in the departments that maintain such records or documents. The updates need to have at least three minimum items for any action plan: (a) what, (b) who, and (c) when due.

Furthermore, it should be recognized that

- D6 requires detailed, hands-on people to be involved in planning D6 actions.
- The 8D process does not supersede an organization's procedures to implement process changes or customer approval processes.
- Some changes may require customer approval before the changes can be made.
- No changes should be made without the champion's direct approval.
- Some people may require training before you can implement the D6 change because their procedures may be affected.
- During D6, the D3 action should be stopped because the ICA is no longer necessary. Continued D3 actions may mask a portion of the defect that is still present.
- D6 is not completed until the PCA is validated (the customer no longer experiences the problem). Prudence dictates that validation should occur at some point before the customer can experience the problem.
- Long-term monitoring is appropriate after the D6 PCA is implemented.
- Failure to document the correction created by the D6 PCA will eventually reintroduce the root cause. Failure to document the PCA typically results in overstocking obsolete parts and stockouts of needed parts.
- Consider adopting the following statement as a normal D6: Flawless implementation of the PCA, where things work right the first time.

D6: Evaluating Questions

When the team feels comfortable with having completed the D6 step, it is a good practice to review the discussion with at least the following basic questions:

- What departments are needed to implement the PCAs?
- Are representatives of those departments on our team to plan and implement their parts?

- What customer and/or supplier involvement is needed?
- Who will do the planning for the customer? For the supplier?
- Has an action plan been defined (responsibilities assigned, timing established, required support determined)?
- Do we have the necessary resources to implement this plan? What is needed?
- At what point is this plan vulnerable to things gone wrong? What can be done to prevent these?
- What are appropriate contingent actions?
- What will trigger our contingent actions?
- How are we monitoring completion of the plan?
- When will the ICA be removed?
- How will we communicate this plan to those who have a need to know? What training will be required?
- What measurable will be used to validate the outcome of the PCAs (both short term and long term)?
- Has the ICA been discontinued?
- Has the unwanted effect been eliminated?
- How can we conclusively prove this?
- How are we continuing to monitor long-term results? What is the measurable? Is this the best way to prove the root cause is eliminated?
- How have we confirmed the findings with the customer?
- Have all systems, practices, procedures, documents, and so on been updated? Do they accurately reflect what we want to be done from here on?

D7: Prevent Recurrence

The purpose of D7 is to modify the necessary systems including policies, practices, and procedures *to prevent recurrence* of this problem and similar ones. It also helps management to infuse the changes into the organization as applicable. D7 is frequently cited as perhaps the most important step of the 8D process. During D7, the root cause of the D4 root cause is fixed. However, behind every change that creates the need to do an 8D is at least one system, practice, or procedure that introduced the change. If the system, practice, or procedure remains the same, the same or a similar problem causing change will be reintroduced. This can and should be prevented.

Therefore, even though D7 is a very substantial step in the process, many practitioners do not complete it for many reasons. We will identify some of the reasons and hopefully making the reader aware of these reasons will help the organization to develop a program to combat them.

- D7 is not even attempted: After the problem is corrected, other priorities get more attention. Infusion is forgotten.
- D7 develops into a blame session: Rather than focus objectively on what happened, the attention drifts to who did or did not do something. Defensiveness sets in and objective review stops.
- The 8D process stopped at D3 containment: The root cause was never found. Prevention of root cause is impossible because it remains unknown.
- The 8D process was never done: Someone penciled in a form that became an 8D report.
- Politics: Fixing systems is a low priority because it requires risking (forcing) a change.
- Fear: People perceive/know that spotlighting defective/inadequate system practices and procedures will result in reprisal.
- Not enough structure/authority: The champion is too low in the organization's power hierarchy to implement necessary system revisions. Further, this champion is unable to solicit involvement of those who have the authority.
- Those who own the system do not have to live with the consequences: The organization is unwilling to create consequences at the proper points of systems management.
- Systems maintenance and management is a low priority: Activity is more important than prevention. Process control is not understood.

Some of these issues obviously exceed the scope of one champion's authority. However, they can be corrected. Some take a lot of effort and time. The first step of any D7 objective is to accurately identify (at least) the cause of the cause—the system, process, and so on that introduced the root cause. This much the champion can prompt with the team's understanding of the first six objectives. D7 often requires bringing all former team members together to answer two questions:

1. What went wrong that introduced the root cause in the first place?
2. What needs to be fixed, modified, or reinforced to prevent this from happening again?

D7 also implies some opportunities. These include the chance to suggest what a good system should look like and to identify places where a defect is about to be created.

The improvements that may follow beyond the process of original concern depend heavily on the rank authority and power of the champion. Generally, three paths may be followed after the systemic issues are identified:

1. The champion uses his or her authority to fix the system, practice, or procedure.
2. The champion carries the recommendation for systemic changes to those who have the authority to change the system, and so on.
3. The need for systemic changes is acknowledged but the organization chooses to actively monitor the day-to-day operation of the system. This is done to better understand the whole system, rather than overreact to a single incident. Other 8Ds might also be reviewed for their recommendations and observations of the system's shortcomings.

On the other hand, we must not forget that the 8D process asks for input at D7. The champion retains 51% of the vote. D7 does not dictate that the champion implement every recommendation. However, the champion works against his or her own best interest if ideas are discounted without serious consideration.

D7: Process Guidelines

As already mentioned, the purpose of D7 is *to prevent the recurrence* of this kind of root cause and escape point as well as infuse the solution to the problem in the organization as appropriate. In addition to the fundamental purpose of the D7, the following items should be recognized and considered for review as well as discussed as needed:

- Approximately 95% of all root causes are introduced because of a procedure, practice, instruction, or management system malfunction. These systems may have been ignored, they may have never worked initially, or they may be too cumbersome or obsolete. If the system is not fixed, it will create a similar kind of problem and/or another root cause.
- Typically, the team is best able to identify the details of the systemic problem.
- The team is chartered with both the responsibility and the authority (empowered) to provide feedback to management (through the champion) as to which systems contributed to the root cause and escape point.

- The team members will draw from all the knowledge and experience gained from the first six steps. The champion may have to take this information to other areas within management to implement systemic recommendations because he or she may not have full authority to make changes as recommended.
- Not all team-identified prevent actions and systemic recommendations will be the best and they should not be implemented. However, in fairness to the team, reasons why these will not be implemented should be communicated. It is very unfortunate that this is hardly done by practitioners.
- If the team has the skill, experience, and knowledge, the champion can ask members to design and implement their prevent actions. If the team does not have the skill, and so on, then the champion will request that other areas of the organization design and implement the changes.
- D7 works best when the people who use the systems have had a hand in the changes. Think simple, easy, and quick in system redesign. Usually, adding more words or instructions adds to confusion as opposed to removing confusion. Practice the keep it simple Stephanie (KISS) principle.
- D7 is best achieved by a face-to-face review between the champion and the team.
- The team members should recognize the champion might have to reflect upon their recommendations before being able to suggest or approve changes.
- The team should follow the format of observation first, recommendation second in the presentation to the champion.
- The team and champion should avoid creating a blame session.
- Sometimes the systems that contributed to the root cause are so big that not one individual in an organization or company owns all parts of that system. For example, cost accounting systems, order entry systems, and so on may require a separate task force by the company to redesign them.
- The champion and team require cooperation and objectivity from each other and from the rest of the organization to make D7 a success.
- The champion's reputation and actions often precede him or her in the D7 face-to-face phase. Most teams will only put as much effort into the D7 steps as they believe the champion will support.
- It is not uncommon for team members to feel threatened at D7. Therefore, the champion must communicate, before D7, his or her interest in receiving legitimate feedback on systems, practices, and procedures. The champion must be careful not to create unrealistic expectations.

- The team needs to realize champions often do not know about every detail and consequence of all the systems, practices, and procedures they are expected to manage. This is the team's opportunity to objectively (not emotionally) point out flaws in day-to-day operations of procedures.
- The leader may have to point out that most D7 actions make life easier in the workplace, not more difficult.
- Use of the Repeated Why method is an effective technique during D7. However, it can also be very frustrating for those trying to answer it. Use it carefully.

D7: Evaluating Questions

When the team feels comfortable with having completed the D7 step, it is a good practice to review the discussion with at least the following basic questions:

- How and where did this problem enter our process?
- Why did the problem occur there and how did it escape? (Why was it not detected?)
- Did confusion or lack of knowledge contribute to the creation of this root cause? Did it contribute to the escape?
- What policies, methods, procedures, and/or systems allowed this problem to occur and escape?
- Did you uncover Band-Aid© fixes in our processes? Where? What are they compensating for?
- Have the affected parties been identified?
- What needs to be done differently to prevent recurrence of the root cause? Of the escape?
- Is a service action required as part of the prevent actions?
- What evidence exists that indicates the need for a process improvement approach (i.e., focused improvement, reengineering)?
- Who is best able to design improvements in any of the systems, policies, methods, and/or procedures that resulted in this root cause and escape?
- What is the best way to perform a trial run with these improvements?
- What practices need standardization?
- What plans have been written to coordinate prevent actions and standardize the practices—who, what, and when?
- Does the champion concur with the identified prevent actions and plans?

- How will these new practices be communicated to those affected by the change?
- Have the practices been standardized?
- What progress checkpoints have been defined to assess system improvements?
- What management policy, system, or procedure allowed this problem to occur or escape?
- Are these practices beyond the scope of the current champion?
- Who has responsibility for these practices?
- Does the current champion agree with the systemic prevention recommendations of the team?
- What data has been submitted to the organization's lessons learned database?

D8: Recognize Team and Individual Contributions

The purpose of D8 is to complete the team experience, sincerely recognize both *team* and *individual* contributions, and celebrate. Therefore, the D8 is the closure of the team effort to resolve a problem. Because of this closure, D8 is a very important step even though many practitioners do not do it. At the very least, it is a clear message for everyone to return to his or her full-time job. D8 says to everyone that the task team is now disbanded and no further study of this concern is justified. More importantly, D8 is one way for the organization to provide feedback to those who did the work (D1 through D7). However, how the message is conveyed to the team is nearly as important as the message itself. It should be fit, focused, and timely. Recognition at D8 includes noting the team's contribution and efforts at all earlier steps. If recognition is omitted, team members interpret that their contributions were not valued. If the 8D process was not implemented using people, D8 could be left out.

To be sure, volumes have been written on recognition—the need for it, how to do it, and how to do it well. This brief section is not an attempt to repeat or condense that. A more complete study on the subject is well advised, but a few highlights are helpful.

- Encourages a repeat of the positive (wanted) behavior.
- Works best when it is perceived to be sincere.
- Is expected when extra effort has been expended.
- Does not have to be expensive in dollars, but does require investment in time; time to find out exactly what was done, time to express the recognition (personal time).

- Is situational.
- Must be on a timely basis.
- Works best when the method of recognition is unique, a bit of a surprise, and considered valuable. Value does not mean money. Often, value means personalized, customized.
- Recharges a person's battery. It feels good. People tend to do things for which they are recognized.
- Several people can benefit from recognition: team members, the team's leader, former team members who were released from the team, and the recorder and facilitator. Each provided important services and contributed. Recognition is a way to encourage future participation. The more effectively recognition is used, the better the organization's results.

D8: Process Guidelines

As mentioned, the purpose of D8 is to recognize the contributions made to the company or department by the 8D team. There is a difference, however, between recognition and reward. Either or both can be used at D8. The responsibility for recognition is the champion's. However, legitimate recognition opportunities exist between team members, the leader and the team, the team and the champion, and so on. Messages about organizational values will be communicated to the rest of the organization by the D8 action. Failure to conduct D8 will also send messages throughout the organization. Additional consideration and opportunity for either review or open discussion should be encouraged for the team participants. Some typical items may be the following:

- Care must be taken that the D8 action is commensurate with contribution of the 8D team.
- Care must be taken that the D8 action is perceived by the team members to be a legitimate and sincere acknowledgment of their efforts.
- It is recommended that variety and creativity be applied in identifying what the champion's D8 actions are. Additional guidelines for success include
    - Fit
    - Focus
    - Timeliness
- The action should fit the contribution of the team. Verbal recognition should focus (accurately describe) the team's actions and contributions. The actions should also be completed in a timely manner.
- Make certain all members, both core and extended, have the opportunity to participate.

- D8 is an opportunity to encourage certain behaviors and actions. If the team's contributions at Steps D5, D6, and D7 are not acknowledged, then the team would rightly presume that these actions do not matter.
- If D8 actions are consistently the same and predictable, the 8D process will be perceived as another lip service program. People look for some legitimate, personal involvement of the champion for the D8 step to be interpreted as sincere recognition.

D8: Evaluating Questions

When the team feels comfortable with having completed the D8 step, it is a good practice to review the discussion with at least the following basic questions:

- Has the 8D report been updated and published?
- Is the 8D paperwork completed and have other members of the organization (who have a need to know), and the customer, been informed of the status of this 8D?
- Is the 8D report and its attachments retained in your historical file system?
- Is there a complete list of all team members, current and past?
- Were there significant contributions by individual team members? What were they?
- Are there opportunities to provide recognition from leader to team, team member to team member, team to leader, and team to champion?
- What are some different ways to communicate the recognition message?
- Are there any nonteam members whose contributions to the 8D process justify inclusion at recognition time?
- Are all current and past team members being recognized?
- Do the results achieved by the team warrant some publicity (e.g., company newsletter)?
- Does the recognition satisfy the fit, focus, and timely criteria?
- D8 is intended to be positive. What are the chances that it might backfire and turn into a negative?
- What have you learned as individuals and as a team? About yourselves? About problem solving? About teamwork?
- How did the organization benefit by the completion of this 8D process?
- Review each 8D objective. What was done well? What was not done so well? Could it be improved? If yes, how?

- What sort of things should be repeated if conditions were applicable so that other escape points could be identified in other 8Ds?
- Are there changes to the business practices that should be considered, based on the learning in this 8D?
- Has all unfinished business now been finalized?
- Has each person had an opportunity to express appreciation to other team members?
- Has each person made a final statement?
- Have you celebrated appropriately? How?

8D Summary Evaluation

When the 8D is complete, it is a good practice to review the entire process. A good review summary for any 8D process is the following. Please notice the tools used for each step. Obviously, these are simple but common and very powerful. There are many more available—see Appendix V—and the reader may want to use others as appropriate.

Prepare for the 8D Process

- The symptom has not been defined.
- The symptom has not been quantified.
- The need for an ERA has not been evaluated.
- The ERA has not been validated.
- Typical quality tools: Trend chart, Pareto chart, Paynter chart.

Establish the Team

- The champion has not been named.
- The subsupplier is not on the team.
- The customer is not on the team.
- Other pertinent personnel are not on the team.
- No indication of a team brainstorming activity.

Describe the Problem

- The problem is not stated in the customer's terms.
- The start/end date of the problem is not listed.
- The percentage defective is not listed.

- The frequency of the problem is not listed.
- The actual reported measurement is not listed.
- The requirements (specifications) are not listed.
- Typical Quality Tools: Is/Is Not analysis, 5 Whys, 3×5 Whys, Pareto chart, process flow diagram, cause and effect diagram.

Develop the Interim Containment Action

- All stock locations were not purged of suspect stock
    - Supplier's site
    - Customer's site
    - In transit
    - Warehouse
    - Next customer's site
    - Subsupplier's site
    - Other
- The percentage of effectiveness of the containment is not listed
- The percentage of effectiveness of the containment is unsatisfactory
- The dates of the containment actions are not listed
- Sort results do not list the _____# Sorted, _____# Found, _____# Defective
- The method of sort is
    - Not stated
    - Unsatisfactory
- Containment action is compromised solely of an audit
- No short-term corrective actions are listed
- Typical quality tools: Paynter chart, SPC data, FMEA

Define and Verify Root Cause and Escape Point

- No fishbone analysis
- No Is/Is Not analysis
- Unacceptable root cause:
    - Operator error
    - Setup piece
- Stated root cause is not a root cause (either a symptom or effect)
- Percent contribution of each root cause is not listed
- Failure to recognize process root cause
- Failure to recognize system root cause

- No verification of root cause
- Process flow diagram missing when nature of the problem clearly requires it
- The 5 Whys have not been explored
- An escape point has not been identified
- Whether a control system exists to detect the problem has not been identified
- The capability of the control system has not been evaluated
- Typical Quality Tools: cause and effect diagram, FMEA, Is/Is Not analysis, 5 Whys, 3×5 Whys, DOE, robust design

Choose and Verify Permanent Corrective Actions for Root Cause and Escape Point

- The dates and actions are not indicated
- The stated actions do not cover all process root causes
- The stated actions do not cover all system root causes
- The actions do not adequately address the timing of the problem/condition/countermeasure requirements
- The actions do not adequately address the magnitude of the problem
- Typical quality tools: Paynter chart, cause and effect diagram, DVP&R, 5 Whys, 3×5 Whys, FMEA, robust design

Implement and Verify Permanent Corrective Actions

- The percentage effectiveness of the corrective actions have not been listed
- The Paynter Chart, or components of it, used to verify the effectiveness is missing
  - Throughput
  - Return sales quantity
  - Scrap
  - Return sales sort results
  - Dock audit
  - Warranty
  - Incoming rejects (IRs)
- The actions listed are detection actions
- Typical quality tools: FMEA, Paynter chart, Is/Is Not analysis, 5 Whys, 3×5 Whys SPC, PPAP

Prevent Recurrence

- The PFMEA was not reviewed
- The PFMEA was not adequately revised
- Control plans we not revised
- Process sheets were not revised
- Prevention action is composed solely of an audit
- Containment action is comprised of an added in-process inspection
- Typical quality tools: FMEA, control plans, ISO/QS procedures, fault tree analysis, process flow diagram

Recognize Team and Individual Contributions

- The 8D report has not been updated
- All parties who need to know have not been advised of the status of the 8D
- All unfinished business has not been finalized

8D Success Factors

To be sure, 8D is a very powerful methodology to resolve problems. However, its success depends on the commitment of the management of the organization. This means in no uncertain terms that

1. Management must create the right climate. This means that they must
    a. Display patience
    b. Provide practice time
    c. Demonstrate interest, even when failure is experienced
    d. State expectations clearly
2. Management's recognition of need to prioritize problems. This means that they must
    a. Continuously improve the system
    b. Have realistic expectations in every respect
    c. Provide the right resources
    d. Communicate the significance of the problem with the team and everyone concerned with the specific problem
3. Management's active support for problem-solving process. This means that they have to demonstrate interest by

   a. Giving authority and responsibility to the appropriate individuals to carry out the assigned task
   b. Always reviewing progress of the project and asking coaching questions
   c. Making sure that the process defined is adhered to
   d. Being an implementation advocate
   e. Providing a system to retain the gained knowledge
   f. Providing the appropriate resources
4. Management's commitment for a system to carry out corrective actions. This means that management must practice empowerment with both authority and responsibility for those individuals that have to carry out the task assigned.
5. Management must encourage prevention of problems. To correct a problem is okay, but it is after the fact. To prevent a problem needs a commitment to advance planning so that problems do not occur.

References

Abudi, G. (2010). The five stages of project team development. The Project Management Hut. http://www.pmhut.com/the-five-stages-of-project-team-development. Retrieved August 30, 2014.

Blanchard, K. and E. Parisi-Carew. (2009). *The One Minute Manager Builds High Performing Teams*. New York: William Morrow.

Ford Motor Company. (1998). *Global 8D Participants Guide*. Version 3.0. Dearborn, MI: Ford Technical Education Program, Ford Motor Company, Ford Design Institute.

Stamatis, D. (2003). *Six Sigma and Beyond: Problem Solving*. Boca Raton, FL: St. Lucie Press.

Tuckman, B. (1965). Developmental sequence in small groups. Psychological Bulletin 63(6):384–399.

Selected Bibliography

McClelland, D., J. Atkinson, R. Clark, and E. Lowell. (1953). *The Achievement Motive*. Princeton, NJ: Van Nostrand.

Rickards, T. and S.T. Moger. (1999). *Handbook for Creative Team Leaders*. Aldershot, Hants: Gower.

Rickards, T. and S. Moger. (2000). Creative leadership processes in project team development: An alternative to Tuckman's stage model. *Br J Manage* 4: 273–283.

White, A. (2009). *From Comfort Zone to Performance Management*. Hoeilaart, Belgium: White and MacLean Publishing.

22

Miscellaneous

Overview

Thus far, we have discussed critical and essential approaches to continual improvement with the goal of doing the right job on time the first time. In this chapter, we address some additional methodologies that organizations may follow toward this end. They are (1) production system, (2) supplier request for engineering approval (SREA), (3) warranty, (4) Weibull distribution, (5) failure mode and effect analysis (FMEA), (6) Six Sigma, (7) systems engineering, and (8) value engineering. We do not want to imply that these are not essential to continual improvement (CI); however, an organization may choose many other methodologies on the way to improvement. The reason we have included these topics as separate is that in most organizations they are part of the topics that we have chosen to discuss at length. On the other hand, they are important enough for any organization to be aware of their existence and apply them as necessary.

Typical Total Production System (TPS)

In today's competitive world of quality, an integrated assessment tool must exist to evaluate and drive deeper into specific elements and processes the notion of CI in the entire organization (vertical and horizontal awareness by management and operator personnel).

Ford Motor Company developed one of the oldest and most comprehensive TPSs. However, one of the most talked about TPSs is the one introduced by Toyota Corporation. Other companies have developed their own. Here we present a generic summary of a typical TPS based on both Ford's and Toyota's systems. This outline is based on 10 fundamental principles. They are

1. Total productive maintenance
2. Training
3. In-station process control
4. Manufacturing engineering
5. Environmental
6. Industrial materials
7. Safety and health achievement recognition program (SHARP): SHARP is a program that recognizes small employers who operate an exemplary safety and health management system. Acceptance into SHARP by OSHA is an achievement of status that will single you out among your business peers as a model for worksite safety and health and will reap rewards for your business. Upon receiving SHARP recognition, your worksite will be exempt from programmed inspections during the period that your SHARP certification is valid.
8. Work groups
9. Managing
10. Synchronous material flow (SMF)

Ford's Approach to TPS

The Ford production system (FPS) was created to standardize and add stability to the existing manufacturing processes. In essence, this program is to focus on zero defects and 100% throughput. More specifically, by utilizing policy deployment, visual management, process confirmation, and time and data management, this process will deliver a more aligned and capable organization, one that will be continually improving and ultimately building the best manufacturing environments in the world. The metrics for this program are based on **S**afety, **Q**uality, **D**elivery, **C**ost, **P**eople, **M**aintenance, and **E**nvironment (SQDCPME).

By measuring these metrics, Ford motor company has recognized that they are the pillars of their CI process, especially in their manufacturing operations. These pillars are aiming to continually improve their processes through standardization and recognition. In addition, these metrics seek to create stability to the inputs, normalize the processes, and seek change (if necessary) for the outputs. Specifically, the FPS is based on the following categories:

- CI board: It is used to visually manage the review and follow-up of elevated issues of the work group. It is located in the start-up communication meeting room. Each manufacturing area is represented on the board. Problems that are not solved at the lowest levels are escalated to the team manager and CI cards are assigned to the appropriate support manager/team for resolution. There are two types of CI cards: yellow, which are items that are expected to be solved in seven days, and red, which are expected to be resolved in 24 hours due to urgency. CI cards are assigned a tracking number for follow-up as necessary and placed on the board for daily review.

- Start-up confirmation: It provides a time where the team leader makes sure that all operators are OK to start to build. If an operator identifies a problem, it is immediately solved or escalated to the supervisor. The team leader completes a start-up confirmation check sheet and reviews it with his or her supervisor. It is imperative that first hours targets are achieved and problems are corrected at the lowest levels.
- Results process/support process: These are working meetings to analyze and diagnose workstation and process results. The intention is to make problems evident and make improvements.
- Time and data management: It is designed to assist the plant in identifying and eliminating waste through improved planning and alignment.
- Basic administration: It is another tool to help reduce waste by clearly defining individual roles and responsibilities, allowing time to change the culture and focus on teaching and coaching of the standards necessary for improvement.
- Kaizen: As problems are escalated to the CI board and the solution is not readily evident, they can be further escalated into a Kaizen project. If the problems are complicated, then the Six Sigma methodology may be employed. In complex issues, a subject matter team is created to deep dive the issues. They collect and analyze data to systematically reduce variation and ultimately resolve the issue.
- Standardized work: The focus of standardized work is on standard process adherence and creating an ideal workstation, process, and method for each operation. There is a three-phase process to accomplish it: (a) job observation, (b) deep knowledge operator review, and (c) identification and understanding of abnormal conditions. During the last two phases of this process, the team leader will replace the operator, who will have an in-depth discussion with the supervisor offline to review the job standards. Through these conversations, process improvements are noted and the working relationship is enhanced.

- Star points: They are policies and procedures for new processes that establish approaches to further promote work team involvement based on things learned from current practices.

The FPS in the final analysis

- IS
  - People working together using FPS tools
  - Continual improvement in SQDCPME
  - Focus on eliminating waste
- IS NOT
  - Another modern operating agreement (MOA)
  - In conflict with the master labor agreement
  - In conflict with local labor agreements
  - Focused on eliminating people

Toyota Production System (TPS)

The TPS is focused on a philosophy that encourages people participation and customer satisfaction. Even though there are no rigid requirements, the philosophy is based on three fundamental output ideas:

1. To provide the customer with the highest quality products, at the lowest possible cost, in a timely manner with the shortest possible lead times
2. To provide employees with work satisfaction, job security, and fair treatment
3. It gives the company flexibility to respond to the market, achieve profit through cost reduction activities, and long-term prosperity

In other words, TPS strives for the absolute elimination of waste, overburden, and unevenness in all areas to allow employees to work smoothly and efficiently. The foundations of TPS are built on standardization (1) to ensure a safe method of operation and (2) to provide a consistent approach to quality. Toyota employees seek to continually improve their standard processes and procedures in order to ensure maximum quality, improve efficiency, and eliminate waste. To do this, the TPS approach emphasizes three fundamental concepts:

1. Kaizen: The heart of the TPS. It is engrained in the employees' minds that continual improvement in all that people and machines do can be improved continually in small steps.
2. Just-in-time (JIT): It consists of allowing the entire production process to be regulated by the natural laws of supply and demand. Customer demand stimulates production of a product. In turn, the production of the product stimulates production and delivery of the necessary parts, and so on. The result is that the right parts and materials are manufactured and provided in the exact amount needed and when and where they are needed. This process is regulated with the Kanban method. The Kanban method depends on a card which is attached to every component that is removed from a process and returned when the component is used. The return of the Kanban to its source stimulates the automatic re-ordering of the component in question. This Kanban process minimizes paperwork, maximizes efficiency and allows the employees of the process to be completely in charge
3. Jidoka: A Japanese word that means automation with a human touch. The idea here is to design machinery in such a way that if a problem arises, the machine will stop and call attention to the problem immediately. The Jidoka principle is a central concept in any application of TPS.

Suppliers and TPS

Suppliers who participate in TPS enjoy the same benefits that Toyota does from the system. JIT manufacturing can dissolve inventories at parts suppliers just as readily and effectively as it does at Toyota's assembly plants. Product quality improves, too. That is because TPS includes measures for illuminating defects whenever and wherever they occur. Suppliers who adopt TPS also report improvements in employee-management relations. That is mainly because the system provides for an expanded role for employees in designing and managing their own work. It brings together employees and management in the joint pursuit of improvements in productivity, quality, and working conditions.

It must be noted here that TPS has adopted (inherited) the principle originated by Henry Ford of breaking down work into simple steps and distributing those steps among employees on the line. The difference is that employees in the Toyota system are in charge of their own jobs. Through their teams, they run their own worksites. They identify opportunities for making improvements and take the initiative in implementing those improvements in cooperation with management.

Suggested Implementation of a Production System

A production system by definition is a lean, flexible, and disciplined common production system defined by a set of principles and processes that employs groups of capable and empowered people learning and working safely together in the production and delivery of products that consistently exceed customers' expectations in quality, cost, and time. What makes this system so powerful is that it focuses on customer satisfaction through some very important concepts such as

- Stability, which indicates consistency, repeatability, and predictability
- Continuous flow, which indicates production without interruptions
- Synchronized production, which indicates parallel production
- Pull system, which indicates a dynamic inventory system such as Kanban
- Level production, which indicates a predictable production

To implement these concepts, there are many options because every organization has its own culture and expectations. However, a good start is to begin with thorough plant visits and reviews to identify the *what is now*. Typical things to consider are

- Visit application areas. Talk with employees about what their role is in the implementation of a production system. Explain the expectations.
- Take a waste walk. Observe areas where waste is prevalent by walking the processes.
- Conduct frequent (at least monthly) meetings between senior leadership and plant managers to review work plans, progress/timing, and provide coaching on how to deal with the implementation issues. A good start is the quality operating system (QOS) meeting.
- Reinforce the importance of union involvement and support. Make sure that employees do not become suspicious of the organization's motives to eliminate them.
- Develop a material flow plan. For example, for current issues, produce the number of scheduled units each day at the lowest assembly variable cost with the highest quality. For future issues, build a customer-driven product sequence in a predictable and stable manner at the lowest total cost and time with the highest quality. Understand and communicate to everyone that this new approach, new vision, perspective, goal, or system requires less

of everything—time, space, effort, capital—to produce goods and services with fewer mistakes in smaller production volumes. It is another way of looking at Lean manufacturing.

These five considerations will result in

- Defining the direction and support to the plants on implementation issues
- Defining measurement and reward for success
- Defining appropriate and applicable measurables
- Making sure that the appropriate frequency for meetings is established
- Reassuring employees that their jobs are safe and not in jeopardy because of the implementation of the new system
- Making the production system a priority in the plant and not another flavor of the month initiative; offer encouragement as well as support whenever needed
- Making the total cost a key measurable in the plant's performance review
- Defining appropriate milestones for reviews and assessments
- Assigning a committee to lead the implementation process

Production System Model Based on an American Engine Plant

The following example demonstrates the approach that management took at an American automotive engine plant to implement and sustain a production system. In the example, we identify the key component of the implementation process without editorializing or comments. We focus on the people, what the plant did to increase value, how JIT was applied, and how the value to the customer was measured.

- People
  - Modern operating concepts
  - Rewards and recognition
  - Development of subordinates
  - Involved employees—self managed
  - Ergonomics
  - Health and safety

- Policy deployment
- Mission, values, guiding principles
- Added value
    - 5 Ss
    - Total productive maintenance
    - Quality process system (QPS)
    - Reliability and maintainability
    - Dimensional control planning
    - 5 Whys
    - Concept of zero
    - Benchmark
- JIT
    - Schedule stability
    - Modern materials handling systems
    - Plant layout
    - Throughput
    - Kanban
    - Visual factory
    - Quick change over
- Value to the customer
    - Eliminate waste
    - CI

Supplier Request for Engineering Approval (SREA)

We all know that what is designed is not always what the manufacturer can deliver. It is possible and, in fact, it happens quite often that the supplier finds a problem with the design. It is imperative at this point not to change the design on your own—even though you may have the best intentions—but notify the customer. The process of notification is called the SREA.

An SREA is an external supplier request to obtain a customer approval for a *plan to implement* a proposed external supplier-initiated manufacturing process or site change. It is worth noting here that an SREA request for approval does not mean that the supplier can ship parts/product to the customer. For shipment, a part submission warrant (PSW) approval is required per production part approval process (PPAP) requirements. The SREA is a way to prevent external supplier-initiated changes adversely affecting the product quality or manufacturing operations of the customer without its knowledge.

In its simplest form, the SREA is the official communication tool used by the customer to meet the external supplier-initiated change notification requirement of PPAP. Changes made without a customer's review and approval have the potential to impact product quality, customer's plant processes, and, more importantly, customer satisfaction. In essence then, the SREA is a method of making sure that the customers do not receive any surprises from their supplier. If they do, the unauthorized changes can result in unanticipated repair, replacement, quality cost, and late deliveries.

An SREA may be initiated everywhere the business is conducted and when an external supplier needs to make a change to a production manufacturing process, site/location, or part design after PPAP approval or when an external supplier needs to make any change to a service-unique part/product or process after PPAP approval.

Generally, an SREA is required when a manufacturing process and/or site location changes to the original PPAP-approved process affecting

- Site location
- Site construction affecting process or process flow
- Part/product processing
- Equipment changes
- Process flow
- Subsupplier changes affecting site location, processing, process flow, or materials
- Material changes within the existing design specification
- Material source changes
- Production

Specifically, an SREA is required for design changes affecting

- Part/product specifications
- Part/product cost
- Appearance, quality, or performance of end-item characteristics
- Compliance with government regulations
- Function, fit, or appearance of component/subsystem/system within a total vehicle system
- Interchangeability of end items
- Serviceability of the part/product
- Subsystem design specifications, worldwide standards and specifications (if applicable)

- Product assembly plant operations
- Warranty and reliability

Service Unique Products

Service unique products are parts/product released by the customer that are not common with production parts/product (i.e., remanufactured components, service chemical, etc.), and past model parts/product no longer used in production or for original equipment warranty repairs, but still produced for service. The SREA requirements for service unique manufacturing and site location changes are the same as the previous list under the general requirements. Sometimes different customers will have different departments handling current model service unique parts/product from the older models.

Service unique design changes that require an SREA are any supplier-initiated changes outlined in the previous list of general requirements, including

- Change components of an assembly due to availability issues (obsolete, not currently available, tooling down, etc.)
- Ship without a component of an assembly due to availability issues
- Change the content of a kit due to availability issues
- Substitute essentially identical final assemblies
- Expand to coverage of one part/product and drop another for complexity reduction
- Material changes or substitution resulting from obsolete or prohibited material revisions
- Worn tooling dimensional and tolerance change approval
- Request to produce part/product to meet last run samples because of
  - Loss of check fixtures
  - Loss of tools
  - Loss of master sample parts/product
  - Design record not up to date due to changes made with no record, changes made with incomplete record, or changes made with inaccurate record
  - Drawing not up to date for all of the above
  - No drawing
  - No formal design record

Temporary SREA

Sometimes the supplier will request to ship parts/product made to a temporary engineering specification. This request is the same as the regular SREA.

The only difference is that the regular one is permanent and this one is temporary. Reasons for consideration for a temporary SREA may be

- Part/product temporarily made using unapproved manufacturing process or site/location (at Tier 1 or subsupplier)
- Temporary emergency change to the part/product or manufacturing process or site/location
- Part/product temporarily not meeting customer's specifications

Supplier Risks for Not Following an SREA

By not following the correct approach for an SREA, the risk to the supplier may be at least in the following areas:

- Loss of customer quality system, that is, loss of Q1 status through violation of trust
- Increase field warranty or customer plant containment costs (through third-party inspection or some other method of segregating good from bad parts)
- Loss of production and reinstatement of prior process
- Loss of future business

Warranty

Our society has become very litigious. We look for perfection and any deviation from that perfection can cause trouble for the OEM customer, but also for the supplier. To mitigate the failure of a product, organizations all over the world have initiated warranties that cover their specific products or services. These warranties may involve the ultimate customer (consumer) or the supplier and its subsuppliers.

For any warranty program to work effectively, there are two approaches that have to be considered. They are

1. Issues that involve common cause processes. Here the focus is on (a) warranty reduction programs, and (b) warranty chargebacks.
2. Issues that involve special cause processes. Here the focus is on (a) warranty spikes, (b) field service action cost recovery, and (c) warranty cost reduction processes once the product is in production.

The idea of a warranty program for the customer is to establish direct responsibility for failures to any external supplier for all or a portion of any costs associated and validated with warranty claims. For the supplier, it is important to negotiate an appropriate and applicable contract to pay claims under the rubric of a warranty claim. This negotiation is best conducted as early as possible, preferably at sourcing.

The terms and conditions as well as the definitions of a warranty spike must be in definitive terms for special cause conditions. Furthermore, it is important to define the product where system interaction or a high possibility of trouble may be present and the responsibility of the supplier may be very difficult to assess or assign.

The idea of chargebacks is based on a refund dollar amount for *bad* parts delivered. It is usually a default process for common cause process and it works best when parts can be analyzed and supplier responsibility is easily determined. Obviously, when the supplier is responsible for the design of the part and system, the application of chargeback and responsibility is very easily assigned.

The idea of a spike is based on the notion of an unusual activity of defect based on a set period, for example, a defect for the current month is 20% higher than the last month. This is called a point-to-point comparison. Obviously, there might be an assignable cause for the unusual behavior and it should be investigated. If a spike is an issue on a regular basis for different parts, then warranty recovery methods should be instituted by the customer to the supplier. Although point-to-point comparisons are very common in many industries, a better way to identify and evaluate warranty issues is the shift of the average by at least one standard deviation in either a positive or a negative way. The baseline for the comparison should be at least six months. Another good analysis tool is the evaluation of the *bathtub* curve. Here, one can see the manufacturing issues early on, which is the left side of the curve—a negative exponential curve. Design issues are the middle of the curve—the steady curve, and the exponential curve to the right indicates wearout situations.

For an effective chargeback and spike recovery, six fundamental elements should be followed. They are

1. Technical analysis for recovery documentation: Failure modes, root causes, characteristics of occurrence, contractual obligations, and prevailing area of recovery must be understood, documented, and discussed with the customer stakeholders prior to supplier recovery notification.
2. Agreement on supplier responsibility: Supplier is engaged in discussion of recovery technical analysis. Documented agreement with supplier on scope and level of responsibility is required for certain areas and circumstances of recovery.
3. Financial calculation: Calculation method or formula for specific area of recovery is applied to scope and level of technical responsibility.

Supplier review and concurrence is required for certain areas and circumstances of recovery.

4. Settlement with supplier: Supplier is engaged in discussion of financial calculation, based on technical responsibility. Format for supplier settlement or concurrence differs by area of recovery, but is required for all supplier warranty recovery collections.
5. Financial collection: Preferred form of recovery is cash transaction by way of debit memo, but deviation is appropriate under certain business conditions with the customer's senior management approval. Finance group differs by area of recovery and region.
6. Escalation within matched pair: Particularly with respect to agreement on supplier responsibility and settlement with supplier, the customer's stakeholders make use of product development (engineers) and purchasing matched pairs to escalate disagreements according to decision makers within the customer and the supplier to teach a feasible outcome.

Weibull Distribution

The Weibull distribution is a good distribution to understand and evaluate *life data* analysis. Weibull and log normal analysis are particularly good for failure analysis. Because of this ability to evaluate failures, there are new applications of this distribution in medical and dental implants, warranty analysis, life cycle cost, materials properties, and production process control. Of course, the use in the automotive, defense, aerospace, and other mature industries continues to be high. Other methods similar to Weibull exist, but they are for specific applications. Some related quantitative models in which the reader may be interested are the binomial, Poisson, Kaplan-Meier, Gumbel extreme value and the Crow-AMSAA. A very good detailed introduction to Weibull is Abernethy (2009), and a summary of it with several examples is Stamatis (2010, chapter 15).

The primary advantage of Weibull analysis is the ability to provide reasonably accurate failure analysis and failure forecasts with extremely small samples. Solutions are possible at the earliest indications of a problem without having to *crash a few more*. Small samples also allow cost-effective component testing. For example, *sudden death* Weibull tests are completed when the first failure occurs in each group of components (say, groups of four bearings). If all the bearings are tested to failure, the cost and time required is much greater. Another advantage of Weibull analysis is that it provides a simple and useful graphical plot of the failure data. There are several software packages in the market to calculate appropriate reliability values as well as plotting the distributions. One of those packages is ReliaSoft's Weibull++ software, which provides a complete array of life data analysis tools and is very easy to use. Minitab® is also good for defining and plotting the distribution functions via linear regression, which can also be used to numerically assess goodness of fit and estimate the parameters of the Weibull distribution. The gradient informs one directly about the shape parameter k and the scale parameter λ can also be inferred. The simplest of all methods to identify a distribution is to construct a histogram because it will show the shape of the distribution. Using the Minitab® software to identify the distribution, go to Stat > Quality Tools > Individual Distribution Identification. This handy tool allows you to compare easily how well your data fit 16 different distributions. It produces a lot of output in both the session window and graphs, but do not be intimidated. To appreciate the output of the Minitab®, there are three measures you need to know:

1. Anderson-Darling statistic (AD): Lower AD values indicate a better fit. It is generally valid to compare AD values between distributions and go with the lowest.
2. P-value: You want a high *p*-value. A low *p*-value (e.g., <.05) indicates that the data do not follow that distribution. For some three-parameter distributions, the *p*-value is impossible to calculate and is represented by asterisks.
3. LRT P: For three-parameter distributions only, a low value indicates that adding the third parameter is a significant improvement over the two-parameter version. A higher value suggests that you may want to stick with the two-parameter version.

The term *life data* refers to measurements of product life. Product life can be measured in hours, miles, cycles, or any other metric that applies to the period of successful operation of a particular product. Since time is a common measure of life, life data points are often called *times-to-failure* and product life is described in terms of time throughout the life of the part/product. There are different types of life data and because each type provides different information about the life of the product, the analysis method will vary depending on the data type. For example, with *complete data,* the exact time-to-failure for the unit is known (e.g., the unit failed at 100 hours of operation). With *suspended* or *right censored* data, the unit operated successfully for a known period of time and then continued (or could have continued) to operate for an additional unknown period of time (e.g., the unit was still operating at 100 hours of operation). With *interval* or *left censored* data, the exact time to failure is unknown but it falls within a known time range (e.g., the unit failed between 100 and 150 h [interval censored] or between 0 and 100 h [left censored]).

Life data analysis requires the following steps to be effective:

- Gather life data for the product.
- Select a lifetime distribution that will fit the data and model the life of the product.
- Estimate the parameters that will fit the distribution to the data. Once this information is collected, then the analysis begins. Typical outcomes are to (a) generate plots and results that estimate the life characteristics of the product, such as the reliability or mean life, and (b) compare suppliers or designs based on reliability.
- Demonstrate that an item meets the specified reliability.
- Make predictions about performance during the useful life (or warranty) period.
- Use plots and other reports to effectively communicate performance estimates to management.

Once you have calculated the parameters to fit a life distribution to a particular data set, you can obtain a variety of plots and calculated results from the analysis, including

- Reliability given time: The probability that a unit will operate successfully at a particular point in time. For example, there is an 80% chance that the product will operate successfully after 3 years of operation.
- Probability of failure given time: The probability that a unit will be failed at a particular point in time. Probability of failure is also known as *unreliability* and it is the reciprocal of reliability. For example, there is a 10% chance that the unit will be failed after 3 years of operation (probability of failure or unreliability) and an 80% chance that it will operate successfully (reliability).
- Mean life: The average time that the units in the population are expected to operate before failure. This metric is often referred to as mean time to failure (MTTF) or mean time before failure (MTBF).
- Failure rate: The number of failures per unit time that can be expected to occur for the product.
- Warranty time: The estimated time when the reliability will be equal to a specified goal. For example, the estimated time of operation is 3 years for a reliability of 90%.
- B_X life: The estimated time when the probability of failure will reach a specified point (X%). For example, if 10% of the products are expected to fail by 3 years of operation, then the B_{10} life is 3 years. (Note that this is equivalent to a warranty time of 3 years for a 90%

reliability.) The B_{10} is usually the common statistic used; however, the "X" can be any number.

- Probability plot: A plot of the probability of failure over time. (Note that probability plots are based on the linearization of a specific distribution. Consequently, the form of a probability plot for one distribution will be different from the form for another. For example, an exponential distribution probability plot has different axes than those of a normal distribution probability plot.)
- Reliability versus time plot: A plot of the reliability over time.
- pdf plot: A plot of the probability density function (pdf).
- Failure rate versus time plot: A plot of the failure rate over time.
- Contour plot: A graphical representation of the possible solutions to the likelihood ratio equation. This is used to make comparisons between two different data sets.

Is Weibull always the best choice? The answer is definitely not. However, it is appropriate for survival analysis, reliability engineering, and failure analysis, industrial engineering to represent manufacturing and delivery times, and in evaluating extreme value theory. On the other hand, it is not appropriate for product failures caused by chemical reactions or corrosion, which are usually modeled with the lognormal distribution.

The following are typical examples of engineering problems solved with Weibull analysis:

- A project engineer reports three failures of a component in service operations during a 3-month period. The program manager asks, "How many failures will we have in the next quarter, six months, and year? What will it cost? What is the best corrective action to reduce the risk and losses?"
- To order spare parts and schedule maintenance labor, how many units will be returned to the depot for overhaul for each failure mode month-by-month next year? The program manager wants to be 95% confident that he or she will have enough spare parts and labor available to support the overall program.
- A state air resources board requires a fleet recall when any part in the emissions system exceeds a 4% failure rate during the warranty period. Based on the warranty data, which parts will exceed the 4% rate and on what date?
- After an engineering change, how many units must be tested for how long, without any failures, to verify that the old failure mode is eliminated or significantly improved with 90% confidence?
- An electric utility is plagued with outages from superheater tube failures. Based on inspection, data forecast the life of the boiler

based on plugging failed tubes. The boiler is replaced when 10% of the tubes have been plugged due to failure.

- The cost of an unplanned failure for a component, subject to a wearout failure mode, is 20 times the cost of a planned replacement. What is the optimal replacement interval?

FMEA

FMEA is a bottom-up method of analyzing a design or a process, and is widely used in many industries. In the United States, automotive companies such as Chrysler (now Fiat), Ford, and General Motors require that this type of analysis be carried out. There are many different company and industry standards, but one of the most widely used is the Automotive Industry Action Group (AIAG) model. Using this standard, you start by considering each component or functional block in the system and how it can fail. This is referred to as failure modes. You then determine the effect of each failure mode, and the severity on the function of the system. Then you determine the likelihood of occurrence and of detecting the failure. The procedure is to calculate the risk priority number (RPN) using the following formula:

$$\text{RPN} = \text{Severity} \times \text{Occurrence} \times \text{Detection}$$

Although this RPN has been used and continues to be used by many organizations, a better way to address risk priority is based on the numerical value of Severity (Priority 1), followed by Severity × Occurrence (Priority 2), followed by the value of the RPN. Some organizations define priority as Severity (9–10), followed by Severity (5–8), and followed by the RPN. In all cases, there should never be a threshold of an RPN value.

The second stage is to consider corrective actions, which can reduce the severity or occurrence or increase detection. Traditionally, you start with the higher RPN values, which indicate the most severe problems, and work downward. The RPN is then recalculated after the corrective actions have been determined. The intention is to get the RPN to the lowest value. If, on the other hand, the organization has chosen to evaluate the FMEA based on Severity, Severity × Occurrence, RPN, the focus will be to make sure that severity is controlled or changed with an appropriate design change. This will be followed by redundant systems to minimize occurrence followed by appropriate controls to catch the failure before it reaches the customer. For detailed information on FMEAs, see Stamatis (2003, 2015).

Six Sigma

Developed by Motorola in the 1980s, this improvement methodology was created to reduce errors, waste, and variations, and increase quality and efficiency in manufacturing. It has since been adapted for use in other types of business processes, and today is in practice in some of the top companies around the globe, including General Electric, Siemens, Honeywell, Ford, Allied Steel, and many more. Six Sigma-driven companies use data to examine, manage, and enhance operational performance by eliminating and preventing flaws in goods and related processes, such as design, management, production, consumer satisfaction, and service delivery (Harry and Schroeder 2006, Stamatis 2002–2003, Pande, Neuman, and Cavanagh 2000).

The bulk of the work in a Six Sigma project is defining failures, measuring deviations, and other activities, which ultimately lead to product quality. In fact, Six Sigma is used as a term for a management style, with the ultimate goal of high levels of customer satisfaction (Stamatis, 2004).

Eckes (2001) and Pyzdek (2001) have suggested that perhaps the most objective way of looking at Six Sigma is to recognize that the Six Sigma methodology essentially provides a framework, and most importantly a strongly branded corporate initiative, for an organization to

- Train its people to focus on key performance areas
- Understand where the organization wants to go (its strategy, related to its marketplace)
- Understand the services that the organization's customers need most
- Understand and better organize main business processes that deliver these customer requirements
- Measure (in considerable detail) and improve the effectiveness of these processes

In specific terms, the Six Sigma methodology at its core centers on the following concepts:

- Critical to quality: Attributes most important to the customer
- Defect: Failing to deliver what the customer wants
- Process capability: What your process can deliver
- Variation: What the customer sees and feels
- Stable operations: Ensuring consistent, predictable processes to improve what the customer sees and feels
- Design for Six Sigma: Designing to meet customer needs and process capability

One of the guiding principles behind the methodology is that variation in a process creates waste and errors. Eliminating variation, then, will make that process more efficient, cost-effective, and error-free. This may sound like a relatively straightforward concept, but its application in a complex and highly integrated business environment can be far from simple. The term *sigma* refers to a scale of measurement of quality in processes such as manufacturing. When using this particular scale, *six sigma* equates to just under 3.4 defects per million opportunities (DPMO). Table 22.1 shows the relationship between sigma and defects.

Six Sigma Methodology

Whether you are trying to implement Six Sigma throughout the entire organization or using it to improve the performance of your own department, you need to understand the two main Six Sigma categories. The first of these is DMAIC, which stands for define, measure, analyze, improve, and control. The second is DMADV, which represents define, measure, analyze, design, and verify.

DMAIC

DMAIC is used to apply the principles of Six Sigma to existing business processes. For instance, if you were trying to find out how to make a particular process more effective, you would use DMAIC to break down the process into its component parts. Using this Six Sigma model, you would start by defining the problems and project goals, measuring data relating to the current process and analyzing your findings to identify cause-and-effect relationships. The next step involves improving existing processes based on your data analysis. Finally, you need to implement controls to avoid variation in the process going forward. Typical tools used in each step are shown in Table 22.2.

TABLE 22.1

Relationship of Sigma and Defect

Sigma Level	Defects (per Million)	Defects (Percentage)
1	691,462	69%
2	308,538	31%
3	66,807	6.7%
4	6,210	0.62%
5	233	0.023%
6	**3.4**	**0.00034%**
7	0.019	0.0000019%

TABLE 22.2
Typical Tools Used in the DMAIC Model

Define Tools	Measure Tools	Analyze Tools	Improve Tools	Control Tools
QFD surveys	Blueprints	Multivariate analysis	Designed experiments	Control plan
GQRS warranty	Process sheets	SPC and charting	Full factorial	Operator illustrations
In-plant data	Metrics	Basic stats	Fractionals	Operator instructions
Pareto	Flowcharts	Hypothesis	Comparison	Error proofing
	Linkage	Basic DOE	Measurement system analysis	Automated control
	Matrices	Check sheets	Capability	QPS
	FMEAs	Histograms	Statistical tolerancing of process	Variation monitoring
	Measurement system analysis	Scatter diagrams		
	Capability analysis	Regression		

DMADV

The classical design for Six Sigma (DFSS) follows the DMADV model, which shares some steps in common with the DMAIC model. It is used for Six Sigma projects that create new product or process designs. Using the DMADV model, you would start by defining your design goals. The next step is to measure the required quality characteristics, product or production process capabilities, and associated risks. You would then conduct an analysis of your findings to develop an appropriate solution. After that, you are ready for the design phase of the new product or process. Once the design is complete, you must test it and verify that it works. Following these steps will result in a successful Six Sigma implementation. Each of the acronyms is defined very precisely in the following manner:

- Define: Address customer needs in relation to a product or service
- Measure: Involve the use of electronic, mechanical, or physical data collection to measure customer needs, response to product, or review of services
- Analyze: Utilize metrics to evaluate areas where product or service can be better aligned to customer goals and needs
- Design: Overlap the improvement of business processes that streamline corporate goals to best meet client and customer needs

- Verify: Build a system of tests and models to check that customer specifications are being met through ongoing improvements

In the evolution of the classical DFSS process, there is also the DCOV model, which has gained in popularity and stands for

- Define: Address customer needs in relation to the product or service desired.
- Characterize: Mathematically transform the customer wants and needs into specifications or requirements.
- Optimize: Out of the possible outcomes, select the best alternative that will satisfy the customer. This is an issue of measuring strengths and weaknesses or risk and benefit.
- Verify: Verify that what you have selected is what the customer wants and is willing to pay for.

Levels of Six Sigma Mastery

The implementation of the Six Sigma methodology is driven by five different levels of experience and knowledge. Three are the actual practitioners and two are administrative in nature. They are

- Green Belts, which are the basic levels in the structure of the Six Sigma methodology. Individuals with this classification generally have about a week's training on the content and philosophy of the Six Sigma methodology and actively participate in projects directed by Black Belts. Generally, the projects in which they are involved are within their primary job duties.
- Black Belts are trained for about 5 weeks in the philosophy and advance statistical tools and methodologies in resolving major problems. Generally, they are dedicated full time to solving specific problems in the organization. They are working under Master Black Belts to apply Six Sigma methodology to designated projects.
- Master Black Belts mentor Black Belts and Green Belts. Like Black Belts, they concentrate on Six Sigma implementation. In addition to spending time on statistical duties, Master Black Belts help ensure that Six Sigma processes are applied consistently throughout an organization's numerous departments and functions. They have an additional 2-week concentrated training in how to teach, how to make presentations, selection of problems, and advanced methodologies for resolving conflicts.
- Six Sigma Champions are individuals chosen from upper management by executive management. Champions are also concerned with organization-wide Six Sigma implementation as well as mentoring

lower-level Six Sigma practitioners. They are the individuals who *remove bottlenecks* in the implementation of the Six Sigma process. Generally, their training of Six Sigma is a week of overview of the entire process.

- At the top of the Six Sigma structure is *executive management*. This includes a company's CEO and other members of top management. Executive leadership determines the overall strategy for the organization's Six Sigma implementation. They also set the parameters for duties of more junior practitioners. They have generally 2–3 days of overall training in the Six Sigma process with emphasis on the philosophy and expectations rather than on statistical tools and other specific problem-solving methodologies.

Systems Engineering

Systems engineering is an interdisciplinary approach to design, implementation, and evaluation that holds the key to the successful development of complex human-made systems. It can be used by both OEMs and suppliers to have effective designs.

The classic model follows the letter V (although there are variations of it) and is shown in Figure 22.1. The V-model provides guidance for the planning and realization of projects. The following objectives are intended to be achieved by a project execution:

- Minimization of project risks: The V-model improves project transparency and project control by specifying standardized approaches

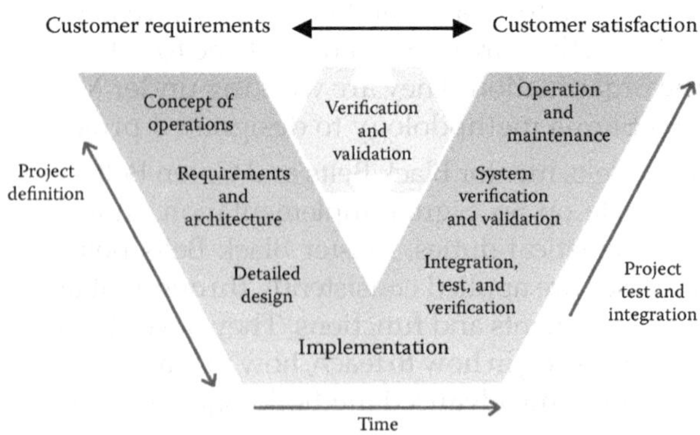

FIGURE 22.1
Typical systems engineering V-model.

and describing the corresponding results and responsible roles. It permits an early recognition of planning deviations and risks and improves process management, thus reducing the project risk.

- Improvement and guarantee of quality: As a standardized process model, the V-model ensures that the results to be provided are complete and have the desired quality. Defined interim results can be checked at an early stage. Uniform product contents will improve readability, understandability, and verifiability.
- Reduction of total cost over the entire project and system life cycle: The effort for the development, production, operation, and maintenance of a system can be calculated, estimated, and controlled in a transparent manner by applying a standardized process model. The results obtained are uniform and easily retraced. This reduces the requirements of dependency on the supplier and the effort for subsequent activities and projects.
- Improvement of communication between all stakeholders: The standardized and uniform description of all relevant elements and terms is the basis for the mutual understanding between all stakeholders. Thus, the frictional loss between user, supplier, and developer is reduced.

It is very important to note the left leg of the V, which is design oriented, and the right leg, which is manufacturing oriented. Generally, the model is depicted as a wide left leg with the arrow going back and forth to indicate adjustments and changes in the design. The right leg is much narrower to indicate the imbedded changes in the design and, consequently, much easier manufacturing. The left side of the V-model begins with customer requirements and the right side finishes with customer satisfaction. What is very important about these two characterizations is that there is a double arrow that connects the two to signify continual sharing of information for both compliance as to what the customer is asking (requirements) and whether the customer is satisfied with what is received and paid for. Finally, it is also interesting to note that validation and verification are in the middle and generally are shown with a double arrow going back and forth between design and manufacturing.

IEEE (2011, p. 452) defines validation and verification in very specific terms as follows:

- Validation. The assurance that a product, service, or system meets the needs of the customer and other identified stakeholders. It often involves acceptance and suitability with external customers.
- Verification. The evaluation of whether or not a product, service, or system complies with a regulation, requirement, specification, or imposed condition. It is often an internal process.

However, in simple terms, the essence of validation can be expressed by the question, "Are you building the right thing?" and verification by, "Are you building it right?"

To improve the design and manufacturing, systems engineering encourages the use of tools and methods to better comprehend and manage complexity in systems. Some examples of these tools are

- System model, modeling, and simulation
- System architecture
- Optimization
- System dynamics
- Systems analysis
- Statistical analysis
- Reliability analysis
- Decision making

On the other hand, early in the development stage of the design, when the primary purpose of a systems engineer is to comprehend a complex problem, graphic representations of a system are used to communicate a system's functional and data requirements. Common graphical representations include

- Functional flow block diagram (FFBD)
- Model-based design, for example Simulink, VisSim, and so on
- Data flow diagram (DFD)
- N2 chart
- IDEF0 diagram
- Use case diagram
- Sequence diagram
- USL function maps and type maps
- Enterprise architecture frameworks, like TOGAF, MODAF, Zachman Frameworks, and so on

Value Engineering

Value engineering (VE) is a systematic method to improve the *value* of goods or products and services by using an examination of function. Value, as defined, is the ratio of function to cost. Therefore, value can be increased by either improving the function or reducing the cost. It is a primary tenet of VE that basic functions be preserved and not be reduced because of pursuing value improvements.

The reasoning behind VE is as follows: If marketers expect a product to become practically or stylistically obsolete within a specific length of time, they can design it to only last for that specific lifetime. The products could be built with higher-grade components, but with VE they are not because this would impose an unnecessary cost on the manufacturer, and, to a limited extent, also an increased cost on the purchaser. VE will reduce these costs. A company will typically use the least expensive components that satisfy the product's lifetime projections.

VE is often done by systematically following a multistage job plan. Larry Miles' original system was a six-step procedure, which he called the "value analysis job plan." Others have varied the job plan to fit their constraints. Depending on the application, there may be four, five, six, or more stages. One modern version has the following eight functions:

1. Preparation
2. Information
3. Analysis
4. Creation
5. Evaluation
6. Development
7. Presentation
8. Follow-up

These eight functions may be formally presented in a four basic steps in the job plan as

- Information gathering: This asks what the requirements are for the object. Function analysis, an important technique in value engineering, is usually done in this initial stage. It tries to determine what functions or performance characteristics are important. It asks questions like, what does the object do? What must it do? What should it do? What could it do? What must it not do?
- Alternative generation (creation): In this stage, value engineers ask what are the various alternative ways of meeting requirements? What else will perform the desired function?
- Evaluation: In this stage, all the alternatives are assessed by evaluating how well they meet the required functions and how great the cost savings will be.
- Presentation: In the final stage, the best alternative will be chosen and presented to the client for a final decision.

These four steps of VE follow a structured thought process to evaluate options as follows.

- Gather information
  - What is being done now?
  - Who is doing it?
  - What could it do?
  - What must it not do?
- Measure
  - How will the alternatives be measured?
  - What are the alternate ways of meeting requirements?
  - What else can perform the desired function?
- Analyze
  - What must be done?
  - What does it cost?
- Generate
  - What else will do the job? (Identify alternatives)
- Evaluate
  - Which ideas are the best? (Must have appropriate and applicable criteria)
- Develop and expand ideas
  - What are the impacts?
  - What is the cost?
  - What is the performance? (Are the proposals meeting and/or exceeding customer needs? Wants? Expectations? If so, to what degree?)
- Present ideas
  - Sell alternatives (focus on the most successful alternative)

References

Abernethy, R. (2009). *The New Weibull Handbook: Reliability and Statistical Analysis for Predicting Life, Safety, Supportability, Risk, Cost and Warranty Claims*, 5th ed. North Palm Beach, FL: Robert B. Abernethy.

AIAG (2008). *Failure Mode and Effect Analysis*, 4th ed. Southfield, MI: AIAG.

Eckes, G. (2001). *Making Six Sigma Last*. New York: John Wiley & Sons.

Harry, M. and R. Schroeder. (2006). *Six Sigma*. New York: Random House.

IEEE. (2011). *IEEE Guide: Adoption of the Project Management Institute (PMI) Standard. A Guide to the Project Management Body of Knowledge (PMBOK Guide)*, 4th ed. Retrieved October 30, 2014.
Pande, P., R. Neuman, and R. Cavanagh. (2000). *The Six Sigma Way*. New York: McGraw-Hill.
Pyzdek, T. (2001). *The Six Sigma Handbook*. New York: McGraw-Hill.
Stamatis, D. (2002–2003). *Six Sigma and Beyond*. Volumes I–VII. Boca Raton, FL: CRC Press.
Stamatis, D. (2003). *Failure Mode and Effect Analysis: FMEA from Theory to Execution*, 2nd ed. Milwaukee, WI: Quality Press.
Stamatis, D. (2004). *Six Sigma Fundamentals*. New York: Productivity Press.
Stamatis, D. (2010). *OEE Primer*. Boca Raton, FL: St. Lucie Press.
Stamatis, D. (2015): *The ASQ Pocket Guide to Failure Mode and Effect Analysis (FMEA)*. Milwaukee, WI: Quality Press.

Selected Bibliography

http://www.barringer1.com/pdf/Chpt1-5th-edition.pdf.
http://www.weibull.com/basics/lifedata.htm.
Bangert, M. (2014). Take these steps to six sigma success. *Quality*. 44–45.
Breyfogle III, F. (1999). *Implementing Six Sigma*. New York: John Wiley & Sons.
Breyfogle III, F. (2003). *Implementing Six Sigma*, 2nd ed. New York: John Wiley & Sons.
Chatfield, C. and G. Goodhardt. (1973). A consumer purchasing model with Erlang Interpurchase Times. *Journal of the American Statistical Association*. 68:828–835.
Montgomery, D. (2009). *Introduction to Statistical Quality Control*. New York: John Wiley & Sons.
Nelson, W. (2004). *Applied Life Data Analysis*. New York: Wiley-Blackwell.
Papoulis, A. and S. Pillai. (2004). *Probability, Random Variables, and Stochastic Processes*, 4th ed. New York: John Wiley & Sons.
Sobek, D., A. Ward, and J. Liker. (2000). Toyota's principles of set-based concurrent engineering. Sloan Management Review. Cambridge, MA. Reprinted from Winter 1999 original.

Epilogue

This book is essentially about how to achieve excellence in customer–supplier relationships. It is not about excellence in manufacturing or service, but achieving excellence in the entire organization for both customer and supplier. Organizations must realize and implement a win–win relationship for a lifetime relationship. This means that the relationship of excellence is measured in loyalty and respect to each other, and repeat business. In this relationship, quality is important; however, this quality must be at all levels of the organization. After all, quality is achieved when *customers return but products do not*! This means in the simplest definition that quality is about *survival* in the future among competitors. So, what sets an organization apart?

We hope that we have demonstrated throughout this book that quality is not about standards, measurement, special characteristics, or special techniques. Of course, these are important, but most important are the *commitment to process excellence* and an overall desire *to do more* than just deliver *conformance to requirements* or products/services that are *fit for use.*

The start of differentiation begins with a flawless launch. To have a successful launch for any product and/or service, both the organization and the supplier must have focus and determination for a mutual engagement that delivers supplier components/systems on time with correct designs, proven performance, best-in-class quality, competitive cost, and required capacity. This means that a simple model must be followed consistently with the following five characteristics:

1. Planning
2. Implementation
3. Measurement
4. Improvement
5. Recognition and award

In practice, these five characteristics may be summarized as

- Early involvement: Work with suppliers more *consistently*, earlier in the product development process
- Teams: Utilize cross-functional teams to work together with suppliers at their facilities
- Standardization: Adopt a single, global, disciplined process to minimize rework and redundancy

- Governance and reporting: Use common, clear deliverables and a structured management review cadence
- Priority: Have a system that prioritizes the needs for an intervention based on criticality, delivery, performance, capacity, or some other factor

An excellent example of these five characteristics is the relationship between Ford Motor Company and two of their suppliers for the 2015 model of the F-150 truck. The first is the Alcoa Company, which provides aluminum for the truck body, and the second one is OSRAM Automotive Lighting, which provides a complete LED forward lighting system.

The relationship of both customer and suppliers in this case demonstrates the growth in quality perception, which is summarized in Table E.1.

Alcoa and OSRAM were committed to satisfying the customer, and they both demonstrated this commitment through four key principles.

1. Quality: We all know how much consumers rely on the products they buy to make their lives easier and more comfortable, so both suppliers pushed themselves to make cars lighter, last longer (aluminum), and perform better (lighting system).
2. Safety: Everyone knows that as long as human beings design anything, there is going to be a failure because nothing is perfect and absolute. In fact, from a reliability perspective, we know this from the $R(t) = 1 - F(t)$ [reliability at time t is equal to 1 minus the failure rate]. Therefore, the focus was to produce products that assure the customers were indeed buying a product that met and/or exceeded the requirements of safety and that the new innovations were part of the philosophy of continual improvement in both cost and safety.
3. Support and satisfaction: Both customer and suppliers worked together in educating each other about the new innovation to optimize product performance and value to consumers.
4. Innovation: Although aluminum usage in the automotive industry is not new, it is the first truck that has annual sales greater than any other vehicle on the market to be made of anything other than steel. The old model worked so well that the F-150 has been awarded the Number 1 status in its category for many years. However, the Ford management was not satisfied with that. They wanted to improve. So, they introduced the concept of aluminum and LED technology. Indeed, the F-150 for the 2015 model year will incorporate both of these innovations. The aluminum will contribute to reduction of weight, thereby increasing fuel efficiency (miles per gallon) and the LED technology will contribute to much greater safety for the high/low beam, light driver module, turn module, and park module.

TABLE E.1

Summary of Growth in Quality Perception

	Old Approach	New Approach
Quality services	• Service quality and performance • Customer satisfaction • Service quality applications	• Service management principles • Customer choices • Service failures and guarantees • e-service
Quality management	• Theory building • Cost of quality • Total quality management (TQM) • Infrastructure element • Externalities • Global issues	• Usage of models • Context and contingencies • Global contingencies • Six Sigma • Supply chain quality • Environmental concerns
Quality product	• Design based on attributes and performance • Product strategy • Process strategy	• Strategic product choices • New product introductions
Quality process	• Project management • Process models	• Improvement models based on data • Use statistics for optimization
Quality performance	• Quality practices • Quality awards	• TQM and performance • Accountability and measurement • Supply chain development • Usage of information technology • Focus on strategic benefits
Quality practice	• Theoretical models • Change management • Baldrige practices • Learning contingencies	• Build on knowledge management (things learned—corporate knowledge) • Accumulating knowledge (historical TGW and TGR) • Baldrige practices • SPC • Six Sigma • ISO standards • Lean practices

The Five Critical Items for the Customer

It all starts with customer *needs* and *wants*. After all, if there is no customer, there is no need for a supplier. We emphasize the word *supplier* because unless the relationship between customer and supplier is based on a win–win philosophy, neither one will prosper. This win–win philosophy is based on a long-term relationship and is the complete opposite of the *vendor* concept, which is based on a short-term relationship and strictly price. Therefore, in order for the customers to receive what they need and want, the following five critical items must be understood and implemented for a flawless launch of their product or service.

1. Develop a formal enterprise risk: This means that it is the customer's responsibility to identify supply chain risk and plan accordingly.
2. Evaluate the supplier base for vulnerability and risk exposure: Based on the results of the development of a formal enterprise risk, it is the customer's responsibility to analyze the individual supplier for its ability to perform to the needed requirements. Special consideration should be given to historical performance, capability, and capacity.
3. Evaluate logistical issues: It is profoundly important for the customer to make sure that logistics will not interfere with delivery of the product and/or service. Issues of concerns here may be availability of utilities, qualified employee pool, proximity to the facility of the customer, transportation issues (availability of rail, airports, and highway system), and so on.
4. Develop a contingency plan: Things happen and sometimes the results are devastating. Therefore, it is the customer's responsibility to develop contingency plans in case of emergencies, such as acts of God, power failures, employee strikes, sudden increases in demand for the product or service, and so on.
5. Develop a symbiotic relationship in the name of continual improvement (CI) and win–win philosophy: Quite often, the supplier may need help in financing to support its business and/or develop its business with new innovations. The customer should have planned for such occasions and be prepared to help. (In the case of the F-150, Ford helped the suppliers in the development of both grants and research.) In addition to direct help, the customer may want to encourage the supplier to pursue innovation and implementation of newer technologies. This is what supplier development is all about.

The 10 Indicators for a Responsible Supplier

No one will deny that suppliers are a key component to every customer's success. After all, they are an integral part of the customer's quality, cost, and ultimate consumer satisfaction. Therefore, due to the significance of their contribution, we have identified some key indicators for a supplier who has the potential of being an *excellent* partner in the customer-supplier relationship. The indicators are

1. Does it meet the standards and specific requirements?
2. Will it make it difficult for us to create more and better quality products/services?
3. Does it use accounting tricks to mask more realistic spending or revenue figures (scrap, rework, capacity, etc.)?
4. Does it increase the price of the product/service?
5. Does it increase the cost of doing business with this particular organization?
6. Does it increase the cost of doing business directly or due to heavy regulations/specifications or policies?
7. Does it secure the safety and sustainability attitude in what they do?
8. Does it support CI for its own improvement or is it viewed as a requirement of the customer?
9. Is the entire organization on board with the concept of continual improvement?
10. Are the policies and procedures in tandem with the philosophy of the customer?

We believe these 10 items will signal to a customer whether the supplier is committed to improvement for its own sake and whether it is willing to perform on the path of excellence.

An analysis of the 10 items will confirm early on in the selection process the attitude of the management and the future relationship. Specifically, a cursory overview of these items will lead to a deeper discussion of the following:

- Financial numbers: Is the supplier financially sound to undertake the responsibility of supplying products and services to the customer?
- Commitment to the customer: Is the supplier committed to CI for all that it does for the customer? Does the supplier have enough capacity to satisfy the customer's needs?
- Measurement ability: Does the supplier have the ability to confirm the performance needed by the customer? How are the key performance indicators (KPI) identified? Measured? Recorded? Are they relevant to the customer needs? Are they actionable?
- Plans and priorities: Does the supplier have plans for CI? How are priorities set? Are the plans and priorities reviewed on a regular basis for changes and/or adjustments?
- Management plans: Does the supplier cascade the plans and priorities to the appropriate level for action? Does management review the quality operating system (QOS) on a monthly basis?

- Tracking the progress: How does the supplier track the progress? How is progress measured? Are measurement data used or are most of the measurements of the attribute type? Is statistical process control (SPC) used by the operators? Reviewed by engineers, supervisors, and appropriate action taken based on the data?

A final thought about the win–win relationship of the customer and supplier may be summarized in seven principles that, if followed, will create a fruitful relationship that is ready to complete a flawless launch. These principles are

1. Make the products easy to find. That is the sourcing advantage. The closer the suppliers are, the better the flow and delivery will be to the customer.
2. Get good advice. Research within and outside your organization for the best alternatives available. Benchmarking and quality function deployment (QFD) is appropriate.
3. Pick the right suppliers. A customer depends on suppliers for the quality of their products. Make sure that the suppliers are able and willing to perform to your expectations. They are your partners in quality. Treat them as equals.
4. Understand supplier rules and regulations. Suppliers are also in business to make money. Therefore, respect their rules, culture, and regulations under which they have to work. If they do not make money in the relationship, they will not be your supplier for long.
5. Communicate regularly and efficiently with your suppliers. Two-way communication is imperative. The information should be on requirements and expectations as well as feedback. The feedback should be objective and data driven.
6. Watch out for channel conflict. To be sure, conflicts will arise. It is imperative that both have a system of conflict resolution that is amicable to both parties.
7. Look for continual improvement. The improvement should be on several fronts, that is, communications, expectations, specifications, capability, capacity, costs, safety, quality, delivery, and in any area of concern.

Glossary

advanced quality planning: a structured method of defining and establishing the required steps necessary to assure that a product satisfies the customer. Advanced quality planning embodies the concepts of defect prevention and continuous improvement as contrasted with defect detection.

andons: special lamps that illuminate problems in the factory. Paging andons are lists to request parts supplies; emergency andons alert supervisors to abnormalities on the line; operation andons indicate the machines' current operation rates; progress andons help us monitor the progress of line operations (Hirano 2009).

angle plate: a precise measurement device used to establish an accurate 90° vertical surface.

angularity: a three-dimensional geometric tolerance that controls how much a surface, axis, or plane can deviate from the angle described in the design specifications.

ASME: The American Society of Mechanical Engineers. ASME is an organization that publishes technical materials and sets industrial and manufacturing standards.

attitude surveys: production system measurable. Surveys measure how employee attitudes help guide Ford production system (FPS) efforts toward 90% employee satisfaction.

bonus tolerance: additional tolerance that applies to a feature as its size shifts from a stated material condition. Both maximum material condition (MMC) and least material condition (LMC) allow bonus tolerance.

build-to-schedule (BTS): production system measurable. BTS is the percentage of units scheduled for a given day that are produced on the correct day and in the correct sequence. BTS reflects a plant's ability to produce what customers want, when they want it. BTS = Volume Performance × Mix Performance × Sequence Performance.

buzz-squeak-rattle (BSR): objectionable vehicle attributes in any product especially in the automotive domain. All BSR issues are tracked by engineering development and subsequently assigned to the most responsible party for resolution, that is, engineering, manufacturing, or the supplier. Although somewhat subjective, BSR issues should be eliminated to achieve customer satisfaction.

capability study: a statistical analysis of the output from a machine or a process to determine its capability. This analysis is especially important when a new product is introduced so product quality and variation reduction can be assessed.

CATIA: a computer-aided three-dimensional interactive applications design and manufacturing system. This software was developed and is maintained by Dassault Systemes of France and is marketed in the United States by IBM. It is known as a CAD/CAM system.

cellular manufacturing: a production approach that uses groupings of manufacturing equipment, tools, and people organized to perform an entire sequence of manufacturing operations in one contiguous physical location (cell) (Duncan 1995).

changeover: the installation of a new type of tool in a metal working machine, a different paint in a painting system, a new plastic resin and/or a new mold in an injection molding machine, new software in a computer, and so on. The term applies whenever a production device is assigned to perform a different operation. Changeovers directly impact the availability of equipment. Whenever a machine goes down for changeover, it is unavailable for production. Improving changeover time increases equipment availability. See also **quick changeover**.

changeover reductions: the eight steps in the classic changeover improvement process are (1) distinguish between internal and external activities; (2) analyze what is being done in the changeover process; (3) list ideas for shifting internal elements to external changeover; (4) list ideas for streamlining internal activities; (5) list ideas for streamlining external activities; (6) prioritize and select ideas to test and adopt; (7) plan next changeover using new ideas; and (8) document procedure on a QPS sheet.

changeover time: time between the last good piece off one production run and the first good piece off the next run.

characteristic: a unique feature of a product or process, or its output, from which variable or attribute data can be collected.

circular runout: a two-dimensional geometric tolerance that controls the form, orientation, and location of multiple cross-sections of a cylindrical part as it rotates (ASME 2009).

circularity: a two-dimensional geometric tolerance that controls how much a feature can deviate from a perfect circle (ASME 2009).

compliance report random audit: an audit performed during a process sign-off (PSO) on-site visit to determine whether arbitrarily selected characteristics of a part or process meet their specification requirements. The part or process samples should be randomly selected.

concentric: sharing the same center.

concentricity: a three-dimensional geometric tolerance that controls how much the median points of multiple diameters may deviate from the specified datum axis (ASME 2009).

constraint: the operation that is preventing the plant's capacity from increasing.

continual improvement: the improvement of products, processes, and/or services on an ongoing basis. The gains made through continuous improvement activities are generally incremental, small-step improvements, as contrasted with more dramatic and sweeping improvements typically associated with initiatives such as policy deployment. In Japan, the continual improvement process is often called Kaizen (Duncan 1995).

continuing conformance tests (CCT): tests and evaluations performed during production to monitor the effects of processing and to assure continued conformance to engineering requirements.

control part: a single part which follows the manufacturing process from beginning to end. Ideally, the control part is an important part of the end product.

control plans: written descriptions of the systems for controlling parts and processes. They are written by suppliers to address the important characteristics and engineering requirements of the product. Each part will have a control plan, but in many cases, *family* control plans can cover a number of parts produced using a common process. Customer approval of control plans may be required prior to production part submission. Control plans must be linked to process failure mode and effect analysis (PFMEA) and process flow chart.

coordinate tolerancing: a system for describing the design of a part that compares its features to distances along three linear axes. These axes create an imaginary rectangular grid.

COVISINT: a web-based system that grants qualified supplier users access to selected portions of the DaimlerChrysler, Ford, and GM networks.

cross-section: a section of a feature that is formed by an intersecting imaginary plane.

cycle time: **(1)** the time required to complete one cycle of an operation. If cycle time for every operation in a complete process can be reduced to equal **target time**, products can be made in **single-piece flow**. **(2)** In cycle reduction activities, the elapsed time of a process from beginning to end. This elapsed time includes both value-added and non-value-added activity. In machine operations, cycle time is more accurately defined as the time required for the machine to cycle through one recurring sequence of activities (Duncan 1995).

cylindricity: a three-dimensional geometric tolerance that controls how much a feature can deviate from a perfect cylinder.

datum: a point of reference. An imaginary, perfect geometric shape or form. A perfect point, line, flat plane, circle, and cylinder are all examples of possible datums.

datum feature: a physical feature that acts as an acceptable substitute for a datum. Datum features relate the various features of the part to each other.

datum reference frame: three imaginary planes perpendicular to one another that are mapped onto the part to relate features to each other.

defect: a product/part that deviates from specifications or does not meet internal/external customer expectations. All defects are created by errors.

design failure mode and effects analysis (DFMEA): a formalized method for quantifying the risk associated with identified potential design failure modes. It is used to identify corrective actions required to prevent failures from reaching the customer. The DFMEA is a living document that will be updated as the design changes and progresses through design verification and production. The first iteration should be done before program release.

design of experiments (DOE): a statistical approach to efficiently plan and structure a series of controlled tests. In its broadest sense, it encompasses planning, setting up, and running the tests, and analyzing the data. The goal is to identify the sources of product variation that should be optimized or avoided for consistent product performance with a minimum number of tests.

design verification plan and report (DVP&R): a formalized test planning and reporting tool. It itemizes all testing necessary to assure that functional and reliability criteria and target requirements are defined in specific measurable terms. It provides a convenient reporting format for engineering development (ED), design verification (DV), production validation (PV) and continuing conformance testing (CCT).

design verification tests (DV): demonstrates that the final released component or assembly meets design intent. The tests, sample sizes, and performance requirements are detailed in engineering performance standards.

dock to dock (DTD): a Lean and production system measurable. DTD measures the elapsed time between the unloading of material and the release of finished goods for shipment. Using tools like visual factory and error proofing improves DTD by moving material through the system efficiently. DTD = Total Number of Units of the Control Part /End of Line Rate.

engineering change level, DCL: The drawing change level (DCL) is the latest design level of the released drawing.

engineering change level, PCL: The part change level (PCL) is the latest design level of the part (not necessarily the drawing).

engineering standards: written requirements that describe materials, processes, performance, reliability, quality, and/or design requirements for a material, process, or part, or a family of materials, parts, or systems.

equipment availability: affected by equipment failure, changeovers, idling/ short stops, reduced speed, process defects, and reduced yield.

error: (1) any deviation from a specified manufacturing process. Errors can be made by machines or people and can be caused by previous errors that have occurred. While an error may not produce a defect, all defects are created by errors. When errors are eliminated, defects will not be created. (2) Act or instance of deviation from unexpected procedures or methods of work that results in defects, scrap, rework, waste, injuries, or any non-value-adding natural consequence (Duncan 1995).

error and mistake proofing: Error proofing is a method used to identify potential process errors and either design them out of the product or process or eliminate the possibility that the error could produce a defect. Mistake proofing is an approach used to develop methods to clearly identify errors that may occur and prevent them from becoming nonconformances.

error proofing: a group-based improvement strategy that is targeted at eliminating defects, errors, and equipment abnormalities in production processes before they occur. Error proofing is part of in-station process control. Using error-proofing techniques to stabilize equipment positively impacts measurables. For instance, error proofing reduces scrap, reruns, retests, and returns. Error proofing improves first time through and customer satisfaction.

error-proofing device: simple and inexpensive methods used to prevent errors about to occur, or detect errors that have already occurred. Some examples of error-proofing devices are guide or reference rod, template, limit or microswitch, counter, odd-part-out method, sequence restriction, standardization, critical condition detector, delivery chute, stopper/gate, and sensors.

error-proofing steps: to do an effective error proofing, the following steps should be taken:

1. Identify and describe the defect
2. Determine where the defect is made/discovered
3. Detail the current standard procedure
4. Identify deviations from standard
5. Identify the red flag conditions where the defect occurs
6. Identify the type of error-proofing device required to prevent the error or defect
7. Create devices and tests for effectiveness

extended enterprise™: a DaimlerChrysler coordinated process that unifies and extends the business relationships of suppliers and supplier tiers to maximize the effectiveness of vehicle development, minimize total system costs, and improve quality and customer acceptance.

external activities: activities that can be performed while the machine is running or producing. External time determines how frequently you can change over. If the external time is longer than your shortest run, waiting, or waste, occurs.

feature: a physical feature of a part that naturally contains variation and imperfections. A corner, edge, flat surface, and hole are all examples of possible features.

feature control frame: a series of compartments containing symbols and values that describe the tolerance of a feature. The order and purpose of these compartments follow a consistent standard.

first production shipment certification (FPSC): ensures that the initial production shipments meet all customer drawing and engineering standards requirements. FPSC is intended to identify part problems and contain the parts at the supplier during first production shipments to ensure a successful launch. The FPSC program requires the supplier to provide statistical evidence of conformance to special product and/or process characteristics prior to the shipment of components to the customer's receiving plants. Certification is for a minimum of 2000 parts. One hundred percent of the FPSC parts are to be inspected for conformance to special product and/or process characteristics.

first tier: a term used for a supplier who is responsible for providing components, services, or raw material directly to a customer.

first time capability (FTC): part of a process capability study that identifies the capability of a process (how well it produces quality parts) by operation. FTC can also be extended from just one operation to the overall production process (plant FTC).

first time through (FTT): A Lean, Six Sigma, and production system measurable. FTT measures the percentage of quality units that go through the system without being scrapped, rerun, retested, returned, or diverted into an offline repair area. FTT improves with the elimination of waste and defects.

FTT = Units entering process – (scrap + reruns + retests + repaired offline + returns)/ Units entering process

five key steps: a phrase resulting from five Japanese words used to describe workplace organization of the visual factories within the Toyota Production System:

1. *Sort* means to separate needed tools, parts, and instructions from unneeded materials and to remove the latter
2. *Stabilize* means to neatly arrange and identify parts and tools for ease of use
3. *Shine* means to conduct a cleanup campaign

4. *Standardize* means setting guidelines regarding work procedures, organization, cleaning, and communications and making sure they are applied consistently throughout the workplace
5. *Sustain* means to form the habit of always following the first four Ss. See also **visual factory, five keys to workplace organization**.

five keys to workplace organization: While there are a number of tools available for creating healthy workplaces at the enterprise level, most focus on specific occupational hazards, industries, or sectors, but no comprehensive scheme for good practice exists. To provide companies with such a scheme, the World Health Organization (WHO), on the basis of the WHO Global Plan of Action on Workers' Health, 2008–2017, has identified the following:

Key 1: Leadership commitment and engagement

Key 2: Involve workers and their representatives

Key 3: Business ethics and legality

Key 4: Use a systematic, comprehensive process to ensure effectiveness and continual improvement

Key 5: Sustainability and integration

five Ss: five Japanese terms perceived by many to represent the fundamental elements of a total quality management approach:

1. *Seiri* (organization)
2. *Seiton* (neatness)
3. *Seiso* (cleaning)
4. *Seiketsu* (standardization)
5. *Shitsuke* (discipline) (Duncan 1995)

Five Whys: a problem-solving method attributed to Taiichi Ohno. The question "why" is asked and answered until the root cause of a problem is identified. Corrections are then put in place.

fixture: a dedicated work holding device used to locate and hold a part during machining or inspection.

flatness: a three-dimensional geometric tolerance that controls how much a feature can deviate from a flat plane.

Ford production system (FPS): a unified system that integrates Ford's worldwide manufacturing, design and development, order and delivery, supply, and management functions. The part of FPS that encompasses what plants do is called the plant operating system (POS). There are seven FPS components contributing to the POS. They are

1. Human resources
2. Industrial materials
3. Material flow
4. In-station process control
5. Ford total productive maintenance (FTPM)
6. Engineering
7. Quality operating system (QOS)

Each step has resources to help plants implement and sustain the production system.

Ford production system (FPS) measurables: seven common measurements that support the vision, principles, and stretch objectives of FPS. The measurables are aligned with POS processes, drive continuous improvement, foster the elimination of waste, and link management objectives to the shop floor. They are first time through, total dock to dock, build-to-schedule, overall equipment effectiveness, total cost, attitude surveys, and SHARP.

form tolerance: a group of geometric tolerances that limit the amount of error in the shape of a feature. Form tolerances are independent tolerances.

functional gauge: a gauge for a specific part that quickly checks its form and fit in a manner similar to its intended use.

gauge inspection instructions: documented instructions for a specifically identified gauge indicating step-by-step procedures for its use or layout of a part.

gauge R&R (repeatability and reproducibility): a method used to rate the appropriateness of a measurement system as a percentage of the product tolerance or process variation. Repeatability is a measure of the ability of the gauge to repeat a given measurement, and is evaluated by repeatedly measuring the same parts (equipment variation). Reproducibility is a measure of the ability of people to reproduce a series of measurements using the same gauge, and is evaluated by two or more operators independently repeating the inspection of a set of parts (appraiser variation).

geometric dimensioning and tolerancing (GD&T): an international standard for communicating instructions about the design and manufacturing of parts. GD&T uses universal symbols and emphasizes the function of the part.

granite surface plate: a precise, flat plate made of granite that is used to establish a datum plane for inspection. Granite surface plates are available in standardized grades.

home line: supplier's assembly line located in the manufacturing facility where volume production of the part takes place using production tooling and processes.

Hoshin planning: a planning technique that derives its name from the Japanese term *hoshin kanri,* which means "shining metal" and "pointing direction." This planning method uses relatively short-term, high-impact goals to achieve breakthrough improvements in company performance (Duncan 1995).

improvement cycle: a step-by-step approach used to prevent defects. The cycle starts with symptom identification, goes to problem description, and ends with improvement implementation. Components in the cycle include adherence, analysis, improvement, and standardization. When the cycle is systematically adhered to, we have created a reliable method.

individual tolerance: a tolerance that does not require a specified datum.

inspection: comparing parts coming out of a process with a standard and removing those that do not conform. There are two types of inspection: self-check and source inspection. In self-inspection, an operator checks his or her work before passing it to the next process. In source inspection, preventive action is taken at the error stage. This prevents errors from becoming defects in the first place. Source inspection involves looking for the cause of defects in the five elements of production: operator, materials, method, machine, and information.

interim approval authorization (IAA): documentation and approval to temporarily use a part that does not meet production part approval process (PPAP) approval requirements for pilot builds or launch.

internal activities: activities that must be performed while the machine is shut down.

ISO: Greek word for *equal.* It was adopted by The International Organization for Standardization in Switzerland to reflect a standardized system for quality. It is considered a basic system of quality for any organization.

ISO/TS 16949: a technical specification that describes, in conjunction with ISO 9001:2008, the quality system requirements for automotive suppliers. ISO/TS 16949 has been prepared by the International Automotive Task Force (IATF) for all participating worldwide OEMs and may become the quality management standard of the future.

just-in-time (JIT): (1) a system for producing and delivering the right items at the right time in the right amounts. JIT approaches just-on-time when upstream activities occur minutes or seconds before downstream activities. At this point, *single-piece flow* is possible. The key elements of JIT are *flow, pull, standard work* (with standard in-process inventories), and *target time* (Hirano 1990, 2009). **(2)** A philosophy of waste elimination. In a JIT environment, the underlying operating principle is that anything that does not add value is eliminated or minimized to the greatest possible extent. In manufacturing, all inventory, queue time, and labor that are not required are targeted for reduction (Duncan 1995).

Kaizen: a Japanese term for continual improvement, founded on the principles of doing little things better and setting, working toward, and achieving increasingly higher standards (Duncan 1995).

Kanban: (1) see **smart card. (2)** A Japanese term that is the cornerstone of the JIT pull system. Kanban actually means "to put away and to bring out." In pull systems, it often refers to a card or other physical device used to signal the previous operation that it is authorized to produce the next unit (Duncan 1995).

key characteristics: features that are applicable to a component, material, manufacturing, or vehicle assembly operation, which are designated by customer's engineering as being critical to part function, fit, or appearance, and having particular quality, reliability, durability, or functional significance. Depending on the customer, they have a specific ID. The most frequent ID is <D>.

launch vehicles (V1): the start of volume production and assembly of vehicles. V1 will result in quality vehicles that meet all customer requirements and corporate objectives. All regulatory requirements and approvals will have been obtained. Part supplier and assembly plant facilities will be complete and ready for full-volume production.

lead time: the moment raw materials enter the manufacturing process until the moment the finished product is delivered to the customer. Lead time includes run-time, downtime, queuing, and transport.

Lean manufacturing: maximizes the efficient use of resources. An inherent principle is involving people to eliminate waste, standardize work, produce zero defects, and institute one-piece flow. It also includes material movement triggered by downstream operations (pull and inventory reductions). More than anything, it is a matter of a cultural change to improve the organization and eliminate waste.

least material condition (LMC): the point at which a feature contains the least amount of material within its acceptable size limit. The largest acceptable hole and the smallest acceptable shaft are examples of LMC.

life expectancy: a time-established service or operating life objective of a system, component, or material.

line speed: the contracted maximum tool capacity rate determined by dividing the quoted customer capacity (pieces per hour) by the supplier's quoted net operating time for the customer parts (h/day). The line speed will be verified during the production demonstration run while process performance studies are maintained at a level greater than or equal to 1.67 P_{pk}.

location tolerance: a group of geometric tolerances that limit the location or placement of features. Location tolerances are related tolerances.

lot: In manufacturing production, it is not to exceed 8 h or 1 day's production, whichever is smaller.

manufacturing floor plan: a pictorial layout of the entire manufacturing facility drawn to scale. It is to include all machinery, office space, lab space, shipping/receiving areas, quarantine areas, hold areas, and so on. The layout will also indicate any proposed areas for expansion and development.

mapping: a method used to look at the complexity of the extended enterprise value chain. Mapping details the manufacturing process, transportation, and other factors that go into component production.

material condition modifier: one of three modifiers that further define the tolerance of a feature in relation to its acceptable size limits.

material requirement planning (MRP): a computerized system used to determine the quantity and timing requirements for materials used in a production operation. MRP systems use: (1) a master production schedule; (2) a bill of materials listing every item needed for each product being made; and (3) information on current inventories of these items in order to schedule their production and delivery. Manufacturing resource planning (often called MRPII) expands MRP to include capacity planning tools, a financial interface to translate operations planning into financial terms, and a simulation tool to assess alternative production plans.

maximum material condition (MMC): the point at which a feature contains the greatest amount of material within its acceptable size limit. The smallest acceptable hole and the largest acceptable shaft are examples of MMC.

median point: a point that is exactly the same distance between two outer points.

net operating time: the number of hours (normal work pattern) less scheduled downtime (breaks, startup, previous maintenance, shared production, tool changes, shut down) and unscheduled downtime (estimated, based on historical or projected information). In case of multiple customer part numbers, the changeover time has to be considered.

noise-vibration-harshness (NVH): measurable perceptions of vehicle attributes. These items are usually discovered and measured by the customer's NVH laboratory facilities. NVH attributes are typically tuned to a certain frequency and amplitude in order to achieve an overall vehicle signature frequency. This signature is optimized by vehicle development for the best customer responses.

non-value-adding: an operation or activity that takes time and resources but does not add value to the product sold to the customer. Non-value-adding activities include work-in-process, inspection, defects, waiting, and inefficiency.

operator instructions: lists of steps that may contain inspection requirements, required tools and gauges, statistical process control (SPC) applications, sample size and frequency, and acceptance/rejection criteria. They will communicate all requirements involved in the process. These are readily available to the operator.

orientation tolerance: a group of geometric tolerances that limit the direction, or orientation, of a feature in relation to other features. Orientation tolerances are related tolerances.

overall equipment effectiveness (OEE): a production system measurable. An indispensable metric in calculating capacity for any organization. OEE measures the availability, performance efficiency, and quality rate of equipment, in particular, the constraint operation. OEE is part of any organization's total productive maintenance program and improves throughput by eliminating downtime. OEE = Availability × Performance Efficiency × Quality Rate.

packaging: material that is used to store and ship parts. It provides protection and containment of parts. It affects ease of handling by manual or mechanical means. Packaging may consist of returnable or non-returnable items including dunnage.

parallelism: a three-dimensional geometric tolerance that controls how much a surface, axis, or plane can deviate from an orientation parallel to the specified datum.

part number: a 10-digit alphanumeric number that represents a production part number that is assigned to a commodity or part. These numbers are assembled to describe the customer's engineering bill of materials used in building a product.

perishable tools: tools that are consumed in the process of producing a product, and are usually a cost item, for example, drill bits, cutters, sockets, driver tips, inserts, hobs, broaches, welding tips, and so on.

perpendicularity: a three-dimensional geometric tolerance that controls how much a surface, axis, or plane can deviate from a 90° angle.

plant operating systems (POS): the part of FPS that encompasses what plants do. Implementation of the POS involves using five sequential phases that take each plant from current state conditions to the improvements envisioned in FPS. The five phases are stability, continuous flow, synchronous production, pull system, and level production.

poka-yoke: **(1)** an error-proofing device or procedure used to prevent a defect during order taking or manufacture. An order-taking example is a screen for order input developed from traditional ordering patterns that questions orders falling outside the pattern. The suspect orders are then examined, often leading to discovery of inputting errors or buying based on misinformation. A manufacturing example is a set of photocells in part containers along an assembly line to prevent components from processing to the next stage with missing parts. The poka-yoke in this case is designed to stop the movement of the component to the next station if the light beam has been broken by the operator's hand in each bin containing a part for the product under assembly at that moment. See also **error proofing**. **(2)** A Japanese term that refers to fool proofing a design such that all ambiguity is removed and it becomes virtually impossible to set up a machine or produce a part or an assembly incorrectly. This is often accomplished through designing the components of the tools and the assemblies so they will fit together only in the proper orientation and sequence (Duncan 1995).

position: a three-dimensional geometric tolerance that controls how much the location of a feature can deviate from its true position.

preproduction sample report (PPSR): a dimensional and performance verification document, submitted with all part samples used prior to S1 build.

pre-PSO documentation review: a meeting conducted by the PSO team, chaired and coordinated by the supplier quality specialist with the supplier and their advanced quality planning project team. The meeting is conducted at a customer's selected location prior to the planned PSO. The purpose is to review all of the supplier's advanced quality planning and system plan documents in a sequential way and determine the supplier's readiness to demonstrate a first production demonstration run (300 pieces) in the form of a PSO.

preventive maintenance (PM): a planned system of actions performed to prevent breakdown of a machine or equipment as a result of normal use.

primary datum: the datum feature that first situates the part within the datum reference frame. The primary datum is the first feature to contact a fixture or surface during assembly.

problem-solving techniques (PST): methods that are used to identify the root cause of a problem. The most common forms of PST are corrective action teams, 8D, quality circles, multivari charts and analysis, design of experiments (DOE), benchmarking, root-cause analysis, fishbone diagramming, quality function deployment (QFD) teams, Shainan Consultants Inc.'s trademarked PST techniques (Isoplot, Red"X", Rank Order ANOVA, etc.), Pareto charts, and many more. The most successful techniques have employee teams contributing in some special way to perform root-cause analysis and then permanently solve the issue or problem.

process: a series of individual operations required to create a design, completed order, or product. It is also a combination of people, equipment, methods, materials, and environment that produces a product or service. Any aspect of a business may be involved.

process capability: the total range of inherent variation in a stable process. It is determined using data from control charts. The control charts will indicate stability before capability calculations can be made. Histograms are to be used to examine the distribution pattern of individual values and verify a normal distribution. When analysis indicates a stable process and a normal distribution, the indices P_p and P_{pk} (or C_p and C_{pk}) can be calculated. If analysis indicates a nonnormal distribution, advanced statistical tools such as transformation of data analysis will be required to determine capability.

process failure mode and effects analysis (PFMEA): a structured method of documenting historical or potential process failure modes, their effects, and the controls taken to ensure their prevention. The PFMEA rating system allows quantification of the risk presented by each failure mode. The risk rating will be used to prioritize failure prevention. The PFMEA will be developed by a cross-functional team and will be uniquely developed for each process/product. It has to be linked to the control plan and process flow diagram.

process flow diagram/chart: illustrates the flow of materials through the process. Usually starting at receiving, material in process, through shipping including any special processing which may result from a nonconformance requiring disposition, rework, or repair operations.

process performance and process capability indices (P_p, P_{pk}, C_p, and C_{pk}): defined in the statistical process control manual published by AIAG (2005) as measures of capability. P_p and P_{pk} are preferred.

process sign-off (PSO) on-site visit: a systematic and sequential review of the supplier's manufacturing process conducted by a PSO team at the supplier's production facilities. Supplier quality is responsible for team leadership.

processing time: the time a product is worked on in design or production, as well as the time an order is actually being processed. Typically, processing time is a small fraction of *throughput time* and *lead time*. In addition, processing time includes *value-adding* and *non-value-adding* activities.

product assurance plan (PAP): a structured method of defining and establishing the required steps necessary to assure that a product satisfies the customer. It is a checklist to track all engineering, supplier, and quality planning events required for new products. It lists all the important tasks to be completed, the persons responsible, the estimated/actual completion dates, and how the individual tasks contribute to the progress of the phases of the product creation process.

product assurance planning manual: a DaimlerChrysler published *blue dot* manual that provides requirements to assure that specific activities occur at the proper time in the product creation process. It also gives specific instructions on how to develop a product assurance plan. It involves assurance activities that focus on the areas of quality, reliability, producibility, maintainability, and serviceability.

product family: a range of related products that can be produced interchangeably in a production cell. The term is often analogous to *platforms*.

product specifications: the sum total of engineering drawing requirements including standards referenced on those drawings and within the contract. Some of these requirements are determined by team consensus during advanced quality planning meetings and design reviews and can take the form of safety standards, material standards, temporary standards, process standards, laboratory standards, reliability standards, or performance standards.

production capacity sheet: a job aid used by team members establishing and maintaining a quality process system. The form is used to determine the processing capacity of each machine in the work sequence.

production demonstration run: a production run consisting of 300 or more parts (2 or more hours of production), from the production line of record, using trained operators, production-approved piece parts or materials from production subsuppliers, and run at a rate equal to the quoted maximum tooling capacity. All special characteristics will meet or exceed a process performance of $P_p \geq P_{pk} \geq 1.67$. FTC shall be calculated for each station and overall. A customer's representative will be on the manufacturing floor during the demonstration to observe and verify the documented production system. PV test parts are taken from this run to demonstrate that the production process has not degraded the design performance that was verified by DV testing. (The customer representative may change the amount of samples to less than 300 pieces if required.)

production part approval process (PPAP): process for ensuring that production parts are manufactured at the production site using the production tooling, gauging, process, materials, operators, environment, and process settings (e.g., feed, speed, cycle time, pressure, temperature). Parts for production part approval will be taken from a significant production run. Customers usually require that this run is a minimum of 2 h or 300 pieces, whichever is more stringent, unless otherwise stated in writing by the customer. Parts from each position of a multiple cavity die, mold, tool, or pattern are to be measured and representative parts tested.

production validation testing (PV): demonstration tests performed to validate design conformance of initial production parts manufactured with production tools and processes.

profile: the outline of a part feature within a given plane.

profile of a line: a two-dimensional geometric tolerance that controls how much the outline of a feature can deviate from the true profile.

profile of a surface: a three-dimensional geometric tolerance that controls how much a surface can deviate from the true profile.

profile tolerance: a group of powerful geometric tolerances that control the size, location, orientation, and form of a feature. Profile tolerances can be either independent or related.

program: the identification of a specific product line and model year.

pull: a system of cascading production and delivery instructions from downstream to upstream activities. The upstream supplier only produces when the downstream customer signals a need. Pull is the opposite of push.

pull system: a way of managing shop floor (or even office-level) activity that minimizes work-in-process and dramatically improves throughput time by eliminating interoperation queues. A pull system requires two things: a pull signal and a fixed upper volume limit. A fixed upper volume limit means the operators must stop producing parts whenever they have not received their (pull signal) authorization to produce more. Pull signals are sometimes called Kanbans (Duncan 1995).

push system: the antithesis of the pull system. Push systems allow production to continue based on a predetermined schedule. Push systems launch orders into the production system on a scheduled interval and assume that they will come out the end of the process at the end of the designated throughput times. Even the best closed-loop push systems are much less responsive to in-process variation, and therefore much less effective for controlling production and work-in-process than pull systems (Duncan 1995).

QPS combination sheet: a job aid used by team members establishing and maintaining a quality process system. The form graphically represents the operation time, which aids work-sequence analysis and the elimination of wasted time.

quality process system (QPS): a group-based method for documenting and standardizing work activities in a production sequence. The QPS supports continual improvement and is a keystone of both the production and plant operating systems. The standard work sequence resulting from QPS eliminates waste and helps improve all customer's measurables.

quality process system sheet (QPSS): a job aid used by team members establishing and maintaining a quality process system. QPS sheets outline the safest and most efficient production procedures. The sheets define the tasks, process flow, and human movements required to complete specific activities. They also include cycle and target time, the work sequence of specific tasks, and the minimum inventory needed to conduct the activity. Quality process sheets become a visual reference that represents the standard practices, safety, and quality at workstations. The sheets do not remain fixed. Work group members frequently modify the sheets to reflect improvements at an operation.

quick-change devices: devices used in manufacturing setup reduction activities to eliminate the adjustments required during setup changeovers. Among the devices most commonly used on manufacturing equipment are quick-change couplings, clamps, locator pins, and fixed-position stops for dies and fixtures (Duncan 1995).

quick changeover: a team-based improvement activity that reduces setup and changeover times by 50% or more. Reducing changeover time reduces downtime and is critical to the success of any production system. For example, shorter changeovers make it possible to produce parts and products in smaller lots. This reduces inventory and shortens lead time. With the resulting flexibility, you can improve on-time delivery rates and respond to schedule changes more easily.

A keystone in quick changeover is shifting internal to external activities. In other words, quick changeover performs the maximum amount of work while the machine is running. A typical process for a quick changeover is the following eight-step approach:

1. Documented current changeover element
2. Separate internal and external activities
3. Shift internal activities to external
4. Locate parallel activities
5. Streamline internal and/or external activities
6. Implement the plan
7. Validate procedure/verify results
8. Document procedure on a QPS sheet

red flag conditions: conditions in the manufacturing process that often lead to errors. The most common red flag conditions are adjustments, tooling/tooling changes, dimensions/specifications/critical conditions, many/mixed parts, multiple steps, infrequent production, lack of effective standards, symmetry/asymmetry, rapid repetition, high/extremely high volume, and environmental conditions.

regardless of feature size (RFS): a modifier indicating that the stated tolerance for a feature applies regardless of its actual size within an acceptable size limit. RFS does not permit bonus tolerance.

related tolerance: a tolerance that requires a specified datum.

reliable method: a method or procedure that always produces the desired results.

root cause: the assignable source of variation that affects all the individual values of the process output/phenomenon being studied. Root cause is the reason for the primary nonconformance, which has induced other failures and for which effective permanent corrective action can be implemented.

roundness measuring machine: a sophisticated inspection device with a precision spindle that measures various circular or cylindrical features.

runout tolerance: a group of geometric tolerances that simultaneously limit the form, location, and orientation of cylindrical parts. Runout tolerances are related tolerances requiring a datum axis.

S/E/N: a characteristic that will affect compliance with customer and/or governmental safety/emissions/noise requirements.

S0 vehicles: prototype vehicles built to provide verification that the production design and the intended production process for components, systems, and vehicles meet objectives. *Production* means that the product and process design release is complete for the product and process design complete (PPDC) milestone. These components may be built using hard or soft tools, as determined to be appropriate based on part or process knowledge required. Design verification testing will be successfully completed before parts are submitted for S0 build. PPSRs are required for all non-PPAP-approved parts.

S1 vehicles: vehicles built to confirm production design intent, validate parts, and prove vehicle and assembly processes (at the assembly plants when feasible). S1 vehicles provide early detection, identification, and resolution of production process issues prior to S2 vehicle build. Production control systems are used to order, track, and supply parts. Production intent containers/racks will be tested using S1 level parts. Grained and new color-mastered materials are required and will be accompanied by a part inspection report for each supplier shipment.

S2 vehicles*:* sellable vehicles built by manufacturing at the assembly plants to verify corporate and supplier production tooling, equipment, and processes, and to confirm successful resolution of any CNs written to correct problems found in the S0/S1 build or testing. All parts, including power trains, supplied by the production suppliers will be manufactured using production tools (on their home lines) and processes. All parts used are required to have full PSO and PPAP approval.

safety and health assessment review process (SHARP): a critical production system measurable. SHARP measures the health and safety impact of production processes. SHARP includes 21 points of reference. The measurable emphasizes improvements in health and safety processes by seeking zero fatalities, zero amputations, and zero permanent disabilities in all customer facilities.

safety characteristics: product features or functions that require special manufacturing control to assure compliance with customer or government safety requirements.

sampling plan: defines the minimum number of parts that will be inspected from a given population.

secondary datum: the datum feature that situates the part within the datum reference frame after the primary datum. The secondary datum is the second feature to contact a fixture or surface during assembly.

serpentine cells: a popular layout for cellular manufacturing operations that deploy operators inside U-shaped work cells that are connected end-to-end in a serpentine manner. See also **U-shaped cells** (Duncan 1995).

setup time reduction: an important part of reducing manufacturing lot sizes in order to reduce cycle times and work-in-process inventory. Setup time reductions were made popular as part of the JIT manufacturing movement in the late 1970s and early 1980s. There are several approaches to setup time reduction, such as dedicated setups, quick-change devices, and fixed-position mechanisms to reduce or eliminate adjustments (Duncan 1995).

seven diamond approach: a preliminary problem-solving technique used primarily by GM to respond to internal quality issues. The first four steps deal with special cause—if it exists (manufacturing)—and the last three deal with whether engineering intervention is required. The steps are as follows:

1. Is this the right process?
2. Is this the right tool?
3. Is this the correct part?
4. What is wrong with the quality characteristic?
5. Will a process change resolve the issue?
6. Will a product change resolve the issue?
7. Is it an extremely complicated problem?

seven recognition success factors: Much has been written in the literature as to how to recognize employees for excellent work. A basic approach is to follow some standard steps for any organization, recognizing that they may be modified to facilitate specific purposes. A generic approach is the following. Step 1: Establish an employee recognition committee. Step 2: Identify recognition program objectives. (It is imperative that this be thoughtful and practical based on appropriate timing, relevance, sincerity, and adaptability, and be meaningful in relation to the event that prompted the recognition.) Step 3: Identify award selection criteria. Step 4: Identify award eligibility criteria such as frequency, eligibility, and selection steps. Step 5: Have a consistent award nomination and selection process. Step 6: Let everyone in the organization know about the award program. Step 7: Evaluate effectiveness of the program (and improve as necessary).

shine: one of the five key steps in visual factory organization. Shine means keeping the workplace clean.

single minute exchange of dies (SMED): a series of techniques pioneered by Shigeo Shingo. Changeover of production machinery is accomplished in less than 10 min. The long-term objective is zero setup for changeovers, which are instantaneous and do not interfere with continual flow.

single-piece flow: Products move through various operations in design, order taking, and production, one piece at a time and without interruptions, backflows, or scrap.

smart card: a small card attached to boxes of parts. The card regulates *pull* in the Toyota Production System by signaling upstream production and delivery. Also known as *Kanban*.

sort: one of the five key steps in visual factory organization. Sorting is a visible way of identifying items that are not needed or are in the wrong place in the workplace. Red, or other color, tags are used to mark sorted goods so they can be temporarily stored and then removed from the workplace.

special characteristics: product and process characteristics designated by the customer and the supplier, including governmental regulatory and safety characteristics as well as characteristics selected through knowledge of the product or process. These include characteristics identified by key: <CC>, <SC>, <OS>, <HI>, <D>, safety <S>, noise <N>, or theft or as determined by the PSO team.

standard work: a precise description of each work activity.

standardize: one of the five key steps in visual factory organization. Standardize means devising methods to make sure the first three keys (Sort, Stabilize, and Shine) are monitored and maintained consistently.

statistical control: the condition describing a process from which special causes of variation have been eliminated. Only common causes remain; for example, observed variation could be attributed to chance causes. On a control chart, statistical control is evident by the absence of points beyond the control limits and by the absence of nonrandom patterns or trends within the control limits.

statistical problem solving (SPS): the organized use of special problem-solving tools (e.g., fishbone diagrams, Pareto charts, ANOVA, etc.) to better understand problems and their causes, and determine solutions.

statistical process control (SPC): the use of statistical techniques such as control charts to analyze a process or its outputs to take appropriate actions to achieve and maintain a state of statistical control and to improve the process capability.

straightness: a two-dimensional geometric tolerance that controls how much a feature can deviate from a straight line.

subtier supplier: provider of production materials, or production or service parts, directly to a Tier 1 supplier to the customer. Also included are providers of heat-treating, painting, plating, or other finishing services.

supplier readiness evaluation run: the first documented build attempt by the supplier following installation, setup, and tryout of the production equipment and tools. The supplier, with the concurrence of the customer's supplier quality specialist, will determine the quantity of parts that are completed during this run, but 30 pieces is a typical build size. The purpose of this run is to provide preproduction data that confirms the proper setup, operation, and performance of the installed production line. Data collected during this run typically include measurement data, machine cycle times, operation times, quantity attempted, quantity accepted, and so on. From this data, estimated production rates, first time capability (FTC), yields, and P_p and P_{pk} for special characteristics are calculated. There are no minimum requirements for process performance, FTC, yields, or other indicators at this time. The customer will evaluate the analysis of this run when deciding whether to schedule the process sign-off (PSO) on-site visit.

supplier tool record (STR): a form generated by the creation of a new tool purchase order (TPO) requiring the identification of supplier, or customer-owned tooling, affected by the TPO.

supply chain: includes all suppliers used in producing a given part; from Tier 1 to the source of production materials.

sustain: one of the five key steps in visual factory organization. Sustain means that visual factory organization has been assimilated into plant practices.

symmetry: a three-dimensional geometric tolerance that controls how much the median points between two features may deviate from a specified axis or center plane.

synchronous material flow: process or system that produces a continuous flow of material and products driven by a fixed, sequenced, and leveled vehicle schedule, utilizing flexibility and Lean manufacturing concepts.

synchronous production: an approach used in just-in-time manufacturing that synchronizes the rate of production from end-item delivery all the way back through supplier production to customer demand. In other words, products are built at the same rate at which they are sold (Duncan 1995).

tact time: the time in which one product, assembly, or part (the item of interest) is to be made. Mathematically, it is calculated as available time for production/required units of production. The calculations are based on theoretical and/or surrogate data.

Takt time: the time in which one product, assembly, or part (the item of interest) is to be made. Mathematically, it is calculated as available time for production/required units of production. The calculations are based on real (actual) data. (Takt is the German word for the baton that an orchestra conductor uses to regulate the speed, beat, or timing at which musicians play. In manufacturing, it indicates the cadence of processing.) Therefore, Takt time is the tool to link production to the customer by matching the pace of production to the pace of actual final sales. Simply put, it is the rate of customer demand. Takt time = available time/customer requirements.

target cost: the development, production, and profit costs that the customer is willing to pay for a quality product.

target time: the available production time divided by the rate of customer demand. For example, if customers demand 240 widgets per day and the factory operates 480 min per day, target time is 2 min. If customers want two new products designed per month, target time is two weeks. Target time matches production to customer demand and is the heart of any Lean system.

tertiary datum: the datum feature that situates the part within the datum reference frame after the secondary datum.

three-dimensional tolerance: a tolerance that controls a shape having a length, width, and depth.

throughput time: the time it takes a product to go from launch, to concept, to order, to delivery, and from raw materials to the product in the customer's hands. This includes both processing and queue time.

tolerance: an unwanted but acceptable deviation from a given dimension. Tolerances indicate the allowable difference between a physical feature and its intended design.

tolerance zone: an imaginary zone in which a part feature must be completely contained for the part to pass inspection.

tooling: refers to any working part of a power-driven machine. Tooling often needs to be changed because a part is not working or is broken, or a part with a different function or specification is needed.

total cost: production system measurable. Total cost is the total per unit cost of material, labor and overhead, freight, inventory, and other related costs. Sometimes it is called lifecycle costing.

total productive maintenance (TPM): a series of methods and processes that ensure production machines receive preventive maintenance so that downtime does not stop production.

total runout: a three-dimensional geometric tolerance that controls the form, orientation, and location of the entire length of a cylindrical part as it rotates.

true position: the imaginary perfect position of a feature described by design specifications.

true profile: the perfect, imaginary profile described by design specifications.

two-dimensional tolerance: a tolerance that controls a shape having only a length and width.

unit load data (ULD) sheet: loading and unloading instructions for a specific durable.

U-shaped cell: in cellular manufacturing, work centers configured in horizontal U shapes, with operators occupying positions within the U. This configuration allows operators to more easily move between positions, adjusting the number of operators in a cell based on the workload present at a given time (Duncan 1995).

value: product characteristics defined by the customer.

value-adding: any operation that changes, converts, or transforms material into the product that is sold to the customer.

value stream: the specific activities required to design, order, and provide a specific product, from concept to launch, order to delivery, and raw materials into the hands of the customer.

value stream mapping: identification of the specific activities occurring along a value stream for a product or product family.

VDA 6.1; VDA 6.4: quality system requirements of VDA ("Verband der Deutschen Automobilindustrie" = German OEMs), the German equivalent to QS 9000 and later to ISO/TS 16949.

visual control: a method of sharing information in the visual factory. Visual controls communicate information and/or build controls into the workplace and process so that activities are performed according to standards.

visual display: a method of sharing information in the visual factory. Visual displays communicate important information, but do not control what people or machines do.

visual factory: the use of controls that will enable an individual to immediately recognize the standard and any deviation from it.

warrant: an industry-standard document required for all newly tooled or revised products in which the supplier confirms that inspections and tests on production parts show conformance to customer requirements.

waste: any expenditure of resources (including time) that does not add value in terms of a customer-satisfying quality characteristic (satisfier) (Duncan 1995).

waste in process (WIP): any additional work in the process that does not add value to the customer. This includes rework and rejects.

work in process (WIP): any inventory against which direct labor has been applied and has not yet been completed. One of three categories of inventory: raw material, work-in-process, and finished goods (Duncan 1995).

zero defects: a quality philosophy that advocates a focus on error-free process performance, resulting in defect-free products and/or services (Duncan 1995).

References

AIAG. (2005). *Statistical Process Control*, 2nd ed. Southfield, MI: Daimler Chrysler Corporation, Ford Motor Company, and General Motors Corporation. Automotive Industry Action Group (AIAG).

ASME. (2009). *ASME Y 14.5–2009: Geometric Dimensioning and Tolerancing*. New York: American Society of Mechanical Engineers.

Duncan, W. (1995). *Total Quality Key Terms and Concepts*. New York: AMACOM.

Hirano, H. (1990). *JIT Implementation Manual*. New York: Productivity Press.

Hirano, H. (2009). *The Complete Guide to Just in Time Manufacturing*, 2nd ed. New York: Productivity Press.

Appendix I
Typical Detailed Expectation List of a Site Assessment by the Customer to the Supplier's Facility

Depending on the organization and type of product, the following two categories are of importance and the supplier must be able to document that it has completed the expectation. The two types are (1) planning for manufacturing process capability and (2) demonstration of manufacturing capability. The items that we have identified here are only examples and may be used as applicable. Obviously, more items may be added as needed. For service organizations, this list may be modified to reflect the specific requirements that the customer needs and wants.

Planning for Manufacturing Process Capability

The supplier should demonstrate and have documented records of the following:

- Quality procedures based on *some* standard. Here we use the ISO/TS16949.
  - An active quality operating system (QOS).
  - Supplier's quality system is third-party certified to ISO/TS 16949 requirements.
  - Supplier's environmental system is certified to ISO 14001.
  - Supplier's quality manual contains processes for all areas of the site assessment. The quality manual is updated as needed to drive statistically valid continual improvements in the supplier's quality system. Supplier's senior management team, including representatives from manufacturing, quality, engineering, and human resources, perform these reviews. The reviews have to be documented.
  - Nonconformances identified in internal audits are not repeated.
  - Supplier has a documented process to identify and regularly update customer expectations, objectives, and requirements.

    - Supplier has a documented process to identify and regularly update internal expectations, objectives, and requirements.
    - Supplier has a documented process to translate internal and external expectations, objectives, and requirements into QOS measurables.
    - Supplier has a documented process to assign targets and objectives to each QOS measurable.
    - Supplier has a documented process to assign chosen measurables to appropriate management levels for review, follow-up, and improvements.
    - Supplier utilizes action plans to drive statistically valid continual improvement of chosen metrics. These are tracked at QOS review meetings.
    - Supplier's management team holds monthly QOS performance review meetings with representatives from manufacturing, quality, engineering, and human resources. Improvement actions are taken by cross-functional teams that include all levels of the organization.
    - QOS reviews include a review of all applicable customer's and supplier's key process metrics.
    - Metrics monitoring customer satisfaction that are related to supplier manufacturing are understood and tracked.
- Failure mode and effect analysis (FMEA)/control plans
    - All customer parts have documented design and process failure mode and effects analysis (DFMEA, PFMEA), and design control plans (or DCPs). There is a robust process in place for developing these documents.
    - FMEAs are reviewed annually or as issues arise. A Pareto is used to focus on (1) severity and (2) the product of severity and occurrence, and (3) to identify high RPNs in order to prioritize actions to drive improvements. All error detection areas are reviewed and, where feasible, plans exist to move to error prevention devices (e.g., poke-yoke).
    - Suppliers, in cooperation with their customers, have identified critical characteristics (CCs), significant characteristics (SCs), and high-impact characteristics (HICs) on their PFMEAs and DCPs for every customer part or family of customer parts.
    - Supplier works with customer to establish final agreement of SCs and HICs. Suppliers obtain specific agreement from customer for any parts not having SCs or HICs.
    - Supplier ensures that all print dimensions and callout notes on the engineering drawing are always met through DCPs, work instructions, job setup instructions, and receiving inspection.

  - Where use of pass-through characteristics is specified, supplier verifies pass-through characteristics and ensures that zero defects are sent to the final customer.
  - DCP is a result of appropriate quality planning and includes clear linkage between DFMEA and PFMEA. A typical DCP contains the following elements: (a) operation number/name, (b) part name/part number, (c) inspection requirements, (d) specifications, (e) control methods for SC, CC, HIC (where applicable), and engineering specifications (ES) requirements are identified and appropriate, (f) SC, CC, HIC (where applicable), and ES control methods (in-process checks, process gauging, and part gauging) are appropriate, (g) error-proofing methods, (h) in-process control methods, (i) sample size and frequency, (j) required tools, gauges, and other equipment, and (k) reaction plans must clearly direct the operator on how to act when a process begins to go out of control or a product fails an inspection or test.
- Employee readiness/training review
  - Supplier is responsible for assessing the skills required to perform all activities that affect product quality.
  - Employees have a skills assessment, training plan, and evidence of training.
  - Supplier ensures that only trained and qualified personnel are involved in all aspects of the manufacturing of customer parts, including rework operations.
- Advanced product quality planning (APQP)/launch/production part approval process (PPAP)
  - Supplier has an APQP tracking mechanism in place.
  - Supplier has a system to manage new product and/or program launches.
  - Supplier has a process in place to support customer's prototype and preproduction build events in a timely manner.
  - Supplier has a process to continuously monitor and improve launch performance.
  - All PPAP documentation must contain the information per the PPAP manual published by the Automotive Industry Action Group (AIAG). This information must also be available for customer review. Suppliers have a process in place to comply with special and specific requirements for any PPAP additions to the AIAG requirements.
  - Suppliers require subsuppliers to have a process in place that effectively satisfies all PPAP requirements regardless of subsupplier business relationship with customer.

    - Run-at-rate must be completed and support the stated program requirements (at full-volume production levels) for all customer parts.
    - Capacity verification must include the following elements and be performed on the entire process or the slowest operation (process bottleneck): machine available hours, shifts, days, machine downtime (planned and unplanned), machine changeover, breaks, and quality or scrap.
    - Supplier monitors capacity and has a process to ensure ongoing sufficient capacity on all customer parts.
- Managing the change
    - Supplier has a documented process to effectively implement post-Job 1 design changes, process changes, or volume changes.
    - Supplier effectively reviews and updates its quality supporting documentation for these changes. Examples of documents that may require changes are process flow diagrams, FMEAs, control plans, operator instructions, visual aids, and PPAP documentation.
    - Supplier follows a documented process to ensure that proper customer approvals are gained prior to making process changes. The approval process will include use of supplier request for engineering approvals (SREAs), change request/concern report (CR/CR), management alerts, and revised PPAP documentation, as appropriate.

Demonstration of Manufacturing Capability

Note: For special or unique processes such as electroplating, fasteners, stamping, welding, plastics molding, painting, electronic PCB and final assembly, fastener torque and machining, and heat-treat processes, the customer may have specific requirements and the supplier must be aware of those. In addition, some processes have regulatory and industry standards in addition to what is identified here. Again, it is the responsibility of the supplier to be aware of them and meet all the expectations.

- Subsupplier quality management
    - Supplier has a defined process for managing its supply base including identification of those with high impact to quality.

- Supplier requires subsuppliers to have a QOS that is similar to ISO/TS 16949 or Verbrand der Automobilindustrie (VDA). Supplier verifies subsupplier compliance by annual on-site assessment of high-impact subsupplier facilities.
- Supplier requires PPAP for all parts from subsuppliers (regardless of subsupplier business relationship with customer).
- Where customer has defined an ES or special characteristics that are the responsibility of the subsupplier, the subsupplier must show compliance to the ES or special characteristics in its PPAP submission to the Tier 1 supplier.
- Where pass-through characteristics are required, supplier requires subsuppliers to take specific actions to ensure that all pass-through characteristics are to customer specifications.
- Supplier must confirm subsupplier compliance by review and approval of subsupplier PPAP submission prior to incorporation of product in supplier manufacturing process.
- Supplier conducts systemic reviews with subsuppliers and assures implementation of corrective actions across supplier facilities for all issues resulting in stop ship or field action.

- Control of incoming quality
  - Supplier has a strategy for receiving inspection, for example, inspection, statistical process control (SPC).
  - Incoming quality requirements are part of a control plan. If incoming quality inspections are made, there is evidence of the specification and compliance to the specification. If no incoming inspections are made, there is a rationale for qualifying subsuppliers.
  - Incoming quality operator follows a reaction plan if raw material or purchased part is found to be out of specification.
  - Supplier has appropriate resources to manage subsupplier quality including on-site assessment of high-impact subsupplier facilities.
- Control plans and operator instructions
  - Standardized operator instructions (including control plans, work instructions, job aids, check sheets, job setup instructions, and illustrations) are developed and available for operators.
  - Standardized operator instructions are legible, specific, and controlled. The instructions are available in the area where the work is being performed and are followed by operators as written.
  - Reaction plans are clearly defined and protect the customer from nonconforming materials.

  - Control plans are revised and updated when products or processes differ from those in current production.
- Process variability monitoring/reduction
  - Supplier has determined appropriate statistical techniques. A key aspect of defect prevention is process variability reduction and the use of SPC or an equivalent process monitoring method to indicate when action is necessary and, conversely, when processes should be left alone.
  - Operator-based SPC control charts are optimized for statistically valid sample frequency, sample size, and appropriate chart type. SPC checks are performed on a timely basis and out-of-control conditions prompt corrective action as appropriate.
  - SPC data is maintained, monitored, and used by the equipment operators to drive process and product improvements by reducing process variability. Evidence of process improvements include reductions in scrap, reductions in rework/repair, and defects per unit (DPU); increases in first-time through capabilities, rolled throughput yield (RTY or, more commonly, YRT), C_{pk}, P_{pk}, process Sigma level, records of error-proofing mechanism implementation; and reductions in machine downtime.
  - Initial P_{pk} and C_{pk} levels are greater than or equal to 1.67 and long term are greater than or equal to 1.33, respectively, for all SCs and HICs (where applicable). For all special characteristics such as SCs and CCs, P_{pk} or C_{pk} trends are tracked over time, and action plans have been put in place to increase these C_{pk} and P_{pk} values by a process of continuously reducing the causes of variability. These trends must be part of the supplier's management routine review such as the monthly QOS meeting. P_{pk} is the preferred method of measuring capability because it uses actual data from processes.
  - CCs require an appropriate control method that prevents shipment of nonconforming product to customer.
  - Machine capability for SCs, CCs, and HICs (where applicable) must be demonstrated on all new equipment and tooling.
- Measurement system capability, calibration, and use
  - Supplier ensures that all gauges are available for use as identified in the control plan to support customer requirements. The gauge calibration and gauge maintenance program ensures that backup gauges or a backup gauging process is available to support inspections required by the control plan.
  - Variable gauges are used for all SCs and CCs, whenever possible. HIC gauges are per the control plan. All gauges (company- and employee-owned) are identified per the control plan.

- Gauge repeatability and reproducibility (R&R) is to conform to the guidelines in the AIAG measurement systems analysis (MSA) manual. Whenever gauge R&R does not meet these guidelines, specific plans are in place that adhere to the guideline. Additionally, the supplier has the appropriate measurement capability for all stages of production. The preferred method for MSA is the ANOVA method with *five or more* number of distinct categories (ndc).
- Supplier has a documented dropped/damaged gauge policy that is followed by all employees. This ensures that only functional, in-specification gauges are used.
- Gauge masters are traceable to a national or international equivalent standard.
- Gauge calibration and maintenance is performed on an appropriate schedule and per specification as well as the individuals who actually do the measuring.
- Supplier conducts regular evaluation of error-proofing devices.

- Control of parts, part identification/packaging/shipping
  - Supplier has a documented process for positive part identification and part control in all stages of production, rework, repair, scrap, testing, laboratories, storage areas, office areas, and so on.
  - Supplier has lot traceability of product shipped to customer, including reworked product.
  - Supplier takes measures to ensure that production processes control handling, storage, and packaging to prevent damage or deterioration and to preserve product quality.
  - Supplier follows customer packaging requirements. At pack-out, the customer packaging guidelines are available for use and followed as required.
- Testing/engineering specifications
  - All inspections, measurements, and tests, including ES and material specifications (MS), are performed according to documented control plans, instructions, and/or procedures to ensure product requirements are met.
  - Appropriate reaction plans address what actions should take place if there is a test or inspection failure, and includes product containment and customer notification policy for all events.
  - Suppliers and subsuppliers providing special processes such as heat-treated components are required to meet the specific customer requirements to reduce the risk of embrittlement.

- The Tier 1 supplier is responsible for ensuring that all tiers of special processes, such as heat-treat suppliers, are annually assessed to the applicable standards and requirements for heat-treat and/or any other special process.

- Preventive maintenance/housekeeping
  - Supplier has instituted practices that include reactive, preventive, and predictive maintenance. The maintenance system supports process capability improvement.
  - Management regularly reviews (at least quarterly) the status of the preventive maintenance (PM) completion to plan.
  - Management reviews corrective action plans to ensure any backlog (past due) maintenance has a plan to become current to the PM schedule.
  - Management undertakes an evaluation to determine if there is a reduction of machine/process downtime.
  - The supplier has continual improvement plans for plant cleanliness, housekeeping, ergonomics, and working conditions, and audits its facility at least monthly. There is evidence that the supplier management team is following continual improvement plans.
  - Supplier takes measures to minimize foreign material, chips, debris, contamination, excessive oil, and so on where part quality can be negatively affected. Active 6S and visual factory policies are followed and documented.
- Manufacturing flow/process engineering/Six Sigma and Lean manufacturing metrics
  - Supplier uses Lean manufacturing and Six Sigma principles for current and new model parts incorporating just-in-time (JIT) production, waste elimination, and team-based structured problem solving to drive continuous improvement for quality, process capability, and manufacturing efficiency.
  - Supplier includes key manufacturing and Six Sigma-type measurables within its QOS. A minimum of two Lean and one Six Sigma-type measurables show trends or history of improvement in the past six months. Lean and Six Sigma measurables include some of the following: first time through (FTT), overall equipment effectiveness (OEE), dock to dock (DTD), percent value add (PVA), and C_{pk} or P_{pk} (P_{pk} is preferred because it actually uses real data as opposed to C_{pk}, which uses approximation).
- Problem solving/corrective actions/preventive actions
  - Supplier uses a structured problem-solving method (e.g., 8D) to address customer concerns and complaints.

- Methods are established to communicate quality concerns to the supplier's production and support personnel.
- Corrective actions are submitted to the supplier's management team for review. The corrective actions are communicated to and replicated in all affected areas in the supplier's organization.
- Supplier has a process in place to address customer plant concerns in a timely and thorough manner.
- Supplier reviews previous launches and incorporates lessons learned in future launch plans to achieve flawless launch.
- Supplier has a structure within its problem-solving approach to separate corrective action and preventive action.
- Supplier has a system to implement preventive methods in the organization.

Appendix II
PPAP Checklist

This appendix provides the reader with a quick check review of a production part approval process (PPAP). The anticipated answers are Yes, No, and Not Applicable or Do Not Know. If the answer is Yes, then there is a good chance that the section of the PPAP is okay. Of course, this is *only* a guide and it is the responsibility of the supplier to make sure that the Automotive Industry Action Group (AIAG) as well as the customer's requirements are followed. Typical concerns that should be addressed (at a minimum) in any PPAP review should be in the following areas:

1. Design record
   a. Released customer drawing in appropriate design system
   b. Ballooned prints (including all dimensions, notes, and engineering specifications [ES] and material specifications [MS])
2. Authorized engineering change documents
   a. Approved supplier request for engineering approval (SREA), if applicable
   b. Approved alert, if applicable
   c. Electronic notification, if applicable
3. Customer engineering approval
   a. When specified on drawing
   b. Part functional trial approval when required
4. Design failure mode and effects analysis (DFMEA)
   a. Customer engineering approved DFMEA provided by customer engineering if customer responsible or from supplier, if supplier is design responsible
   b. DFMEA must align to latest revision of AIAG and customer requirements
5. Process flow diagram
   a. Confirm document linked to process failure mode and effects analysis (PFMEA) and control plan, including part number/revision level or process/part name
   b. Confirm it flows from receiving to shipping encompassing all production process steps (including all production and service steps as applicable)

6. PFMEA
   a. Confirm PFMEA is linked to DFMEA, process flow diagram, and control plan
   b. If ▼ critical characteristics (CC) parts, then customer engineer and supplier engineer must approve
   c. Aligns with AIAG and customer requirements (CC if severity is 9 or 10, or SC if severity is 5–8 and occurrence >3)
   d. Aligns with special characteristics requirements of the customer regardless of the location of the special control in the supply chain (Tier 1–Tier N)
7. Control plans
   a. If ▼ CC parts, then customer engineer and supplier engineer must approve.
   b. Controls for all special part/product (dimensional and performance) and process characteristics are to be documented in the control plan.
   c. Control plan must identify appropriate reaction plan.
8. Measurement system analysis studies
   a. All gauges, both production and service, must have gauge repeatability and reproducibility (R&R) study in accordance with customer requirements and AIAG MSA handbook
   b. ANOVA method preferred, variable study: 10 part, 3 operators, 3 trials; attribute study: 50 parts, 3 operators, 3 trials using customer parts
   c. Reported in % Study Variation and % Tolerance with ≥5 distinct categories
   d. Variable gauge: Gauge repeatability and reproducibility (GR&R) acceptable <10%, >30% is unacceptable. Greater than or equal to 10% but less than or equal to 30% requires supplier engineer approval
   e. Attribute study, kappa value ≥.75
9. Dimensional results
   a. Specified measurement results are recorded on APQP/PPAP evidence workbook, GD&T sheet, checked (ballooned) print, or another method of recording
   b. One part from each cavity/mold/tool/line is measured with a minimum of five parts
10. Material/performance test results
    a. Test results for test specified on the print/engineering specification
    b. Number, date, and change level of specification to which part was tested

   c. Date on which testing took place, quantity tested, actual results, and material supplier's name
11. Initial process studies
   a. Capability studies on all SC, HIC must be documented. Control charts included with capability studies, PPAP requirement of $P_{pk} \geq 1.67$. CCs require controls that *prevent* shipment of nonconforming product.
   b. Recommended sampling of 125 parts with 25 subgroups including full range of expected variation in manufacturing process.
   c. Data available from stable, in-control, and normally distributed process to compile six pack capability chart from Minitab® (p-value included; normal, $p > 0.05$); non-normal distributions require appropriate methods to determine process capability.
12. Qualified laboratory
   a. The internal or external lab will comply with ISO/IEC 17025; however, accreditation is not mandatory.
   b. A laboratory scope is documented.
13. Appearance approval report (AAR), if required
   a. Shall be signed and completed for items requiring AAR
14. Sample production parts
   a. Must be taken from run-at-rate of 300 pieces minimum production run (unless otherwise specified by customer-authorized personnel)
15. Master sample (retained at the supplier plant, properly tagged)
   a. Must be taken from run-at-rate or quality evaluation (all stream) production run and properly stored and tagged at supplier
   b. A master sample must be kept for each position of a multiple cavity die, mold, pattern, or process unless otherwise specified by customer
16. Checking aids (descriptions, drawings, control numbers)
   a. Supplier to certify all aspects of checking aid to agree with dimensional requirements
   b. Measurement system analysis (MSA) to be conducted in accordance with customer requirements and preventive maintenance program
   c. Checking aids can include fixtures, variable and attribute gauges, models, templates, Mylars, and so on
17. Customer specific
   a. Capacity verification study by Tier 1 including subsuppliers

b. All processes must be included in the capacity analysis report (CAR) and any shared processes must have a shared analysis tab completed; this CAR is not to be confused with corrective action report, which also is designated as CAR
c. Capacity calculated at the end of the PPAP process must be entered in the appropriate system per customer request
d. Evidence of part approval/PPAP from subsuppliers for all sub-tier supplier components, as applicable (including production and service)
e. Pass-through characteristics list/matrix, if applicable
f. Supplement K or equivalent must be completed
g. APQP evidence workbook
h. Branding compliance must be validated for both production and service
i. Customer tooling tagged properly per terms and conditions with photo evidence
j. Approved all customer requirements with transmittal date including all evidence showing customer acceptance

18. PPAP submission warrant status (must use customer-specific format)
    a. All fields have been completed (no blank fields). N/A has been inserted if any field is truly nonapplicable.
    b. Confirm the part numbers exactly match the print included in the design record and the PO or note from buyer. Include cross-referenced service part number if applicable.
    c. Include the mold/cavity/processes.
    d. "Customer" is the receiving plant name, not "name of customer." "Application" is the model year and make.
    e. Include PO#. PO# must not be blank and cannot be 111111 or part number.
    f. Include Notice # (e.g., NT00…) under "Engineering Drawing Change Level."
    g. Customer engineering must sign timing priority launch parts.
    h. Capacity section must align with CAR and any other applicable system.
    i. Weight entered to four decimal places from 10-part average measured in kilograms.
    j. Polymeric parts identified with appropriate ISO marking codes.
    k. Part submission warrant (PSW) status must be updated in the customer's electronic system by the supplier, as applicable.

Appendix III
Mistake Proofing

Typical Mistake Proofing Development Process

Step 1: Defect

Describe the defect

Show the defect rate

Step 2: Location

Identify the location at which the defect is discovered or made

Step 3: Operations

Detail the current standard procedure/elements of the operation where the defect is made

Make sure that one element is identified at a time

Step 4: Mistakes/deviations

Mistakes/deviations from standards where defects were made

Step 5: Analysis

Analyze the cause for each mistake/deviation

Identify the red flag conditions

Step 6: Improvement ideas

Identify the ideas to eliminate or detect the mistake

Check mistake proofing device required

Level 1: Eliminate cause of mistake at the source

Level 2: Detect mistake as it is being made

Level 3: Detect defect before it reaches the next operation

Step 7: Implement devices

Create a mistake-proofing device

Mistake-proofing improvement must consider the current and future benefits from a

Cost to install and any special comments, perspective

Time to install and date, perspective

Generic Mistake-Proofing Worksheet and an Example

A Generic Mistake-Proofing Identification Worksheet	**Mistake-Proofing Identification Worksheet (Sample)**
Plant: Plant Name Area: Department Date: Month/Day/Year	Plant: XYZ Engine plant Area: Camshaft machining department Date: 10/15/2014
I. Mistake-Proofing System: A brief description of what is used to mistake proof the process. This may include a device, an operational practice, or a fixture.	I. Mistake-proofing system: Proximity switch on jaws of gantry robot.
II. Type Is the mistake-proofing approach prevention or inspection/correction (refer to Key Card)?	II. Type: Inspection/correction.
III. Location Where is the mistake-proofing system located in the plant?	III. Location: Camshaft machining—CAM unloading station
IV. Description A brief description of the process being mistake proofed and how the mistake-proofing system prevents mistakes.	IV. Description: Camshafts entering the machining process are unloaded by a gantry robot. The coordinates that the robot uses to locate the CAM are fixed. If a camshaft is not seated in the crate correctly, or if the pallet is not in the right place, the robot can secure the CAM incorrectly and damage the CAM or deliver it to the conveyor incorrectly.
V. Purpose What mistake is being addressed and what is the significance of this mistake?	V. Purpose: The proximity switch releases the gripper jaws before the CAM is damaged. The mistake-proofing device protects the CAM if the CAM is seated incorrectly in the pallet, or if the pallet is located incorrectly under the robot.
VI. Condition Is the mistake-proofing system operational, and how is it maintained?	VI. Condition: The proximity switch is in working condition. The robot's most recent calibration and scheduled PM were completed as indicated in the maintenance log.
VII. Preventive Maintenance Schedule What is the PM schedule for the mistake-proofing system? The condition of the error-proofing device and PM schedule can be gathered from the plant floor information system (PFIS) or something similar to that. Maintenance personnel should also keep a maintenance log in the area.	VII. Preventive maintenance: Weekly calibration and monthly maintenance.
VIII. How the team recognized the mistake-proofing system What indicator allowed the team to find the mistake-proofing system?	VIII. How the team recognized the mistake-proofing system: Knew the CAM seating and pallet location were critical. Checked to see if there was a mistake-proofing device.

Mistake proofing helps us achieve zero waste and zero defects. In other words, mistake proofing is a process improvement that is designed to prevent a specific defect from occurring. There are three sides of mistake proofing: (a) Physical: install hardware, (b) operational: enforce procedure, sequence, and/or execution, and (c) philosophical: empowerment of the workforce, and team concept. No matter which of the three is followed, there are two approaches for improvement: (1) prevention: prevents mistakes from creating defects and (2) inspection/correction: detects a defect and immediately initiates a corrective action to prevent multiple defects from forming.

A defect is the result of any deviation from product specifications that may lead to customer dissatisfaction. To classify as a defect, two conditions are necessary: (1) The product has deviated from manufacturing or design specifications, and (2) the product does not meet internal and/or external customer expectations.

Defects and Errors

Mistakes are generally the *cause* of defects. So one may ask, "Can *mistakes* be avoided?" To answer this question requires us to realize that we have to look at errors from two perspectives:

1. Errors are inevitable: People will always make mistakes. Accepting this premise makes one question the rationale of blaming people when mistakes are made. Maintaining this *blame* attitude generally results in defects. The discovery of defects is postponed until the final inspection or, worse yet, until it reaches the customer.
2. Errors can be eliminated: If we utilize a system that supports (a) proper training and education and (b) fostering the belief that mistakes can be prevented, then people will make fewer mistakes. This being true, it is then possible that mistakes by people can be eliminated. Sources of mistakes may be any one of the five basic elements of a process: (a) measurement, (b) material, (c) method, (d) manpower, (e) machinery, and (f) environment.

   Each of these elements may have an effect on quality as well as productivity. To make quality improvements, each element must be investigated for potential mistakes of operation. To reduce defects, we must recognize that defects are a consequence of the interaction of all five elements and the actual work performed in the process. Furthermore, we must recognize that the role of inspection is to *audit*

TABLE III.1

Examples of Mistakes and Defects

Mistake	Resulting Defects
1. Failure to put gasoline in the snow blower.	1. Snow blower will not start.
2. Failure to close window of unit being tested.	2. Seats and carpet are wet.
3. Failure to reset clock for daylight savings time.	3. Late for work.
4. Failure to show operator how to properly assemble components.	4. Defective or wrap product.
5. Proper weld schedule not maintained on welding equipment.	5. Bad welds, rejectable and/or scrap material.
6. Low charged battery placed in griptow.	6. Griptow will not pull racks, resulting in lost production and downtime.

the process and to identify the defects. It is an appraisal system and it does nothing for prevention. Product quality is changed *only* by improving the *quality* of the process.

The first step toward elimination of defects is to understand the difference between defects and errors. That is, defects are the results, and mistakes are the causes of the results. Therefore, the underlying philosophy behind the total elimination of defects begins with distinguishing between mistakes and defects. Typical examples are shown in Table III.1.

It is obvious that mistake-proofing systems is a process that focuses on producing zero defects. That goal is accomplished via a reactive approach (defect detection) or a proactive approach (defect prevention). The *reactive approach* relies on halting production in order to sort out the good from the bad for repair or scrap—see Figure III.1. On the other hand, the *proactive approach* seeks to eliminate mistakes so that no defective product is produced, production downtime is reduced, costs are lowered, and customer satisfaction is increased (see Figure III.2).

Mistake Types and Accompanying Causes

Obviously, there are many types of mistakes and certainly many more causes. Here we provide a list of 10 types with their accompanying potential causes to demonstrate the variability of both:

1. Assembly mistakes (potential causes)
    a. Inadequate training

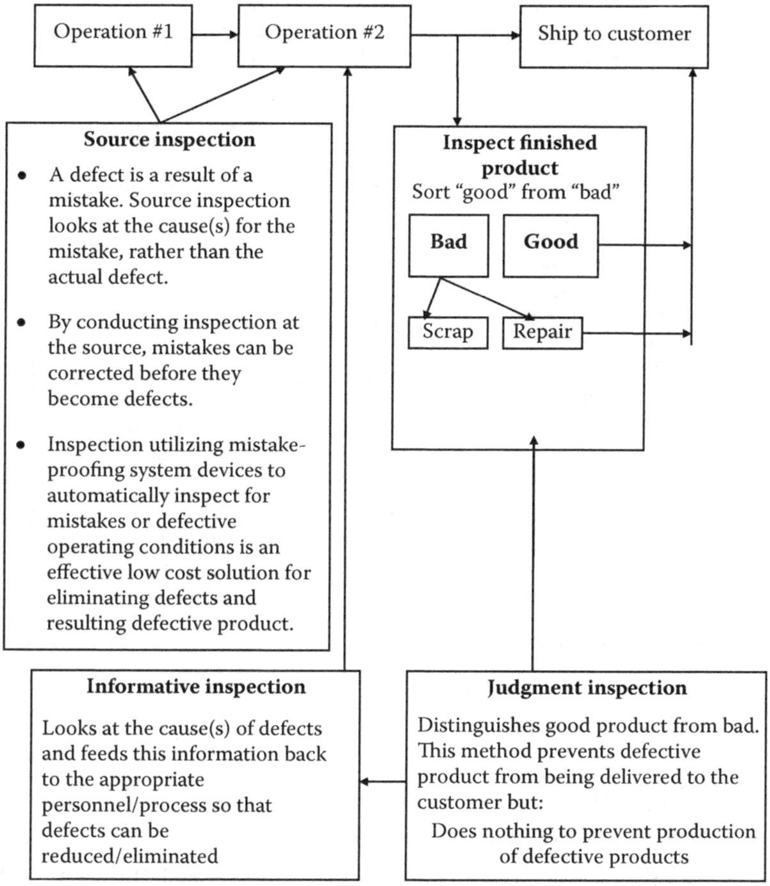

FIGURE III.1
Reactive approach.

b. Symmetry (parts mounted backward)
c. Too many operations to perform
d. Multiple parts to select from with poor or no identification
e. Misread or unfamiliar with parts/products
f. Tooling broken and/or misaligned
g. New operator

2. Processing mistakes (potential causes)
   a. Part of process omitted (inadvertent/deliberate)
   b. Fixture inadequate (resulting in parts being set into incorrectly)
   c. Symmetrical parts (wrong part can be installed)
   d. Irregular shaped/sized part (vendor/supplier defect)

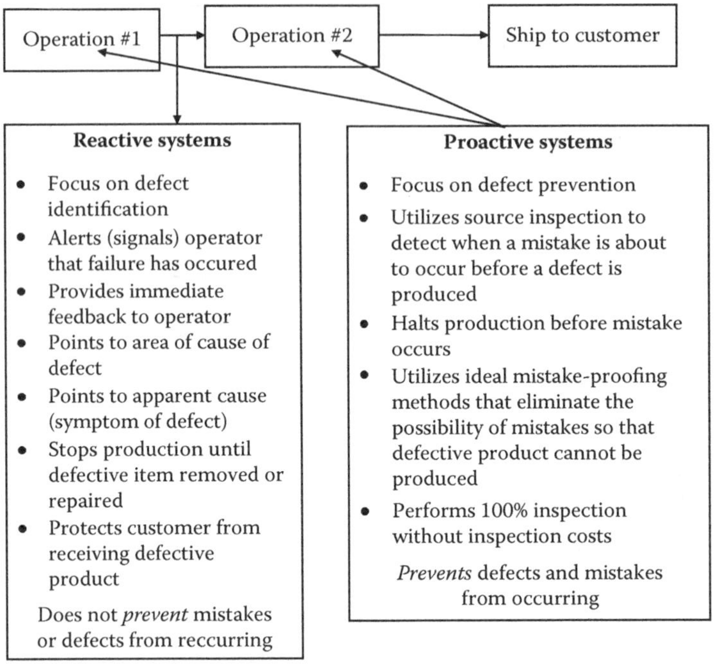

FIGURE III.2
Proactive approach.

   e. Tooling damaging part as it is installed
   f. Carelessness (wrong part or side installed)
   g. Process/product requirements not understood (holes punched in wrong location)
   h. Following instructions for wrong process (multiple parts)
   i. Using incorrect tooling to complete operations (impact vs. torque wrench)
3. Inclusion of wrong part or item (potential causes)
   a. Part codes wrong/missing
   b. Parts for different products/applications mixing together
   c. Similar parts confused
   d. Misreading prints/schedules/bar codes, and so on
4. Operations mistakes (potential causes)
   a. Process elements assigned to too many operators
   b. Operator error: Even though it is used, it is not an acceptable error. This type of error is a definite design issue
   c. Consequential results

5. Setup mistakes (potential causes)
   a. Improper alignment of equipment
   b. Process or instructions for setup not understood or out of date
   c. Jigs and fixtures mislocated or loose
   d. Fixtures or holding devices will accept mislocated components
6. Assembly omissions: missing parts (potential causes)
   a. Special orders (high or low volume parts missing)
   b. No inspection capability (hidden parts omitted)
   c. Substitutions (unexpected deviations from normal production)
   d. Misidentified build parameters (heavy duty vs. standard)
7. Measurement or dimensional mistakes (potential causes)
   a. Flawed measuring device
   b. Operator skill in measuring; lack of appropriate and applicable training
   c. Inadequate system for measuring
   d. Using best guess system
8. Processing omissions (potential causes)
   a. Operator fatigue (part assembled incorrectly/omitted): see Item 4b
   b. Cycle time (incomplete/poor weld)
   c. Equipment breakdown (weld omitted)
   d. New operator
   e. Tooling omitted
   f. Automation malfunction
   g. Instructions for operation incomplete/missing
   h. Job not setup for changeover
   i. Operator not trained/improper training
   j. Sequence violation
9. Mounting mistakes (potential causes)
   a. Symmetry (parts can be installed backward)
   b. Tooling wrong/inadequate
   c. Operator dependency (parts installed upside down); see Item 4b
   d. Fixtures or holding devices accept mispositioned parts
10. Miscellaneous mistakes (potential causes)
   a. Inadequate standards
   b. Material misidentified
   c. No controls on operation

d. Counting system flawed/operating incorrectly
e. Print/specifications incorrect

Signals That Alert Mistakes

Signals that *alert* are conditions present in a process that *commonly* result in mistakes. Typical signals that *alert* are

- Many parts/mixed parts
- Multiple steps needed to perform operation
- Adjustments
- Tooling changes
- Critical conditions
- Lack of or ineffective standards
- Infrequent production
- Extremely high volume
- Part symmetry
- Asymmetry
- Rapid repetition
- Environmental
    - Housekeeping
    - Material handing
    - Poor lighting
    - Foreign matter and debris
    - Other

These alerts are not to be confused with the alerts that we use to stop the process because of problems. They are indicators that something may go wrong.

Common Error-Proofing Devices

Perhaps the most common error-proofing device is the sensor. A sensor is an electrical device that detects and responds to changes in a given characteristic of a part, assembly, and/or fixture. For example, a sensor can verify with a high degree of accuracy the presence and position of a part on an assembly or fixture and can identify damage or wear. Some examples of types of sensors and typical uses are the following:

Welding position indicators: Determine changes in metallic composition, even on joints that are invisible to the surface.

Fiber sensors: Observe linear interruptions utilizing fiber optic beams.

Metal passage detectors: Determine if parts have a metal content or mixed metal content, for example, in resin materials.

Beam sensors: Observe linear interruptions using electronic beams.

Trimetrons: Exclude or detect pre-set measurement values using a dial gauge. Value limits can be set on plus or minus sides, as well as on nominal values.

Tap sensors: Identify incomplete or missing tap screw machining.

Color marking sensors: Identify differences in color or colored marking.

Area sensors: Determine random interruptions over a fixed area.

Double feed sensors: Identify when two products are fed at the same time.

Positioning sensors: Determine correct/incorrect positioning.

Vibration sensors: Identify product passage, weld position, broken wires, loose parts, and so on.

Displacement sensors: Identify thickness, height, warpage, surface irregularities, and so on.

Appendix IV
Preparation of an Audit

As we discussed in Chapter 14, audits are very important in any organization. They can be based on international standards, customer requirements, and/or for internal improvement. In this appendix, we have identified several areas of concern and split them into three categories. The first one is the required items, the second is the criteria of the requirements, and the third is the verification possibilities. Obviously, this is not an exhaustive list and certainly not a universal characterization, but it can serve as a model for any audit in any organization. In any case, the items identified here are a demonstration of what an organization must be able to accomplish with an audit. It is a road path for preparing for excellence.

Required	Criteria of Requirements	Verification Possibilities
Fast Response		
Daily leadership meeting held with cross-functional, multilevel attendees to address significant external and internal concerns.	1. There is a daily fast response (FR) meeting with cross-functional attendees and led by manufacturing. 2. The FR meeting is a communication meeting and a precise timing is respected. 3. All the significant external and internal issues are addressed. 4. Natural owners are assigned to problems; next report out date is assigned. 5. Take into account the production forecast and the quantity produced. 6. Safety and near miss accident issues are reported out.	Attend FR meeting. Observe: Lead by manufacturing with cross-functional attendees How leader controls the FR meeting (keep timing max 10–20 min, focus on subject, not going to the details) Environment is suitable (everyone can hear and see the meeting) How issues are reported out Problem-solving report format is used for report out and document the status of the issue
FR board tracks all major concerns with appropriate timing and exit criteria.	1. FR board is used for tracking of issues, being updated before FR meeting (board could be a dry erase board, laminated poster, Excel sheet projected by Dell projector or something similar, etc.).	Prior to the audit, check last customer complaints and focus on open ones.

(*Continued*)

Required	Criteria of Requirements	Verification Possibilities
	2. Method of communicating problems to all key stakeholders is defined (e.g., quality alert, temporary work instructions, bypass procedure, etc.) 3. Exit criteria represent the core 6 steps of problem solving (1. Define 2. Contain 3. Root cause 4. Correct 5. Validate 6. Institutionalize) including timing for each of the exit criteria. 4. All the exit criteria are identified (red, yellow, green). Red items have a planned date to go green with next steps. Problems are not closed until all criteria are met. Overall status represents the worst condition or overall planned timing has been exceeded.	Prior to the FR meeting, ask if there are any significant internal issues. Check the board if it contains previously described external and internal issues. Follow an issue from FR tracking board through the exit criteria confirming actions are in place and all documentation has been updated. Check a few statuses if they are rated well based on their timing; judge a few N/A items on whether they do not need to close the issue.
System is in place to ensure fast response to operator's concerns and information flow between shifts.	1. System in place to respond to the concerns of the operator. 2. Escalation process is defined in order to ensure that problems are quickly communicated to people who can help. 3. If there is more than one shift, all the information that can affect the next shifts has to be passed across and documented. 4. Manufacturing leadership reviews shift book daily to verify proper containment or corrective action done, decides further escalation to FR meeting.	Ask operators how they can escalate their issues. Andon system or similar in place, if applicable. Test Andon's function and response to request (light boards, lamps, or audio signals work, support arrives soon). Participate at shift change, check shift book and information shared.
A defined process for problem solving including a standard for documenting the tools used for root cause identification.	1. Standard process (practical problem-solving report [PPSR], 8D, or equivalent) used with a format that follows the core 6 steps of problem solving (1. Define, 2. Contain, 3. Root cause, 4. Correct, 5. Validate, 6. Institutionalize). 2. Analysis process is developed and applied for all customer returned parts. 3. Problem-solving forms used across the plant for internal, customer, and supplier issues.	Prior to the audit, check last customer complaints focus issues where root cause was found and corrective action was implemented. Verify that problem solving was used efficiently, all core 6 steps were applied, especially the real main root cause was found and action was implemented against the root cause. If no customer complaint was issued, verify via an internal or subsupplier issue.

Required	Criteria of Requirements	Verification Possibilities
	4. Evidence of cross-functional team approach and team members have the necessary trainings about problem-solving methods applied. 5. Tools for identifying root cause are applied (GM's 7 diamonds, Ford's G8D, 3 × 5 Why to correct systemic issues, fishbone diagram, etc.).	Check a drill deep (5 Whys), main systematic root causes found.
A system to capture and institutionalize lessons learned.	1. Electronic form or database is used to document lessons learned. 2. Method that assures implementation supported by evidence of review dates, distribution lists, or posted lessons learned as well as reviewed by leadership. 3. Institutionalization is applied for all the issues where problem solving was done. 4. Completed lessons learned information, which is easily retrievable by all who need the information (e.g., master FMEA, APQP program checklist reviews). 5. A process to collect suggestions of personnel is put in place and contributed to lessons learned and process improvement.	Ask people for examples how they are using lessons learned system. Check sixth step of problem solving (Institutionalize) via examples of point FR4. Check drill wide or D7 (eighth step) of 8D of last customer complaints.
Targets are defined and followed to ensure effectiveness of fast reaction for external and internal issues.	1. Tracking of safety-related metrics (e.g., accident without lost working days, distinguishing lost time accident and no lost time accident, etc.). Lead time to solve safety issues. 2. Quality Q or other method to track FR performance is posted and up to date. 3. Tracking of total amount of significant internal and customer complaints based on their trend and statuses (not responded, open, or in delay and closed in time) and average closing timing. 4. Tracking of the repetitive issues occurred after corrective action was implemented.	Prior to audit, check number of last 12 month's PRRs. If higher than 24, do not count line accumulation ones, or if trend is significantly negative, a special PRR reduction team has to be established. Check last customer complaints if due dates were kept. If not reasons for delay, actions need to be addressed. Check red items percentage; evaluate actions addressed to eliminate roadblocks. Evidence of periodical review of average closing time for each exit criteria and set action plan for any deviation.

(*Continued*)

Required	Criteria of Requirements	Verification Possibilities
	5. Continuous improvement (CI) data for line stoppage/downtime from Andon. 6. Location with a high level of customer complaints (>24 complaints in last 12 months) shall have a special team created to work on complaints reductions; performance shall be tracked at FR meeting on open issues list.	
Control of Nonconforming Product		
Traceability rules are applied according to the class of traceability of the finished product and FIFO is kept.	1. Development of processes integrates traceability requirement (including tier parts) agreed with customer and applied at each stage of the manufacturing process. 2. Quality status of the product (physical or data processing) and last operation performed are identified in entire process. 3. Equipment and stock areas are designed to facilitate FIFO and allow easy identification of noncompliance with FIFO. 4. When FIFO cannot be strictly respected (derivative flows), rules are defined to minimize the disturbances (e.g., stock rotations). 5. Quantitative approach is taken into account to determine theoretical suspected batch size. Precision Calculations (dilution calculations) are carried out and rules/procedures are defined to comply with threshold of dilution.	Check parts in process and suspected material that last operation performed is identified. When traceability is required, check it via a part. If you have unitary traceability on a parameter or a component, choose a finished product and ask for the component/process value associated with the product. Assure that layered process audit (LPA) check sheet is assuring traceability application. Precision calculation: Check how many finished products must be verified if an incoming part, a process parameter, or a rework process failed.
Nonconforming and suspect material are identified and segregated in order to prevent them from unintended use.	1. Consistent tagging and visual management (footprint, labels, etc.) procedure is defined and kept in entire organization to ensure that identification and handling of not OK or suspected material is in place to avoid mixing with normal parts.	Check in shop floor that identification tagging system is used in all areas including documentation (check incoming, working stations, control station, rework) and including visual management like footprint, color coding, labels, etc.

Required	Criteria of Requirements	Verification Possibilities
	2. If random checks (e.g., sample check, product audit) find not OK (NOK) part, it is ensured that all the parts produced from the last known good parts are handled as suspected. 3. If red tags are used for both scrap and suspect material, tag must have disposition. 4. NOK parts are segregated and recorded and their storage is managed including a quarantine area, which has authorized access and quantities in quarantine are controlled.	Ask operators that everyone in shop floor understands color coding used inside the organization. Scrap boxes size should match with part size. Verify quarantine, access is defined, quantity is controlled. Verify avoidance of mixing NOK part, ensure via layout of workplace, handling, and storing of NC parts.
Alert is issued for significant issues to ensure communication and action for stakeholders.	1. Alert process is defined based on the severity level of nonconformity and applied for both external and internal issues once problem occurred at least as per FR issues. 2. Alert includes minimum problem definition, the standard and deviation from standard, tasks, and timeline. 3. Escalation process is put in place with internal/external distribution and notification (contact list for customer and tier suppliers' exits). 4. The standard defines the criteria of withdrawal of posting.	Review FR board and check open issues has quality alert posted. Check quality alert is clear, understood by team, and posted in stations needed (operation, quality). Customer and tier supplier contact lists are available.
Containment is segregated and standardized; all the potential quantity and location of suspected material are identified.	1. Containment worksheet or equivalent contains all the potential locations and quantities including physical quantities and status (OK/NOK). 2. Containment area is separated from production line; standardized work (SW) is applied. 3. Containment process systematically includes securing of stock and pipeline and application of immediate countermeasures to guarantee conformity of production and deliveries; break point identified to customer.	Verify that containment worksheet is used before initiating any containment. Check containment worksheet contains all the potential locations. Check that countermeasures are put in place for each alert.

(Continued)

Required	Criteria of Requirements	Verification Possibilities
	4. Containment actions are defined for each customer issue and verified in order to prevent further defects; breakpoints are documented and communicated to all involved parties. 5. Exit criteria are defined to stop countermeasures.	Ask team members about containment and its rules. Ask one team member who performed containment in any FR issue, how he or she did the containment, and how the containment worksheet was filled out.
Rework or repair is standardized, performed only with necessary authorization and process of reintroducing parts back to line.	1. Prohibited or authorized rework operation clearly defined. Authorized rework operations have to be part of part/process approval. Parts waiting for rework have to be handled as suspected parts. 2. Failure modes of rework are considered in PFMEA, identified in process flow and control plan (CP) with its reintroduction at or prior to removal point. 3. Reintroduction of reworked part includes all downstream checks and color-coding to ensure that all CP inspections and tests are performed. 4. If a rework needs to be applied, which is not part of the approved process, authorization (if needed, includes customer) and releasing responsibility is defined. 5. Reuse of components is considered a rework operation. The component reused must be traced on the finished product. 6. Each reworked part is traceable via an identifier, documented serial number, etc.	Rework is part of process flow. Potential failure modes of rework are detailed in PFMEA. Check rework station; SW is applied. Check if team members understand rework identification process and follow one rework part, how it is handled and identified, and how it is reintroduced back to line.
Parts that have deviations but approved by customer are managed; traceability of concerned products is guaranteed.	1. All the parts that have deviations are handled as suspect parts. 2. Deviation is approved according to customer requirement; deviations are granted for a limited period or a defined quantity of parts. 3. Traceability of the parts delivered under deviation is guaranteed. Manufacturing batches are identified.	Site history of requests for deviations. Procedures and forms used on site. Check PPAP or appropriate customer-specific requirement. Check request to deliver nonconform product for PSA and initial samples are available for deviations.

Required	Criteria of Requirements	Verification Possibilities
	4. Impact of deviation on process and product is analyzed and based on its result, all responsible parties have to be included on communication and decision about usability of deviated parts.	Multiskill approach for the decision for initiate deviation request. Evidence of traceability.
Targets are defined and followed to ensure effectiveness of fast reaction for external and internal issues.	1. Dilution calculation: affected batch size vs. containment worksheet in case of internal or customer complaint. 2. Performance metrics such as scrap rate (internal ppm), rework percentage, first-time quality (FTQ), voice of the customer (VOC), etc. are established and followed at all levels of the operation. 3. Tracking of customer complaints caused by rework, best practice violation, known NOK parts shipped. 4. Costs of poor quality (including indirect costs: sorts, etc.). 5. Tracking of number of containment at customer and in-house. 6. Tracking time in containment (long lasting).	Prior to audit, check controlled shippings and customer complaints link to control on nonconforming product. Verify that containment is driven back to the source of error. Verification of countermeasure action.
Mistake-Proofing Verification		
A system in place focuses on building quality in station through prevention, detection, and containment of abnormalities. (The order here is very important.)	1. Verification station (VS) strategy is defined on procedure/ instruction level, contains selection of VS place, guideline for alarm limit, and definition of escalation route. 2. Characteristics have to be checked that 100% are defined and VS needs to be developed according to SW requirements. 3. Customer complaints have to be covered in VS. 4. Permanent VS is implemented based on customer requirement. 5. Exit criteria is defined to remove nonpermanent VS and approved by quality.	Ask for setup of a VS for a theoretical problem and check that conditions are defined to establish VS in a short time. Check that VS is clearly identified, developed according to SW: instruction developed, layout defined to avoid bypass and mixing of parts, training and necessary certification done. Check for customer requirements compliance. For presupplier agreement, suppliers final inspection implemented as VS. If needed, VS capacity confirmation via "limited" R&R.

(Continued)

Required	Criteria of Requirements	Verification Possibilities
	6. Impact of VS on capacity of the line is verified.	
Alarm system is defined and performance tracked for defects entering VS.	1. Results of controls are recorded and posted at or near VS. A real-time follow up of results is applied. 2. Escalation procedure is defined and followed when alarm limit is reached. 3. Issues are documented and countermeasures are applied immediately. 4. The decision criteria to stop production are established.	Tally sheet posted at or near the VS, filled in properly. Ask VS operator about escalation; when and who to call, who responded, and when. Check results back, proper escalation was done when alarm limit was reached. Alarm limits are reasonable: e.g., for customer complaint, specific productivity tool, i.e., work flow, containment system.
Problem-solving process is applied, communication is in place for defects leaving the station.	1. A problem resolution management is quickly applied directly at the workstation according to escalation route, following problem-solving process. 2. When alarm limit is reached, an upstream reaction process is defined. 3. Leadership reviews VS results and follow up of action plan, if needed, institutionalize/read across its results. 4. In case of lack of detection of nonconformance part, VS controls have to be re-evaluated and improved.	Check problem solving was applied; corrective actions were defined against main root cause (not retraining of operators). VS is owned by management.
The error-proofing (EP) devices are identified, managed, and regularly verified.	1. A master list of EP devices is available including description of verification (self or with master samples). 2. Each EP device/system and its master samples are clearly identified and managed. 3. EP devices are verified regularly with master sample at least once per day, but at all part number changes and start-ups (including after significant production stop). 4. Verification documented with operator sign off. 5. When applicable, gauge R&R is conducted to confirm EP efficiency.	Participate in an EP verification, check that the process is kept and documented well. Identification of EP devices on shop floor and coherence with the list. Records of verification (CP, start-up work instructions). Check identification, conservation, easy access, calibration of master samples. Work instructions for verification. Is gauge R&R result less than 10%?

Required	Criteria of Requirements	Verification Possibilities
In the event of EP devices malfunction, the suspect products are managed.	1. All EP failures have to be documented and reaction plan includes who is notified and actions to be taken. 2. All parts produced since the last OK verification have to be handle as suspected material and apply containment.	Ask people who make EP verification about their responsibility in case of EP failure and escalation process. Check back records that containment was done for all EP failures.
Reaction to failures, corrective actions, and reverification are documented.	3. Corrective actions to fix EP device failure is documented. 4. In the event of EP device malfunctioning or unavailability, production is stopped until capable substitution processes/control are identified and handled as bypass processes.	
Targets are defined and followed to evaluate VS effectiveness.	1. Performance graphic, Q chart, or report card (i.e., calendar days-red/green, I-chart, etc.) showing feedback from downstream and upstream processes. 2. FTQ or internal scrap metrics showing improvement trend, reduction of events/defects over time. 3. VS applied flexible, alarm limits are reviewed continuously. 4. Tracking of EP failures.	Prior to audit, check customer complaints caused by failed EP or no detection on VS. Any long-lasting processes or temporarily VS. Actions to close them. Review charts, verify action brought expected result. Check how often the alarm limit is reached: no alarm = no improvement. Check that VS selected efficiently: 1. Review data: customer complaints (any major or repetitive issue), FTQ results, high-severity numbers or RPN items from PFMEA, process capability data. 2. Based on data reviewed, evaluate if VS implemented to right place or there is a need to implement a new one. Verify that EP verification frequencies are reasonable.
Standardized Work		
The workplace environment is safe and ergonomic.	1. Relevant safety standards are applied for each workplace.	Check on shop floor potential safety issues, e.g., hidden corners, potential accidents, pedestrian way, color-coding on floor, noise, temperature, etc.

(Continued)

Required	Criteria of Requirements	Verification Possibilities
	2. Design of general and workstation layout considers safety aspects to avoid potential safety risks. 3. Rules related to ergonomics of workstations are defined and applied for each workstation design. They take into account environmental conditions. 4. Operators are involved in workstation design (ergonomic work stream) and major player in the qualification process, which always includes safety and ergonomic aspects.	On the shop floor, appreciate the level of light, the temperature (cold/hot), the level of noise, the loads carried by operators, the level of work (hands up, etc.). Look for result of ergonomics evaluation. Look at a "painful" workstation. Is there a specific control to verify its action plan.
The workplace is clean and orderly, arranged to contribute to higher quality and optimization.	1. A process to define layout including internal stock or buffer is available. 2. Organization of workplaces is suitable to ensure compliance with FIFO including repackaging operations. 3. Systematic approach for all workplace organization (like 6S) is implemented and maintained. 4. Visualization is used in the shop floor. 5. A continual improvement and optimization/waste elimination process is in place related to workplace organization.	Check workplace organization and visualization at several different places (incoming/storage area, workstations, maintenance room). 6S audit records and verify actions implemented for findings. Layout is in coherence with the workstation. Method for waste elimination is applied (VSM, 7 Waste, etc.) and periodical review minutes are available.
Working instructions are standardized and available for each workstation.	1. Standardized work instructions (SWI) are available for each operation. 2. SWI covers the whole of the produced references and line balancing means leveling of the workload across all workstations. 3. Revision history is traced back for process/product changes. 4. SWIs are developed by a cross-functional team and include at a minimum: Work elements including quality controls and their sequence	Work instructions at different workplaces (manufacturing, quality control, material handling/logistic), check: 1. Easy availability. 2. Compare instructions to work performed by operators. Observe three full cycles of the job in station and verify that the major steps and key points are followed and the reason is understood. 3. Controls listed in CP are added to SWI, efficient time allocated for quality checks. Try to perform a manual operation based on SWI, check all necessary information, hint, key points described to perform operation.

Required	Criteria of Requirements	Verification Possibilities
	Operator movement with sketch of work flow Takt time and overall cycle time Standard in-process stock Required PPE and safety requirement, if applicable Support description with pictures, sketches, and images Reference to product/process/control standards 5. For each work element, instruction describes: Major steps (What) Key points (How) Reason (Why) 6. SWI is posted near or at the operation.	Ask a few operators to explain SWI. Do operators understand it? Various balances are managed for different planned production outputs and product mix (e.g., modification of line speed).
The operations to validate the start of production are described and applied.	1. Start-up instructions are available (can be covered in SWI or checklist) for manufacturing operation and tool change/setup. 2. A start-up process is applied at the beginning of the shift, part number change, and after significant production stop (planned or unplanned) and documented. 3. Start-up instruction includes a list of the checking tasks to be carried out and recorded: Availability and functionality of all the manufacturing and control equipment, EP, and personal protective equipment (PPE) Process/product parameters with tolerance limits Availability of components and materials Environmental conditions at the workstation (cleanliness, lighting, etc.) 4. All activities/operations to be carried out prior to tool installation are defined (e.g., preheating of tool). 5. Start-up instruction defines first-off parts validation. Traceability to be ensured until validation is completed and result is documented.	Take 2–3 startup instructions for automatic equipment and verify content. Ask a setup person about roles of setup. Evidence of traceability between parts produced and first-off part. Check back records for startup activities and verify: Exact date of start-up documented Setup parameters recorded and within tolerance First-off parts result In case of any deviation action initiated and results verified

(*Continued*)

Required	Criteria of Requirements	Verification Possibilities
Reference parts and boundary samples are available and managed in order to confirm inspections.	1. Responsibilities for definition of samples are established (including customer, if required). At boundary samples, tolerances for each characteristic and decision criteria are clearly established. 2. Samples (in some cases, they can be replaced by photos) are clearly identified and in accordance with latest design and approval status. 3. Usage of samples is described/referred to in instructions, used in training process. 4. Samples are periodically reviewed and results are documented based on acceptance criteria and customer feedback/complaint.	Verify that boundary samples are available for operators. Check that samples represent typical failures. Ask operators when they are using samples. Verify sample storage condition. Evidence that samples are used for training. They have easy access in an area where they are used, storage preserves original condition.
The capability of the control devices is checked prior to using them in production process and it is periodically verified.	1. All gauges are periodically calibrated and recorded according to procedures, ensuring that only calibrated gauges are in use. 2. If calibration is performed in-house, necessary skilled staff, equipment, and facility are available; if outsourced, external laboratory accreditation is verified. 3. The capability of the measurement systems/equipment (gauge R&R) is periodically reviewed according to procedures. 4. The acceptance criteria are defined for calibration and capability. 5. For each deviation/nonconformity or equipment exceeding calibration due date, containment and corrective actions are defined and validated and followed by quality manager (apply handling of nonconformance). 6. SW is applied for gauge instructions.	Check several gauges in different areas (production, lab, incoming, storage) for: 1. Identification 2. Calibration status and their record 3. Proper usage and storage (ask operators about usage and handling of gauges, are aware about risks of damaged gauge) List of gauges contains identification and calibration period. Check schedule for calibration. Verify a work instruction for a gauge (see the acceptance criteria). Gauge R&R has been conducted to calibrate operators from shift to shift/line to line.
Target defined and followed to optimize processes.	1. Tracking of external and internal issues created by working instruction that is not well defined.	Prior to audit, check customer complaints where root cause is linked to SW (working instruction not detailed, wrong setup, etc.).

Required	Criteria of Requirements	Verification Possibilities
	2. Layered audit result related to noncompliance of SW. 3. Direct labor efficiency (ratio of real number of parts produced against the theoretical number to be produced during the opening time). 4. Cycle times leveling. 5. Capability and calibration follow up indicators (e.g., number of late calibrations). 6. Tracking of downtime and/or scrap rate caused by setup. 7. Ergonomics evaluation rate.	
Training		
Training process is defined. Organization, methods, skills, and facilities are available for training activity.	1. The training process covers the entire staff of manufacturing, logistics, and support functions including temporary employees. 2. Internal trainers are identified and qualified to ensure that only certified trainers train. 3. A graduating step approach is developed and applied for each training activity, such as the 4 Step: (a) prepare, (b) demonstrate, (c) tryout performance, and (d) follow-up. 4. Dedicated infrastructure (outside of the process) for hands-on training simulating production conditions, if possible.	Verify at different workplaces whether their operators received training from a certified trainer. Record of Internal qualified trainers. Ask operators how they were trained. What specific training did they receive? Rules for graduation process are described at training procedure.
Training needs for safety are identified and training is performed for all relevant people.	1. A safety policy is defined and communicated to all staff. A safety handbook is established and regularly updated in coherence with the policy. 2. An organization is defined according to policy and local legislation with suitable equipment and qualified people. 3. Training needs for safety are defined based on the risk assessment linked to the industrial domain and specific workstation. 4. Based on safety training needs, all relevant staff are trained in two levels: (a) basic training about general safety rules, and (b) individual job training, if applicable.	Safety manual and policy. New employee training covers safety requirements. Ask operator who works on station where safety requirements are established, about awareness of safety rules. Organization chart: responsibility for the safety is defined.

(*Continued*)

Required	Criteria of Requirements	Verification Possibilities
	5. Visitors' safety is managed; standards and equipment are available (quick training, suitable safety devices are at disposal, i.e., safety glasses, hearing plugs, appropriate gloves, covers for static electricity, etc.).	Facilities on site: first aid station, etc. Safety standards are kept, e.g., Personal protective equipment (PPE), circulation on the shop floor, risk related to process (stamping, welding, assembly, etc.). Signs, posters on the shop floor, line marking, behavior of logistic employees.
Basic and individual job training needs are identified, their content is formalized, and all the trainings are recorded.	1. Basic training is given to each new employee which includes: - New hire orientation - Knowledge and respect of the product (including management of key characteristics) - Proper record keeping (production/quality) - Understanding of workplace organization responsibility - Quality requirements (containment, red-bins, Andon, etc.) - Escalation procedure (are the right people notified during a problem?) 2. Individual job training is documented for each employee showing training needs, skill or knowledge level of the job, who trained them, timing of operator sign off. 3. Individual job training is held in case of change in training content or need (e.g., working instruction change). Traceability for latest change is ensured. 4. Documentation, scheduling, and tracking forms for employee refresher training if they have not performed that job within a specific duration. 5. All training activities are documented; records are available and easily retrieved.	Check if individual job training has all training content needed to perform the task. Check individual job training records for different job elements (operator at a working station, quality controller, team leader, a support function). Ask team member to explain the documentation regarding quality, workplace organization, and escalation process. Verify that a few operators follow SWI and know the quality and productivity requirements. Check a few working stations and verify that training records are available to the latest working instruction level. Verify if there are employees that go back to the operation after a specific period out of operation without refreshing training.

Required	Criteria of Requirements	Verification Possibilities
Operator qualification process for each job position and workplace is applied, including requalification if needed to ensure that only qualified people perform the job.	1. Qualification levels are established. For each, measurable criteria are defined. 2. A calibration process is in place for all people making check operations where result depends on subjective decision. 3. Flexibility chart or equivalent posted at all operations or work area which Contains numbers of qualified people per workstation as well as workstation per person are targeted; associated action plans are implemented Indicates the steps in training and skill qualification level achieved for each job Has been updated 4. Criteria to revise qualification level are defined; they take into account the operational results at the specific workstation, the result of the layered audit, time off job, etc. 5. If requalification failed actions are implemented to reach required qualification level, include reassessment or degrade qualification level.	Check several operations where result depends on subjective decision. Check the record of calibration process, frequency, results, and action plan. Chart showing cross-training/certification level in a cell or work area such as a flexibility chart. Look for a job rotation plan or log. How often does team rotate? The number of team members certified per station should support the job rotation plan. Check if the training procedure describes the requalification process. Check if a requalification process (employee performance review) is in place. Evaluate if an action plan was generated in case of low performance.
A process of motivation (individual recognition/reward) is implemented. The policy is defined and managed.	1. Process in place to recognize employee's achievements. 2. Individual objectives are in line with the site targets and are regularly reviewed, evaluated with feedback to the employees. Individual interviews are systematic for the all staff. 3. Process in place to collect and address employee grievances with short- and medium-term improvement actions established.	Check how the leadership is promoting the recognition (e.g., 7 recognition success factors; see Glossary). What kind of system does the supplier use to receive feedback from the employees? Check if the action plans regarding employee grievances are on track. Verify if the individual objectives were deployed from the business plan.
Metrics are defined and followed to ensure effectiveness of training process.	1. Flexibility target (number of qualifications per employees and number of employees qualified per workstation).	During audit, check how team members follow general procedures (e.g., handling of nonconforming material), as a confirmation about training effectiveness.

(Continued)

Required	Criteria of Requirements	Verification Possibilities
	2. Layered audit results related to people awareness of procedures (e.g., SW, safety requirements, material handling, workplace organization, etc.). 3. Number of retraining needed in same subject (failure of first training). 4. Training assessment by trainee. 5. Internal and external issues due to the lack of training. 6. Employee satisfaction survey.	Check all findings found during audit that can be linked to lack of training (root cause of deviation is lack of knowledge). It affects the effectiveness of training. Does supplier measure the turnover in its departments (fluctuation)? Why?
Audits		
A generic layered process audit (LPA) is established.	1. Written procedure: Defining the rules of LPA. 2. Layered audit check sheet: Developed and applied for all manufacturing and material handling (logistic) areas. LPA check sheet shall define what to look for. 3. LPA check sheet is comprised of the following elements, at minimum: Safety and ergonomic items Proper safety practices and PPE are being followed. Manufacturing/material handling (logistic) items SW is being strictly followed. Workplace organization standards are maintained, proper tools, gauges and materials are available and used. Operators work according to flexibility chart. Standard in-process stock is according to SW. FIFO is being followed. Quality focused. Required quality controls completed and documented according to PCP. EP verification done and documented properly.	Verify if the LPA procedure defines - Frequency - Who will perform the LPA - How to conduct the LPA (standard method) - How to record and treat issues Examples of fields in the LPA check sheet form: Safety PPE: The team member is using all the posted PPE. Manufacturing/material handling Work instructions (e.g., standard operation sheet, job element sheet) Proper tools, gauges, and materials available and used Quality focused Ensure control of significant process elements that can impact areas such as customer satisfaction, PPM, warranty Ensure control of high-risk elements including operator/process sensitive operations, key process control operations/checks, and mandatory assembly sequence operations Verify that the auditor understood the questions, LPA check list filled in properly; follow a team/group leader in an LPA. Verify if they use the standard method

Required	Criteria of Requirements	Verification Possibilities
	4. Training of auditors: Using a standard method for LPA (choose the workstation, conduct the audit, give feedback, document results, and conduct follow up). 5. LPA is applied for standard processes of supporting functions and internal, process-specific audits are performed (i.e., process/commodity specific audit, CQI audits etc.).	
LPA, which covers all operational activities, is carried out and owned by manufacturing.	1. Each workstation is audited and schedule frequency is defined to ensure: Covering all shifts Each operation is audited a minimum of once per month Containment activities (e.g., sorting, controlled shipping, specific processes, etc.) are covered by LPA 2. LPA schedule for all levels: Showing participation of all levels (from team leader to top management) with established frequency for all manufacturing areas. 3. LPA schedule is tracked by manufacturing.	In the shop floor, select a workstation and verify how the workstation is audited by LPA: Frequency is according to the plan Once a problem is identified, how is the team member informed Check if safety issues are detected by LPA Verify if top level is conducting the audit Check back LPA records, verify that audits were really performed according to schedule, and all operational activities were audited (not only manufacturing operations, but also material handling, storage, shipping, etc.). Verify if containment/rework activities are included in the LPA plan.
A follow up of the LPA and associated action plans are in place. Deviations are treated.	1. LPA records: All LPA results are documented including: No deviation found (Y) Deviation found/not corrected during audit (N) Deviation corrected during audit (NC) Not applicable (NA) 2. Problem solving is applied directly at the workstation to treat deviations; actions are defined and recorded with responsibility and target closing date.	How organizations apply the problem-solving methodology for issues detected during the LPA and how organizations record the countermeasure (e.g., in the countermeasure sheet) and how they do the follow up of actions. Actions defined against root cause (e.g., not only retraining). Due dates kept. Implemented actions were verified. Check the involvement of management (knowledge of the result, of the ongoing action plan).

(*Continued*)

Required	Criteria of Requirements	Verification Possibilities
	3. Countermeasure sheet: Deviation found/not corrected during audit (nonconformances or operator claims as well as safety/ergonomic issue) must be addressed on countermeasure sheet. Corrective action implementation is followed up. Countermeasure sheet is used for CI (e.g., if a good practice is discovered during the LPA, it should be used as a driver to improve the current standard work). 4. LPA results are used for CI.	CI Flexibility chart revised using the results of the LPA. Problem-solving methodology revised using the results of LPA. Workstation organization performed. LPA used to capture a more efficient way to work and lead the standard work revision.
LPA effectiveness is continuously monitored and analyzed via LPA results in order to ensure procedures are kept.	1. Tracking of audit results with visualization to share status on affected area (notice board [nb] of nonconformances per LPA, Pareto of nonconformances, repetitive findings, nb of customer issues caused by nonrespect of SW; it should have been found by LPA). 2. Tracking of audit schedule and effectiveness of action plan implementation (average time to treat deviations/effectiveness).	Perform a layered audit together with team leader. Compare your results with team leader. In case of gap, identify reasons for different evaluation. At shop floor, verify in the visual management board (area or plant): - LPA plan - LPA tracking - LPA results - LPA action plan/effectiveness
Risk Reduction		
PFMEAs shall be developed and maintained by cross-functional teams for all manufacturing processes and support functions.	1. PFMEA is developed by cross-functional team. 2. PFMEA available for all part numbers and all operations 3. Scoring of severity/occurrence/detection is according to customer's guideline; ratings are consistent. 4. Content of PFMEA fields is defined properly in accordance with customer guideline. 5. If supplier is design responsible, DFMEA has to be used to develop the PFMEA.	Check member of PFMEA team and verify that necessary trainings are provided. Evidence that PFMEA was prepared for all "risky" operations even though the base was a generic PFMEA. Scoring is according to predefined standards and customer requirements. Verify content: Effects evaluated from both customer and manufacturing point of view, potential cause of failure defined specifically, ambiguous phrases (e.g., operator error or machine malfunction, etc.) should not be used. Real preventive actions are listed.

Required	Criteria of Requirements	Verification Possibilities
Proactive approach for reduction of PFMEA highest risk items is implemented.	1. For high-severity rankings or high-risk items, FMEA team ensures that the risk is addressed through existing design controls or recommended actions. 2. Prioritization is based on a defined approach (e.g., severity, criticality, and RPN or a combination of severity, occurrence, and detection (risk priority number [RPN]). 3. Recommended actions documented with responsibility and due date. 4. Quality tools (e.g., LPA, VS, etc.) are applied until recommended actions are implemented in order to keep risk under control. 5. Recommended actions re-evaluated after their verification.	Review risk reduction action plan, evaluate that actions are defined against root cause or improve detection, target dates are kept. Evaluate that quality tools implemented are efficient to keep risk under control. Review scoring after recommended action implemented. Where severity 9 or 10, detection is low (visual inspection alone is not acceptable). These items must be identified in the CP.
PFMEA is reviewed periodically as proactive and reactive activities.	1. PFMEA is reviewed and updated for each past quality issue and corrective action that has been implemented within target completion date. 2. After each issue, 5 Whys analysis is performed to understand why PFMEA did not predict failure. Action implemented to avoid in future. 3. Periodic PFMEA reviews are scheduled based on process capability, process/product changes, etc., which cover - All processes and their controls are included - Detection ratings are accurate - Occurrence ratings are analyzed using data (SPC, FTQ, ppm, scrap data, VS results, etc.)	Revision date of PFMEA linked to past failures. Check last customer complaint or quality issues and their update in PFMEA. Review checklists, agendas, or equivalent that assure adequate PFMEA review. Check some operations/processes (material handling, labeling, rework/repair, etc.) are included and accurate. Compare top internal scrap data with occurrence scoring.
There is a reverse PFMEA (proactive approach) process in place to identify new potential failure modes on shop floor.	1. A cross-functional team conducts on-station review. 2. A schedule of reverse PFMEA (fault tree analysis [FTA]) is implemented and regularly updated (timing for review with prioritization of operation and its status/planned/done).	Check updates after reverse PFMEA is performed. Choose one station and perform a quick reverse PFMEA to confirm all current controls are rated properly and all potential failure modes are covered (try to create new ones).

(*Continued*)

Required	Criteria of Requirements	Verification Possibilities
	3. All the findings are driven back into process flow, PFMEA, CP, and work instructions as applicable.	Have the current occurrence/detection numbers been revised?
General assessment is conducted to identify potential risks that could impact plant (normal processes/activities).	1. There is a formalized process of evaluation and control of the risks concerning resources, facilities, tooling, and supplies. 2. The responsibilities of evaluation and management of risks are clearly established, coherent with the typology of the risks. 3. A structured tool to identify, evaluate, and prioritize risk based on a multidisciplinary approach is used (e.g., FMEA approach). 4. Prevention and reaction plans are established and deployed for each identified elementary risk. When appropriate, plans are audited and simulations are carried out. 5. The risks and the associated plans are periodically reviewed based on the plant, corporate and external lessons learned.	Master securing plan, which contains major risks, procedures, and owners in case of flood, fire, etc. Verify that master security plan covers the relevant major risks. How risks are evaluate (which criteria, is there a tool like FMEA, etc.). Who is responsible for the whole process? Who is responsible for a precise risk? An example of risk is a fire. Look at the action plan (preventive: extinguisher; training/corrective: site evacuation plan, sprinkler, firefighters on site, etc.).
Targets are defined and followed to ensure effectiveness of continuous risk reduction activity.	1. Tracking the number of high-risk items (trend chart or a Pareto chart). 2. Tracking of delayed recommended actions by leadership; necessary actions performed in case of delay. 3. Track that PFMEA reviews and reverse PFMEA are performed based on their schedule. 4. Number of new failure modes and root causes covered after complaint (both external and internal).	Make sure that the evaluation process is according to your guidelines as well as the customer's. Start with the highest severity, then criticality, and then the RPN. Try to avoid the RPN analysis as a prioritization of failures unless dictated by the customer. Avoid RPN numerical thresholds. Review actions/implementation dates/delays. Percentage of error proofing/error detection.
Contamination Control		
Sediment is controlled with identification of possible risk areas for contamination as well as continuous measurement and monitoring of sediment level.	1. Identify all areas/operations that could be affected by contamination and failure modes related to contamination are considered in PFMEA.	Verify if the organization has mapped all areas/operations that could be affected by contamination and identify the type of contamination in each area. Verify, if this information was used during the PFMEA and process control plan (PCP) development.

Required	Criteria of Requirements	Verification Possibilities
	2. Procedure developed that includes method defining sediment collection and system to measure cleanliness. 3. Cleanliness results are documented and plotted on control chart. 4. Control limits are utilized to trigger reaction plans. Do not confuse specification limits with control limits. They are not the same. Control limits are calculated based on the *voice of the process*. Specifications are given by the customer.	Check if the cleanliness controls are being done according to frequency established at PCP. Frequencies are reasonable. Check sediment measurement in lab (instruction and process). Application using statistical method to monitor and control (include individual parts, subassemblies, full assemblies). Verify if there is a documented reaction plan for cases where the limit control or unstable condition is reached. Ask team member and/or inspector what they should do when an out of control condition is reached.
Sediment reduction strategy is followed to ensure proper functionality of equipment and processes designed to remove/prevent sediment contamination.	1. Procedure/instruction to define and maintain method and frequency of checks to ensure quality and cleanliness of - Metal working fluid system - Fluid/air probes - Work station cleanliness - Dunnage and part storage - Purchased parts - Washer - Debur	Check workstation cleanliness, daily verification to be established (layered audit, tape test, etc.). Dunnage cleaning process defined and kept. Parts are covered/properly protected. Parts on the assembly lines handled with lint-free gloves/towels. All wash/debur machine parameters captured and monitored/documented. Wash solution and process parameters monitored and analyzed on an ongoing basis.
If clean room/area is required (due to sediment or painting requirement), special rules are utilized in order to minimize risk of contamination.	1. Clean room protective clothing defined and enforced. 2. Positive pressure to stop outside air/contaminants from being drawn into the clean room. 3. Limited access by employee and enter/exit route to clean room are identified. 4. Sticky mats or shoe protection to remove contaminants from footwear. 5. Control of chemicals detrimental to the process.	Check if protective clothes are worn (e.g., hairnets, shoe covers, lint-free lab coats, rags, gloves, etc.) Check how access control to clean room is kept. See if air quality and pressure are monitored. Verify if filters are changed according to planning. Check if controls of restrictions (e.g., no fibrous material, no aerosol sprays, no food, etc.) are in place. See if LPA is in place (if out of condition is identified, there is a reaction plan in place).

(*Continued*)

Required	Criteria of Requirements	Verification Possibilities
Maintaining the processes to control and prevent dirt in paint or parts and foreign material contamination.	1. Sources of dirt and foreign material as potential failure mode are considered in PFMEA. 2. Methods are developed to prevent extra parts or materials that may fall into or stick to products and are verified through EP verification and/or LPAs (e.g., sensor for masking used). 3. Procedure or instruction describes dirt control and prevention for key areas: people, process, facility, and material. 4. Application of statistical method to monitor and control painting process such as dirt count (SPC, U-chart), dirt identification (Pareto), etc. 5. Ensure proper functionality of equipment and processes designed to remove/prevent dirt in paint or part. 6. Behavior instructions are defined and respected.	Check if foreign material/extra part (screw, pin, washer, any small components, etc.) effect (noise, function issue, etc.) are considered in the PFMEA. Verify that dunnage, fixtures, conveyors, washers, and preassemblies are protected to prevent falling material. Check possibility of assembling part without masking. Verify if all stations in the paint area are clean and operators do a self-inspection of the area. Ask a team member to explain what the main sources of dirt are—key issues in the people, process, facility, and material area—and how team member works to avoid them. See if the sources related to process (paint spatter, overspray, towels, etc.) are checked by PM and if there is a reaction plan established. Check if general housekeeping process is established for facility and if it is followed.
Targets are defined and followed to ensure continuous contamination reduction activity.	1. Sediment result monitored, reduction plan defined. 2. Layered audit and process specific audits results. 3. Internal or customer issues related with cleanliness. 4. Pareto of defects for key sensible process (painting, metallization, etc.).	Review sediment results and verify sediment reduction activities. Evidence of site leadership reviews. Audits performed focusing on contamination control.
Supply Chain Management		
Procedures or work instructions to select and evaluate suppliers are implemented.	1. Procedures that require supplier audits to support sourcing decisions, evaluation of their systems. 2. Criteria to become sourceable supplier are defined. 3. Approved supplier list to track tier suppliers' performance (e.g., bid list). Performance indicators are defined with threshold.	If supplier selection is done at manufacturing location, check documentation of a tier supplier selection process. If selection is done centrally, check Tier X status is known by manufacturing location. Evidence of SQ responsibility (SQ organization chart or SQ training records).

Required	Criteria of Requirements	Verification Possibilities
	4. Contractual documents (such as system of record [SOR]) contain all Tier 1 requirements, expectations, and processes, signed by tier supplier. 5. Supplier quality department or organization is fully involved in the sourcing process.	Effect of manufacturing location to tier supplier's performance evaluation.
Customer-specific requirements are cascaded to all tier supply chains and deliverables of Tier X are validated/approved by Tier 1.	1. Customer-specific requirements are implemented at tier suppliers: - Technical specification of product - Product- and process-specific standards - Procedures need to be applied (e.g., PPAP/EI, PFMEA, etc.) - Traceability, FIFO, and labeling requirements 2. Tiered suppliers comply to product and process validation requirements. Approval is available for all the components needed for assembled part. 3. Final product key and pass-through characteristics are identified on component (or material) level as well as shared and controlled (if required with SPC) with tier supplier. 4. All design and process changes are communicated, managing change requirements are applied. Tier X supplier production transfer needs to be managed with specific procedures to all their suppliers. 5. Tier suppliers capacity is confirmed for all the materials/components needed for assembled part.	An example of request for quotation. Technical specifications forwarded to Tier X including list of critical characteristic in their PCP. Supplier quotation file including technical/process feasibility. Check critical components product/process qualification files (PPAP). Check customer drawing whether it contains any key/pass-through characteristics that are produced and controlled by Tier X supplier. If yes, verify that Tier X supplier drawing, PPAP, and CP cover them as well as Tier 1 incoming control includes them. Example of management on a Tier X process change. Check capacity confirmation for critical components.
An escalation process is applied in case of tier supplier issues.	1. For any delivery of parts with deviation, Tier 1 validates the formal deviation request of the Tier X complaint. 2. For single issues, a systematic and disciplined approach to problem solving is implemented, an action plan is set up by Tier X and verified by Tier 1, keep due dates expected by customer.	Prior to audit, check whether there was any customer complaint caused by tier supplier. Example of supplier complaints toward its Tier 2, verify contents of the complaints.

(*Continued*)

Required	Criteria of Requirements	Verification Possibilities
	3. Approved supplier list to track tier suppliers performance (e.g., bid list six panel). Performance indicators are defined with threshold. 4. In the event of nonrespect of a target, an action plan to reach it is systematically defined and followed by the Tier X supplier. 5. When a tier supplier moved from Green ranking, escalation process takes into account the Tier 1 support of the Tier X supplier to improve. Exit criteria are defined. 6. Regular audits are performed at supplier's plants. In case of complaint, verify action implemented and regular audits at key suppliers to improve their system (CQI audits, process-specific audits, PCPA, etc.).	Is there a timely management with milestones? Verify content of 5 Whys prepared for customer complaint with problem resolution report (8D, 5P, Red-x, etc.). Example of action plans given by Tier X. How is it validated and verified? Example of supplier audit, evidence of audits at key suppliers. Escalation criteria (e.g., controlled shipping, top focus, new business hold, etc.). Critical suppliers identified and tracked via top focus process, exit criteria defined. Review Tier X performance matrix with Tier X corrective action plan approved by Tier 1.
Incoming inspection process in place.	1. An incoming inspection CP and the related control standards and sampling rules are established for each key/critical supplier's product. 2. Only approved (PPAP/EI and incoming inspection released) components/material are used for assembled part. 3. At the reception of supplied product, the first controls take into account checking of the quantity, the integrity of the packaging, and the identification of the product. 4. Part inspection requirement including pass through is established and implemented for each supply since design phase and updated in case of nonconformity.	Check identification of material released for production after receiving. Verify that incoming covers key and pass-through characteristics, frequencies are reasonable. See a nonconformance found during incoming. Verify link between incoming frequency and supplier performance (high ppm supplier). Check incoming records for a component.
Tier supplier targets are defined and their performance is tracked via indicators.	1. Tier X performance result (ppm, rate of complaints, delivery, etc.). 2. Mean time to respond to complaints. 3. PPAP curve (planned full PPAP date vs. achieved). 4. Metrics on bid list (number of red suppliers, etc.).	Check customer complaints, ratio of Tier X supplier issues. Red suppliers on bid list, actions for improvement. If any significant concern about Tier X issues had been addressed in FR meeting.

Required	Criteria of Requirements	Verification Possibilities
	5. Tracking results of the audit result at Tier X. 6. Average time to close problem resolution.	
Managing Change		
All product, process, or source changes are monitored and controlled.	1. A procedure to manage product/process/source changes is defined and applied—both planned and emergency changes. 2. A change form is utilized to document all changes and controlled through a document control process (e.g., tracking log sheet, revision numbering system, approval process, etc.). 3. All changes are managed like a project; responsibilities and milestones are defined; planning, activities, and the deliverables are established in agreement with the customer. 4. All changes need to be reviewed and approved by customer (PPAP and/or engineering). Inform customer in early phase to get approval before kickoff.	Check documentation via example: A design change and a process change: - Organization (project team) and milestones - Evidence of customer approvals - Planning and evidence of reviews - Tool to ensure traceability of modifications Evolution of data system such as MRP system, storage management software, EDI server must be considered as major changes.
A risks analysis is applied for any product/process change.	1. For any product/process/source change, a feasibility analysis is carried out. The study takes into account the impacts in terms of costs, technical, performance, quality, timing, and capacity. 2. Potential risks of change are considered via FMEA approach and documented in DFMEA and PFMEA. 3. Based on risk analysis planning and implementation of the change are carried out, a product and process validation plan is defined. 4. Break point is defined when change becomes irreversible; it is communicated to customer. 5. According to risk analysis, a securing approach is put in work in the launch phase (e.g., mixing of old and new design).	Check one of last of modification and verify: - A feasibility analysis including lead time analysis - Formalized impact evaluation and risk management - PFMEA and DFMEA (if applicable) - Action in place to cover risk identified

(*Continued*)

Required	Criteria of Requirements	Verification Possibilities
A production trial run (PTR) process is implemented.	1. A standardized communication procedure and form to control and monitor all PTRs that documents each step of the process and records all approvals and results. 2. Traceability of trial run batch is ensured. 3. Quality reviews documented to release product for PTR shipment and verification process has returned to normal production.	Example of a trial run. Trial run validation criteria and results (records).
Maintenance		
Maintenance organization and strategy are established and deployed.	1. Maintenance process is formalized and covers all the machines, tools, devices, equipment, and facilities on site. It includes preventive and corrective maintenance. 2. A system for managing, planning, organizing, and monitoring maintenance operations is set. 3. Resources are available during all production periods and outside production periods to ensure noncritical maintenance operations (e.g., preventive maintenance). 4. Suitable maintenance facilities and equipment are available. 5. Fast communication between production and maintenance is assured. 6. A process is in place to improve production output. It is based on analysis of operational availabilities, operators' suggestions, Six Sigma techniques, etc. 7. Approach to standardize equipment (e.g., using same filter, same interfaces to tools, etc.).	Preventive maintenance for safety equipment has to be performed on time without exception. Check that a maintenance system is in place to manage all maintenance activity, supported by IT tools like CMMS (computer maintenance management system), Excel, and so on. Available resources by technology including outsourced experts (flexibility chart). Implementation of resources near the manufacturing activities. Are there free resources to manage the corrective maintenance? Facilities available (areas well defined, conditions, 6S level, etc.). Management improvement strategy and periodical reviews. Review major brand types of major equipment; ask about process to purchase new machines. A generic plan to improve a type of equipment (e.g., electrical screwdrivers).
The activities of maintenance are planned, performed. and tracked.	1. The preventive maintenance planning takes into account risk classification of the equipment (safety, constraints and bottleneck equipment, unique process with no substitution process, etc.). It is built together with MPS, which takes into account time allocation for maintenance.	Choose machines (constraint/complex one) and a tool to verify: Maintenance planning: identification of the equipment, task to do, and when Follow up of the maintenance schedule and its visual management

Required	Criteria of Requirements	Verification Possibilities
	2. The planning is regularly followed and updated (including outsourcing maintenance). 3. Corrective maintenance is carried out for any deviation from the nominal (according to manufacturing priorities). The preventive maintenance activity is periodically reviewed based on the corrective maintenance activities. 4. Maintenance activities are considered SW. Technical documentation is available and managed for all equipment. 5. Records of all maintenance activity and results are formalized and filled in. 6. Necessary requalification is done after maintenance activity; its result is recorded to ensure traceability.	Maintenance work instruction (including changes due to lesson learned) Technical documentation for a precise equipment including document management Records of corrective maintenance activity (type of equipment, is there repetitive breakdown, etc.) Records of preventive maintenance on an equipment with a recent breakdown Verify that maintenance activities are fully deployed and covers all equipment (machines, facilities, tools). Different type of preventive maintenance depending on the type of equipment (pure preventive approach and conditional maintenance).
Level 1 maintenance is systematically applied and integrated to workstation instructions.	1. Level 1 (L1) maintenance operations take into account equipment identification, cleaning, self-maintenance, and safety device verification. 2. L1 maintenance is performed under manufacturing responsibility at operator's workstation. L1 operations are integrated in the workstation's work instructions. 3. Any deviation, anomaly, or improvement suggestion is recorded and, if necessary, escalated to a higher-level maintenance activity. 4. Records are analyzed and used as lessons learned to improve maintenance operations.	Ask operators about Level 1 maintenance responsibilities. Verify a few L1 maintenance working instructions and records. Look at the equipment status on the shop floor (is there an identification number, cleanliness, protection in good condition, leakage, etc.?). LPA records. Evidence of activities transfer from preventive maintenance to L1. (The levels of maintenance are: L1–unit; L2–direct support; L3–general support; L4–depot.)
The spare parts and their storage are managed.	1. A list of critical spare parts is determined, managed, and regularly updated. 2. A spare parts stock is available with minimum stock level. Spare parts tracking system is combined with maintenance system in order to control physical inventory.	List of critical spare parts. Stock of spare parts: reception/organization/consumption

(Continued)

Required	Criteria of Requirements	Verification Possibilities
The critical parts are identified.	3. The spare parts are stored in suitable conditions and periodic physical inspections are performed for long-term stored items.	Conditions of storage. Computer-aided system. Inventory (take an example of a critical part and verify the robustness of the inventory).
The customer's product specific tools are managed to preserve tool condition until the end of their lifetime.	1. Each customer-specific tool (CST), including customer-owned, is identified in a single and unalterable way according to customer requirements (customer marking). Identification includes tool change level. 2. Lifetime of tools is strictly followed and documented in diary sheet, which includes lifetime of tool, shoot number, and maintenance activities. 3. In case of subcontracting, a list with the localization of the CST is established and communicated to the customer. 4. The storage of the CST is organized and managed. Activities to be carried out prior to storage are defined. Storage conditions guarantee the safeguarding of the CST until its end-of-life. 5. Tool release process after storage is established. 6. Product/process requalification is carried out for each replacement tool according to customer requirement.	Check a few tools to verify General conditions of the tools (leakage, rust, etc.) Identification and visual management of the tools A diary sheet and records, lifetime followed Communication and customer approval for each change on an example Storage conditions Standardized maintenance operations
Indicators are defined and tracked to ensure effectiveness of all the maintenance activities.	1. Performance and reliability targets are defined based on historical data, and related indicators are tracked. Failure rate, MTBF, MTTR, stop of lines 2. Tracking maintenance performed vs. planning (including service provider activity). 3. Pareto analysis of breakdown. 4. Average rotation of spare parts. 5. Deviations found during spare parts inventory audit. 6. Ratio of corrective maintenance against preventive maintenance.	Customer complaints caused by machine or tool problem (e.g., burrs issue, etc.). Verify quantity of a few spare parts in stock. Check a few maintenance working instructions whether they are standardized.

Required	Criteria of Requirements	Verification Possibilities
Manufacturing and Material Flow Management		
A structured process of manufacturing scheduling is implemented and systematically reviewed.	1. Long-term strategic scheduling is managed via a sales and operating planning (S&OP), which includes a complete forecasting of customer demand. 2. The S&OP is reviewed regularly and is used to define manufacturing capacity, stock level, and investment plans. 3. A master production schedule (MPS), coherent with the S&OP outputs, is managed on site. It provides a complete forecasting of the customer demand at the lowest level (part number) short term. 4. The MPS is reviewed regularly by logistics and manufacturing team (supported by maintenance) and is used to define resources needed (equipment time allocation, human resources, material). 5. S&OP and MPS are transferred to concerned entities on site and to Tier X.	Check long (SOP) and short MPS term scheduling for a product. Verify that Capacity meets long-term customer demand Time allocated to other product is part of review SOP regularly reviewed based on electronic data interchange (EDI) data Long-term demand deployed into short-term scheduling Check a Tier X demand whether it is in line with Tier 1 manufacturing schedule Supplier should have at least a sheet with the list of constraints in the plant and up to date availability information (opening time, preventive maintenance planning).
The daily manufacturing activity is planned in detail and follow up at site level.	1. A process is in place led by manufacturing to generate manufacturing detailed program on a daily basis coherent with MPS outputs. 2. The program takes into account the preparation time batch change, tool change, setup, start-up process, etc.) and up to date OEE rates. 3. Deviations between forecast and real production are followed and controlled on a daily basis at production line level. An escalation process is defined. 4. A process to improve the setup time is in place. Organization should establish a goal, measure the setup time, and define the action plan once the setup time goal is not reach.	Check product chosen in production section and check daily deployment: Manufacturing detailed program Complete preparation time for a batch Different programs from one week to another (is it stable, level of flexibility) The real production level. Planning and minutes of optimization workshops. Hourly board control. A strategy is in place to improve the setup time (e.g., SMED strategy).

(Continued)

Required	Criteria of Requirements	Verification Possibilities
Constrain operation is identified and specifically managed.	1. Based on customer demand, all bottleneck operations are identified. 2. Constraint operation identified among bottlenecks and prioritized concerning qualified operators, training, maintenance, scrap, setup, and fast reaction in case of any deviation. 3. Manufacturing detailed program is optimized taking into account constraint operation. 4. Constraints shall be managed by identifying problems, establishing action plans, and verifying effectiveness of action plans on a regular basis. 5. A backup plan for each bottleneck operation is defined.	List of bottleneck equipment. Check via record that - Constraint is running continuously - Number of qualified people via flexibility chart - Breakdowns and actions against them Check buffers size; ask operators how often they run out.
Product packaging (final product, intermediate, and supplies) availability and conditions are managed to ensure product quality.	1. Loops or flows of full/empty packaging are organized and managed. 2. The stocks of empty packaging are organized and managed. The number of empty packaging available is enough to keep production running. 3. A process to check the packaging is applied; it takes into account the requirements of the customer and the standard definition of the packaging. All activities are considered as SW. 4. Packaging control work instructions are established and applied, which include decision criteria for the use of packaging, the associated countermeasures, and corrective actions (e.g., alternative packaging, cleaning, maintaining, special transportation of empty packaging, etc.). 5. Supplier responsibilities are clearly defined. Deviations are systematically communicated and treated with the customer.	Examine the flow of empty packaging. Try to see the loops are respected. Verify the procedure to initiate the supply of additional packaging (threshold definition, alert process, timely response). Check empty packaging storage area. Verify work instructions for checking and cleaning. What happens if there is no more empty packaging? Look at the conditions of the packaging, particularly the label support. Check actions for critical packaging from availability point of view (e.g., alternative packaging is defined, for final product it is approved by customer).

Required	Criteria of Requirements	Verification Possibilities
Handling and storage conditions of product (final, intermediate, and supplies) are respected in order to protect parts from damage and environmental effects.	1. All storage areas are identified and organized by type of product (incoming, intermediate, work in process, final, etc.). Process flow charts are effectively matching with processes. 2. Potential failure modes related to material storage and handling are considered in PFMEA (e.g., damage by handling equipment, rust caused by storage condition, mixing up of products, etc.) 3. When needed, storage conditions are controlled by devices in real time (e.g., temperature, humidity, etc.). Records of stock condition are kept. Alert procedures and countermeasures are defined. 4. SW is applied, structured approach for organization of storage is defined and applied. System in place allows easy visualization of storage operations and level of stock for each reference. 5. Associated equipment is suitable for stocking and handling. 6. The stock management system takes into account product expiring dates, product change level, and the respect of the FIFO. 7. Stock level is checked via regular quality and quantity audits (e.g., LPA).	Check storage area and condition at several places (incoming, work in process, final product). Ask operators and material handling personnel whether they are aware of and are following instructions. Verify that adequate protection is in place to protect parts from damage and mixing. Condition of storage (temperature, waterproof, etc.). Visual management in place (level mini and maxi are visible, token board for the FIFO). Check expiring dates by reading labels. FIFO: Risky situation to examine, intermediate stocks, double flows/lines (e.g., 2 paint lines), how do they manage these types of situations (specific rules and procedures)? Specific management for the high runner references. Results of audit or inventory. Equipment to manage the stock (barcode reader, informatics systems).
A system is in place that ensures materials needed for production are organized and available at place of their application.	1. Feeding plan is established and tracked. Suitable feeding equipment adapted to layout, to flows configuration, and to available space are used. 2. The feeding process is based on a structured methodology (e.g., pull system) which: Minimizes overflows and non-value-added operations (repacking) Guarantees the availability of materials at the workstation during all production periods	Check if a standardize feed route plan is defined and followed. Check a few workstations and check: Respect of the "pull system" principles Alerts from the lines and their management; management of the risk of stockout Work instructions (line feeding operations, loading, repackaging operations, etc.) FIFO kept (organization of the rack).

(*Continued*)

Required	Criteria of Requirements	Verification Possibilities
	3. All feeding activities are considered as SW and guarantees the respect of FIFO. 4. At the workstation, materials are used in manageable size and material flows are organized and managed following the standard work/ in-stock process. 5. Material flow optimization is in place based on the use of tools such as MIFA or VSM.	Organization of "supermarket" areas (visual management), if existing. Organization of the material flow at the workstation (entry and exit point for each component, packaging, useless movement). Optimization activities (minutes of meetings, action plans, etc.).
Indicators are defined and tracked to ensure effectiveness of material flow management.	1. Indicators concerning availability of packaging. 2. Stock management indicators: Inventory Stock targets fulfillment 3. Lines feed indicators: Feeding tracking Internal service rate Production losses due to feed missing Internal flows lead time 4. Tracking of constrain output.	Prior to audit, check any customer complaint issued related to packaging and material handling. Layout of areas are optimized. Where applicable, the edges of lines are fed automatically or by little trains (no big boxes or full of components). Stock level (min-max visual management). Workplace visual management of stock (min-max).
External Logistics		
Supply process (incoming material) is managed, organized, and tracked.	1. A process to follow deliveries is formalized and applied. Related operations are standardized. Resources are available and properly managed. 2. A master schedule of deliveries is defined and followed up. A visual management system, showing smoothing of deliveries (leveling), allows management to follow them. An escalation process is implemented in case of deviation. 3. A logistic protocol is defined and regularly updated for each Tier X supplier. Transportation organization toward the Tier X suppliers is optimized. 4. The service rate of the Tier X suppliers is followed; actions plans are established for most failing Tier X.	Organization of the incoming. Master schedule of deliveries (stability through weeks). Working instructions. An example of protocol with a Tier X. Service rate metrics. An example of alert in case of deviation. Resources available match incoming and outbound deliveries.

Required	Criteria of Requirements	Verification Possibilities
A process to secure supplies is applied on a basis of risks analysis.	1. Components/materials at risk of supply are identified. Securing plans including safety stocks are established with the Tier X. 2. Securing plans are periodically reviewed based on lessons learned. 3. Safety stock, advance warehouses are contractually defined and managed for the "distant" Tier X (more than 48 h of transportation) and when applicable for the "risky" Tier X. 4. The management of the site is strongly involved and validates the whole process.	Identify a "risky" supply (e.g., distant supplier or failing supplier). Examine the securing plan associated with the identified risk. Emergency procedures with alternate supply process. In case of safety stocks, identify clues of contractual definitions.
Shipping process (outgoing) is organized and tracked. Packaging is properly labeled.	1. Shipping process is described and managed in a timely manner. All operations are standardized. 2. Shipping process integrates milestone operations (e.g., verification of availability of finished products, truck loading completed (i.e., Lear corporation's system AVIEXP, which has a purpose to provide the final consignee of the goods, with detailed information relating to the actual contents of a consignment sent by a consignor) allowing to alert customer in relevant time in case of issue. 3. Shipping planning is visually managed (timing table of the preparations of shipping orders). 4. A preparation list is available for operators. Operations done are recorded. Loading is prepared in "bogus truck" areas. 5. All the packaging is labeled properly and packaging/shipping label EP operation is a best practice in order to prevent part misidentification. PFMEA covers potential failure mode of labeling.	Shipping management board. Verify that the shipping process includes at least the following milestones: Verification of finished products' availability Start of preparation End of preparation; ready to ship Truck loading completed (EDI message sent to customer) Identification of deviations on the board (late supply). Organization of bogus truck areas. Preparation lists. Labels correctly fulfilled (with the right routing code). Equipment and work instructions to check the product and pallet labeling. Hooping of the pallets. Emission of the EDI message AVIEXP.

(*Continued*)

Required	Criteria of Requirements	Verification Possibilities
	6. Specifically for synchronous deliveries, a constant monitoring/follow up of the process is carried out in real time; all deviations/issues are recorded and analyzed, and countermeasures and corrective actions to eradicate issues are started immediately.	
In order to guarantee that shipping process is carried out according to customer needs, the electronic data interchange (EDI) network is operational and a logistic protocol is managed.	1. EDI communication is installed and is validated with the customer. Qualified people are permanently available on site to manage EDI. 2. Back-up solutions are defined, validated with the customer, and periodically tested. 3. Any evolution of EDI communications must be considered as significant change. 4. A logistic protocol is established together with the carrier (transporter) and the customer. It is continuously updated to take into account changes during current production (change of customer site, change of schedule, etc.). Resources responsible for the protocol management on site are identified. 5. Any deviation in the protocol application is treated and managed by action plan (alternative transportation mode, adapting packaging, etc.).	Several logistical electronic protocols (LEPs) and validation status. Coherence between different LEPs and shipping schedules. EDI equipment. Securing equipment (hardware, backup, networks). What happens if EDI network fails? Customer site contact list.
Indicators are defined and tracked to ensure effectiveness of external logistic management.	1. Indicators concerning the supply process: Fill rate of the trucks Service rate of Tier X Suppliers Tracking of logistic issues with Tier X suppliers 2. Indicators concerning the shipping process: Shipping lead time Customer service rate Tracking of customer log issues Pareto of failures (customer line stops/stockout) Rate of mislabeling	Prior to audit, check any customer complaint issued related to reception issues, stockout. Visual management (indicators are followed on boards at the shop floor). How does supplier determine safety stock level, risk analysis?

Required	Criteria of Requirements	Verification Possibilities
	3. Indicators concerning the safety stock (applicable only if a safety stock has been contractually defined with customer) Stock level Rotation index of the safety stock	
Strategy and Project Management		
The industrial strategy vision and the associated roadmap are defined and deployed.	1. There exists an industrial long-term vision (2–5 years) coherent and communicated to the whole staff. 2. Long-term vision is deployed to strategic targets of the site. 3. There is a robust process to transfer targets until the operational level (from the site targets to the operator targets). 4. Based on target improvement master plan developed, which includes the modification of the internal processes and takes into account the organization, the resources, technology, the evolutions of product/process, the evolution of skills, etc. 5. The improvement master plan is managed by the site manager director. Actions are followed by the site top management team and, when applicable, by corporate industrial central department.	Check long-term vision for site and strategic targets deployment for a few departments. Evidence that performance was reviewed and action initiated in case of deviation.
The organization of the manufacturing site is completely described for all key activities.	1. The manufacturing site is involved in new projects since advanced stages; project team includes at least staff from manufacturing site. 2. A project milestone states the transfer of responsibilities between project and production site. After transfer to manufacturing location, process needs to be requalified. 3. An organization chart details the site organization and is systematically updated based on changes in the organization and its missions. 4. A job description is available for each job and systematically updated based on changes in the organization and its missions.	Check site management organization chart. Check project responsibility, site contact part of it. Check a few job descriptions, one of them for quality manager.

(*Continued*)

Required	Criteria of Requirements	Verification Possibilities
	5. Quality is independent of the operational directions. It has authority to stop production in the event of nonconformity.	
Human resources for all operation activities are evaluated and continuously managed.	1. There is a formalized process to identify new resources needed. 2. Rules are defined for its implementation in the event of modification of organization or process, evolution of production program or new product launch, etc. 3. The process integrates management of the rates of new employees including ratio of the temporary staff based on indicators of production schedule and efficiency.	Check whether new project requires new staff. Check process for identification of needs (number of employees, their qualification). Predefined plan based on production needs (new shift/additional operators, etc.).
For all new facility/equipment/tools, etc. (manufacturing, control, logistics, etc.) or in case of revamping, a structured validation approach is applied.	1. Milestones are established and aligned with project timing for installation and validation of new equipment/tools, etc. 2. Based on milestones, detailed timing plan is established and regularly followed up for validation process. In case of any delay/deviation, actions initiated including notification of customer. 3. The qualification process aims to verify compliance with specifications, contract, and performance (productivity, capability, quality) of equipment. In the case of gauges, it also integrates equipment calibration. 4. Dedicated resources assigned for qualification of equipment, connected to maintenance.	Check whether new project requires new facility/equipment/tool, etc. Check process for identification of needs. Check new equipment and its documentation for qualification, technical data, and maintenance.
For every new product/process or significant product/process changes, the product/process qualification is defined, continuously tracked, and finally approved.	1. Milestones are established and aligned with customer project timing for product/process qualification. 2. Based on milestones, detailed timing plan is established and regularly followed up; their results are shared with the customer. 3. Qualification process includes all manufacturing processes (including outsourced), components, tier parts, and raw materials.	Check a timing plan for an advance project. Check a qualification package (PPAP) and crosscheck a few elements (e.g., gauge R&R, dimension and material report, function/durability result, capability study). Check an R&R result.

Required	Criteria of Requirements	Verification Possibilities
	4. Deliverables to the customer are managed and controlled. Customer product approval status is known and tracked for all products. 5. Capacity analysis with its verification, process audit, and capability of key characteristics are part of qualification process. 6. Any deviation is treated; action plans are led by the project manager and/or his or her representative on site. Customer is notified.	Check qualification (PPAP) confirmation from customer.
Service parts (SP) production process is managed and controlled during and after the mass production phase.	1. A process to manage "end of life" of current production of reference part number is deployed. It takes into account the quantities of SP remaining to produce (strategy defined, e.g., prepare stock to cover SP demand, keep tool, production transfer, etc.). 2. In order to maintain efficiency and effectiveness of production and measurement equipment dedicated to SP, specific procedures of storage and maintenance are applied (e.g., storage conditions, 6S, oiling, protection against dust, etc.). 3. Specific operations or processes for SP (e.g., specific marking) production are industrialized, managed, and controlled according to the same methods as mass production process. 4. In the event of storage of SP for a long period before shipping, a process is applied in order to guarantee the safeguarding of the parts (intermediate packaging, new storage areas, etc.).	Business planning + production master plan (verify that SP needs are included). Example of "end of life" production. Process qualification of specific operations. SP storage. Procedure for long storage.
Product packaging (final product, intermediate, and supplies) and handling equipment are designed and validated to guarantee product conformity.	1. Packaging is defined during product design phase (take into account in project milestones). The technical specification is built based on customer needs and is approved by the customer (drawing of packaging).	An example of technical specifications. An example of packaging. Packaging part protection material. PFMEA concerning handling and storage. Storage CP when necessary. Test result (static and dynamic).

(*Continued*)

Required	Criteria of Requirements	Verification Possibilities
	2. Packaging is designed in order to protect product from damages. 3. Test procedures and validation process are shared with the customer. Static and dynamic tests are implemented. Test reports are available. 4. The definition of the storage and handling conditions starts during product/process design phase and are shared with the customer. 5. Handling and transportation equipment design is done taking into account risks of degradation of product quality.	
Part and Process Control		
There is a system to guarantee the availability to users of the documents at the latest version at the right place. All the technical information is available.	1. A standardized process guarantees that all the users have access to the latest versions of the documents. 2. Revision history available for each document. 3. Customer approved drawings are available at production facility with the latest change level. 4. Drawing is completed with all the product characteristics (tolerances, GD&T, correct datums, key process characteristics [KPCs], etc.). 5. All technical regulations are available at production facility.	Prior to audit, check drawing date and engineering change level of latest customer drawing. All technical regulations referred to in drawing are available.
All the key characteristics are identified in order to ensure their proper control.	1. All the key characteristics defined by customer are identified in PFMEA and CP. 2. Key customer characteristics are deployed to process characteristics if applicable. 3. If applicable, they are communicated to tier suppliers, necessary controls are verified and confirmed. 4. Required customer controls (e.g., SPC) applied for key characteristics. 5. Pass-through characteristics are identified and controlled. 6. EP strategy is applied.	Check all the key characteristics deployment into documentation (PFMEA, CP, and working instructions), verify their control effectiveness. Ask operators about awareness of key characteristics.

Required	Criteria of Requirements	Verification Possibilities
All the operations are listed and charted in process flow diagram.	1. All process steps including logistic flow are listed and charted. 2. The flow chart is used as an input for PFMEA. 3. Link ensured between process flow, PFMEA, and CP.	Process flow includes receiving, all the operation steps, rework, scrap, gauging/inspection, shipping, labeling and Part ID at receiving, waste in process (WIP), finished goods, and shipping. Flow is clearly arranged. Workplace properly configured and matches with process flow diagram. Check link/number matching between process flow, PFMEA, and CP.
CP is available for each product in order to define necessary controls needed to be performed.	1. A CP is available for each product and process. 2. All current controls are listed on the PFMEA detailed on the CP. 3. CP defines the characteristics and the parameters to be monitored, the specifications and the tolerances, the control frequencies, the methods, the equipment, the records, and the reaction plans (in the event of nonconformance). 4. Sample size and frequency take into account process and measurement capabilities, phase of project, and how often reaction is needed. 5. CP is reviewed systematically in the event of PFMEA review.	CP addresses print requirements relative to dimensional, material, rust preventative, heat-treat, micro-finish, contamination/sediment, etc. Sample sizes and frequency for each operation are reasonable (e.g., based on internal data: scrap, rework, FTQ, etc.). Evidence that CP is being updated. Checks to be done under the PCP requirements are performed.
Statistical process control (SPC) is applied.	1. Characteristics, where SPC needs to be applied, are defined in CP based on customer and process capability requirements. 2. Sample size and frequency are defined according to the stability of the process. 3. Control limits are calculated for each characteristic according to the capability target and updated based on trends. 4. Results are documented via control charts and used for process capability calculation. Result needs to meet customer requirement. 5. Out of control point is noted with corrective action taken.	Where the PCP calls for SPC, look at actual SPC records and data collection: - All data boxes filled in - Control limits calculated - Trends/patterns are present that would indicate a need to recalculate limits - The limits actually are control limits and not spec limits. - Verify that every out of control point is noted and corrective action defined and followed by leadership. - Ask someone noted on the action plan to explain his or her involvement and/or responsibility.

(*Continued*)

Required	Criteria of Requirements	Verification Possibilities
	6. Leadership periodically reviews SPC results and ensures that out of control conditions are managed.	- Manufacturing process demonstrates the required capability or performance (based on SOR requirements).
Results of control are recorded and archived as well as analyzed to correct deviations and improve processes.	1. Control records are documented (using standard forms or information systems) according to CP requirement. 2. Control records must guarantee product/process traceability and conformity and are archived according to customer traceability requirements. 3. Results are reviewed, out of control point are noted. 4. Actions are implemented in accordance with reaction plan defined in CP. 5. Results are analyzed for CI activities.	All results recorded where CP requires. Records are readable. Out of control points are noted, necessary action implemented. Track back records to check results are archived, traceability ensured.
The information systems are managed in order to secure all the data related to complete material flow of product.	1. A list of critical information systems is established and managed. 2. Resources are identified, qualified, and available during all production periods. The responsibilities are defined for each critical information system: - To guarantee the availability and the integrity of the data - To ensure the breakdown service in case of failure - To implement needed functional evolutions 3. There are procedures of backup and recovery of the data and they are regularly tested. 4. For all main changes concerning critical information systems, a qualification process is carried out prior to full implementation.	List of critical systems (system connected to the material flow management, software, programs, system related to the traceability of the parts, incoming, and shipping). Resources are available onsite to manage IT and experts on short term. Check backup plans for loss of data/computer breakdown/ network breakdown. Specific procedure for degraded modes.
Product audits (including dock audit as well) are defined and performed according to CP.	1. Product audit procedure is defined and performed according to CP and all the results are documented. 2. Product audits completely cover KPC verification. 3. Product audit is considered SW. 4. Decision criteria to alert the customer are defined. 5. In case of outsourced test, labs must be certified.	CP defines product audit clearly. Check product last audit results. Was it performed according to required frequency? In case of any NOK results, action taken.

Required	Criteria of Requirements	Verification Possibilities
Safety stock location, organization, and management comply to customer expectations.	1. Safety stock warehouse (SSF) is separate from the supplier site. The distance between the two is sufficient to prevent against social risks, flood, fire, etc. The location is confidential and unknown to the personnel of the site. 2. The SSF is managed by staff independent from the supplier qualified to storage and shipping supplier standards. Schedules of activity of the SSF warehouse are consistent with customer site activity. 3. A procedure of emergency releasing is defined and applied. The time of provision of the parts (washing, repackaging) is managed. A releasing is possible during the night or the weekend. 4. The rebuilding of the SSF (in case of releasing) and the inventory turnover are organized and tracked. On customer request, the level of safety stock can be communicated and the turnover frequency can be adjusted. 5. Safety stock is fully integrated in the management system of supplier stock.	Stock levels. Results of audit and inventories. Stock conditions. List of contacts with the customer plant. What happens if there is a request for an emergency release?

Appendix V
Miscellaneous Items

I thank Cary D. Stamatis and Stephen D. Stamatis for their contribution in both designing and drawing the forms. In addition, their help with the computer was very valuable, especially with very short notice. Thanks guys!

Six Sigma Helps

General Requirements of Understanding Six Sigma

A typical calculation for a Six Sigma capability
Brainstorming thought starters: general
Brainstorming thought starters: what if
Verb-noun listing: products
Verb-noun listing: process
Function identification example for a product
Function identification example for a process
Cascading the Y = f(X)

Forms

Define

Maslow's Theory of needs

1. Translating the voice of the customer (VOC) to requirements (CTQs)
2. Understanding inputs and outputs
3. Cause and effect matrix
4. Project charter
5. Project plan milestone chart
6. Process and boundary development
7. Input, process, and output measures
8. Computing the cost of quality (COQ)
9. Six key areas to address when improving the cost of quality

Measure

10. Check sheet development
11. Data collection plan
12. FMEA form
13. Gage R&R: long form
14. Gage R&R: short form
15. Attribute gauge study
16. 5S observation summary
17. Control plan
18. Simplified control plan
19. Sources of data

Analyze

20. Main effect and interaction setup

Improve

21. Criteria matrix (evaluating improvements)
22. A form that may be used to direct effect and resources
23. Payoff matrix
24. Understanding systems and structures for the Six Sigma methodology

Control

Some of the previous forms are used here, such as 8, 12, 13, 15, 17, 21, and 23

Miscellaneous forms

25. Problem statement
26. Understanding the operational definition of the problem
27. Gauge control plan
28. Work breakdown structures
29. Who does what in the project
30. How the work is done
31. Exploring the values of the organization

Typical Forms

In addition to the above, these forms are used in design for Six Sigma.

32. Risk identification and mitigation
33. Competition matrix
34. CTQ matrix and assessment of design risk
35. Design scorecard
36. Pugh matrix
37. Score card: critical to satisfaction (CTS) items
38. P-diagram
39. Translating language data to numeric data
40. Brand profiler
41. Status of items critical to satisfaction and relationship to customer satisfaction
42. Customer dimensions
43. Reliability and robustness checklist
44. Design verification (DV) test plan
45. Triple 5 Why analysis worksheet
46. Project proposal worksheet
47. Action plan
48. Cost function
49. Process sequence flow chart
50. Cost index worksheet
51. Manufacturing information gathering worksheet
52. QFD block diagram
53. Function identification worksheet
54. Function diagram
55. Action plan
56. Competitive evaluation
57. CTS scorecard
58. Design and manufacturing or service scorecard (to be used with Microsoft Solver)
59. Reliability and robustness demonstration matrix
60. Assessment of Six Sigma status

General Requirements of Understanding Six Sigma

Typical Calculation for a Sigma Capability

Step	Action	Equations	Your Calculations
1	What process do you want to consider?	N/A	Billing
2	How many units were put through the process?	N/A	2167
3	Of the units that went into the process, how many came out OK?	N/A	2059
4	Compute the yield for the process defined in Step 1	Step 3/Step 2	2059/2167 = 0.9502
5	Compute the defect rate based on Step 4	1-Step 4	1 − 0.9502 = 0.0498
6	Determine the number of potential things that could create a defect	Number of CTQ characteristics	18
7	Compute the defect rate per CTO characteristic	Step 5/Step 6	0.00498/18 = 0.0028
8	Compute the defects per million opportunities (DPMO)	Step 7 × 1,000,000	0.0028 × 1,000,000 = 2800
9	Convert the DPMO (Step 8) into a sigma value. You may use a conversion chart or calculate the number	N/A	4.3
10	Draw conclusions	N/A	Little better than average performance

Brainstorming Thought Starters				What If Questions
Adapt	What else is like this? Does the past offer similarities? What could we copy? What other ideas does this suggest?	**Rearrange**	Can we interchange components? Different layout? Different sequence? Change pace? Different pattern? Different schedule?	You were spending your money? Money was no object? You combined two functions? You were a man from Mars, what questions would you ask about this?

Brainstorming Thought Starters				What If Questions
Combine	Use a blend, alloy, assortment? Can we combine units? Combine purposes, functions? Combine ideas?	**Reverse**	What's the opposite? Can we turn it around, upside down, backward? Can we reverse roles?	You were 8 years old, what questions would you have?
Magnify	What can we add? Thicker? More frequent? Stronger? Duplicate?	**Modify**	Could we change the color, sound, motion, form, shape, meaning, odor? What new twist?	You knew your managers would buy an idea you came up with, what would it be? You were from 1000 years in the future? You didn't perform the function at all?
Minimize	What can we subtract? Smaller? Lighter? Condensed? Omit? Streamline?	**Substitute**	What can we use instead? Who else can? Another approach? Another material? Another ingredient?	You had the perfect material, what would you expect? You performed the function backward? Gravity did not exist, what would you do then? You were trying to prevent the function from being performed? You had an unlimited amount of time? Nothing was sacred?

Verb-Noun Listing: Products

Verbs		Nouns	
Absorb*	Generate	Access	Friction
Access	Guide	Air	Heat
Actuate	Improve	Appearance	Impact
Allow*	Increase	Bending	Light
Apply	Isolate	Circuit	Mass
Attach	Limit	Climate	Material

(*Continued*)

Verbs		Nouns	
Attract	Maintain	Cold	Moisture
Circulate	Pivot	Comfort	Motion
Conduct	Position	Component	Noise
Connect	Prevent	Corrosion	Occupant
Contain	Protect	Current	Parts
Control	Provide*	Deflection	Path
Convert	Reduce	Dirt	Performance
Create	Regulate	Drag	Pressure
Decrease	Resist	Energy	Stability
Direct	Rotate	Entry	Surface
Enclose	Seal	Environment	Torque
Enhance	Sense	Flow	Travel
Extend	Support	Fluid	Vibration
Facilitate*	Transmit	Force	Weight

*Try to avoid use of these verbs.

Verb-Noun Listing: Process

Verbs		Nouns	
Allow*	Join	Alignment	Flash
Apply	Load	Assembly	Gauge
Assemble	Maintain	Burr	Gas
Assure	Make	Casting	Heat
Blend	Move	Cause	Hole
Clean	Position	Cleanliness	Inventory
Control	Prevent	Cold	Length
Convert	Protect	Component	Locator
Create	Provide*	Container	Machine
Decrease	Receive	Correction	Material
Deliver	Release	Damage	Mold
Facilitate*	Remove	Defect	Operation
Fasten	Repair	Device	Part
Fill	Rotate	Die	Priority
Finish	Seal	Dimension	Schedule
Form	**Store**	Dirt	Shape
Identify	Supply	Environment	Surface
Improve	Thread	Equipment	Tool
Increase	Transport	Finish	Uniformity
Inspect	Verify	Fixture	Waste

* Try to avoid use of these verbs.

Function Identification Example for a Product/Service

Team Members: **Date:**

Project

Listing of Functions Performed	
Active Verb	**Measurable Noun**
Improve	Appearance
Position	Parts
Enhance	Stability
Prevent	Vibration
Improve	Assembly
Assure	Location
Reduce	Margin Variation
Enhance	Cooling
Control	Location
Support	Part
Limit	Deflection
Fasten	Parts

Function Identification Example for a Process

Team Members: **Date:**

Project:

Listing of Functions Performed	
Active Verb	**Measurable Noun**
Transfer	Material (parts)
Move	Material
Position	Material (parts)
Apply	Material (sealer)
Actuate	Circuit
Weld	Assembly
Remove	Assembly
Position	Assembly
Store	Assembly
Produce	Assembly
Ship	Assembly
supply	Material
Repair	Assembly

Cascading the $Y = f(X)$

Level 1 $Y = f(X_1, X_2, \ldots X_n)$

↓

Level 2 $y = f(x_1, x_2, \ldots x_n)$

↓

Level 3 $y_i = f(x_i, x_{ii}, x_n)$

where:

Y is influenced by a number of potential Xs

y is the new requirement based on X_1s and so on; X_1 is further examined for more potential x's

y_i is yet another requirement of the customer based on x_1's and so on; x_i is further examined for more potential x's

Example: High-level application of a meal in a restaurant:

Level 1 Y = a good meal in a restaurant is dependent on X_1 = price, X_2 = service, X_3 = satisfaction

Level 2 y = the result in satisfaction (which is the X_3) is dependent on x_1 = service and x_2 = food selection

Level 3 y_i = the result in satisfaction (X_3) with service (which is the x_2) is dependent on x_i = availability of food, x_{ii} = management

Level 4 and so on

Define

To know who and what the customer needs or wants, one must understand the customer through Maslow's theory of motivation and human needs. Furthermore, to extend that understanding into the concept of customer service, we must be cognizant of the PACT principle. PACT is essential in this understanding because customer service is perishable and it depends so much on the moment of service. PACT is an acronym for prompt, professional, personable; accurate, attentive, acknowledging; courteous, caring, concise; and tailored, timely, and thoughtful. Maslow's theory is based on a hierarchical pyramid of needs shown as

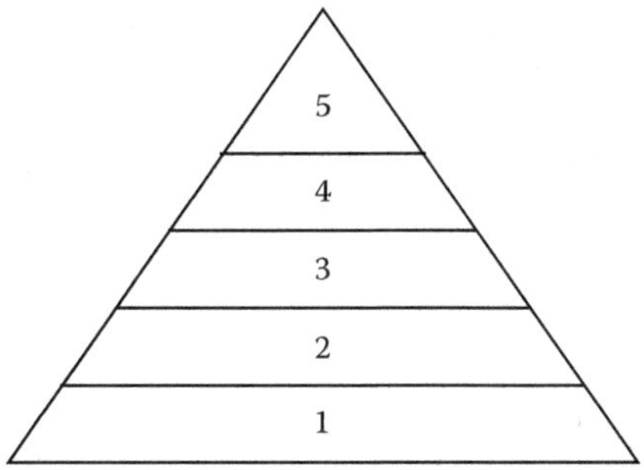

Levels of Needs	Motivation and Behavior
Level 1 = Your body: Physical safety and security Need to stay alive! Biological and cultural imperatives to live. Includes having enough healthy food, air, and water to survive.	Level 1 = Your body: Survival skills Eat, sleep, and take care of your bodily needs, provide for clothing, shelter, comfort, be free from pain.
Level 2 = Your family and work: Social safety and security Need to be safe from physical and psychological harm in the present and future, and trust in a predictable future.	Level 2 = Your family and work: Survival skills Work, save for future, improve skills and talents, be responsible, and want an organized predictable world.
Level 3 = Love and relationships: Communication and response Need to be loved and to love. Includes the desire for affection and belonging.	Level 3 = Love and relationships: Validation Join and be active in clubs and groups, be able to talk to others, contribute to society, marry and have a family.
Level 4 = Self-esteem: Self respect and acceptance Need for reputation, prestige, and recognition from others. Contains the desire to feel important, strong, and significant.	Level 4 = Self-esteem: Brainpower Display your talents and skills, have self-confidence, appreciate attention and recognition from others.
Level 5 = Self-actualization: Fulfillment of goals & dreams Need for self-fulfillment. Desire to realize your full potential and become the best you are capable of becoming.	Level 5 = Self-actualization: Creativity Be a self-starter, have enthusiasm, be creative, be dedicated, enjoy challenges, love to accomplish results.

1. Translating the voice of the customer (VOC) to requirements (CTQs)

Level	VOC	Key Issues	Requirements	
			Good	Bad

2. Understanding inputs and outputs

Process Input Variables	Process Output Variables

3. Cause and effect matrix

Rating of Importance to Customer								
	1	2	3	4	5	6	7	
Process Inputs (Xs) CTPs	Process Outputs (Y's), CTC, CTD, CTQ							Total

4. Project charter

Project Charter			
Project title:		Key roles: Name: Title: Task:	
Problem statement:		Timing: Start: End:	
Company's impact:		ROI: Expected benefit	
Project scope statement:		Team members:	
Project goal:			
Project location			

5. Project plan milestone chart

Cycle of Project	Time Line in Weeks or Months				
Project start					
Define					
Measure					
Analyze					
Improve					
Control					
Project finish					

6. Process and boundary development

Process Name	Starting Point	Ending Point
1		
2		
3		
4		

7. Input, process and output measures

Input Measures	Process Measures	Output Measures

8. Computing the cost of quality (COQ)

Currently Measured	Not Measured At This Time
Scrap:	Increased maintenance:
	Lost sales:
Warranty expense:	Customer dissatisfaction:
	Downtime:
Inspection costs:	Engineering and product:
	Development errors:
Overtime:	Bill of material inaccuracy:
	Rejected raw materials:
Internal Failure	**Appraisal**
Scrap:	Inspection:
Rework:	Testing:
Supplier scrap:	Quality audit:
Supplier rework:	Initial cost and maintenance of test equipment:
External Failure	**Prevention**
Cost to customer:	Quality planning:
Warranty cost:	Process planning:
Complaint adjustments:	Process control:
Returned material:	Training:

9. Six key areas to address when improving the cost of quality

Key Drivers	Basic Issue
1. Basic organizational capabilities	Skills and tools required to implement improvements in business processes are lacking
2. Industrial process variations	Poor industrial process capabilities result in high COPO (rework, scrap, field failure)
	Customer demands are frequently not passed on to engineering
	Inefficient front-end engineering
3. Business process variations	Product cost estimation is often widely off the mark, resulting in poor financial performance and incorrect manufacturing decisions
4. Engineering/design process and documentation	Engineering systems and design processes and documentation are often inadequate and flawed
5. Quality of specifications	Specifications sent to suppliers/subcontractors vary considerably in their quality, resulting in poor-quality parts
6. Supplier capabilities	Lack of quality suppliers, resulting inpoor-quality parts/services, late deliveries, higher parts/service costs, and so on

Measure

10. Check sheet development

Name:		Date:
Defect Type	Frequency	Comments

11. Data collection plan

What			Why	Who	How	When	Where
Needed Measures	Operational Definition	Formula or Formulas	Purpose of Data	Individual Responsible For The Collection	Method of Selection	Date, Time, or Frequency	Source

12. FMEA form

Process or product name:		FMEA Team:	Prepared By:			Original Date:		Revised Date:		Page _ of _		FMEA Number:			
Responsible:															
Process Step	Potential Failure Mode	Effect	SEV	Class	Causes	OCC	Controls	DET	RPN	Actions Rec.	Actions Taken	SEV	OCC	DET	RPN

13. Repeatability and reproducibility study (gauge R&R): Long form

	1	2	3	4	5	6	7	8	9	10	11	12
Appraiser	A-				B-				C-			
Sample #	1st Trial	2nd Trial	3rd Trial	Range	1st Trial	2nd Trial	3rd Trial	Range	1st Trial	2nd Trial	3rd Trial	Range
1												
2												
3												
4												
5												
6												
7												
8												
9												
10												
Totals												

Sum / Xbar → $Rbar_A$ Sum / Xbar → $Rbar_B$ Sum / Xbar → $Rbar_C$

$Rbar_A$	
$Rbar_B$	
$Rbar_C$	
$\bar{\bar{R}}$	

# Trials	D_4
2	3.267
3	2.574

$(\bar{\bar{R}}) \times (D_4) = UCL_R$ *

$(\_) \times (\_\_) = \_\_\_$

Max. Xbar	
Min. Xbar	
Rxbar	

* Limit of individual R's. Circle those that are beyond this limit identify the cause and correct. Repeat these readings using the same appraiser and unit as originally used or discard values and reaverage and recompute R and the limiting value UCLR from the remaining observations.

Notes: ______________________________

14. Repeatability and reproducibility study (gauge R&R): Short form

Parts	Operator A	Operator B	Range
1			
2			
3			
4			
5			
		Sum of percentage	
D_2 values for distribution of R			
Parts	**Two operators**	**Three operators**	
1	1.41	1.91	
2	1.28	1.81	
3	1.23	1.77	
4	1.21	1.75	
5	1.19	1.74	
$R-\text{bar}=\frac{\sum R}{n}$; RR error $\left(\frac{\overline{R}}{d_2}\times 5.15;\ R\,\&\,R=\frac{\text{RRError}}{\text{Tolerance}}\times 100\%\right)$			

15. Attribute gauge study

Operation title:			**Gauge name:**			**Attribute**	**legend**	
Characteristics:			**Gauge number:**			A	Accept	
	Name:			**Data:**		B	Bad	
Operator A						D	Defect	
Operator B						G	Go	
Operator C						G	Good	
						NG	No Go	
Known attribute		**Operator A**		**Operator B**		**Operator C**		**Score**
Sample ID	**Attribute**	**Trial 1**	**Trial 2**	**Trial 1**	**Trial 2**	**Trial 1**	**Trial 2**	
1								
2								
3								
4								
5								
6								
7								
8								
9								
10								

11									
12									
13									
14									
15									
16									
17									
18									
19									
20									
21									
22									
23									
24									
25									
26									
27									
28									
29									
30									
Score via trial via operator									
Score via operator									
		Operator A		**Operator B**		**Operator C**			
		Trial 1	**Trial 2**	**Trial 1**	**Trial 2**	**Trial 1**	**Trial 2**		

Result Summary:		
	% Appraiser effectiveness	
	Operator A	
	Operator B	
	Operator C	
% effectiveness		
	Total	

16. 5S observation summary

5S Summary Report:		Score Results: (X/100)	
Shift:	Area/Department:		
5S Target Objective:			
Historical 5S Success:			
Priority Items that Need Attention	Existing Situation: Score		Future Situation: Score
1			
2			
3			
4			
5			

17. A control plan

Department:			Prepared By:				Page ___of ___		Document Number:	
Process:			Approved By:				Revision Date:		Other:	
Location:							Supersedes:			
Employee Check	Characteristic	CTQ or Critical Level (CL)	Requirement	Measurement Method	Sample Size	Frequency	Who Measures	Where Recorded	Decision Rule or Corrective Action	Reference Number

18. Simplified control plan

Project Title:				Date:	
Task/Action to Be Accomplished	**Specific Steps to be Taken**	**Responsible Person for Task**	**Starting Date**	**Finish Date**	**Deliverables**

19. Sources of data

	Y (CTQ)	X_1	X_2	X_n
Existing data				
Needed data				

Analyze

20. Main effect and interaction setup

	Orthogonal Array (Setup of Experiment with Appropriate Levels)			**Individual Factors and/ or Interactions**				**Response (It May Be a Single or a Multiple Response)**
Level effect +								
Level effect –								
Main effect difference								

Improve

21. Criteria matrix (evaluating improvements)

Desirable Criteria	Alternatives				
		Alternative 1		Alternative 2	
	Weights	Score	Weighted Score	Score	Weighted Score
	Totals				

22. A form that may be used to direct effort and resources

Part Name:

Functional Area	Present Cost	High	Low
X			
Y			
Total cost			

23. Payoff matrix

		Effort	
		Low	High
Benefit	High		
	Low		

24. Understanding systems and structures for the six sigma methodology

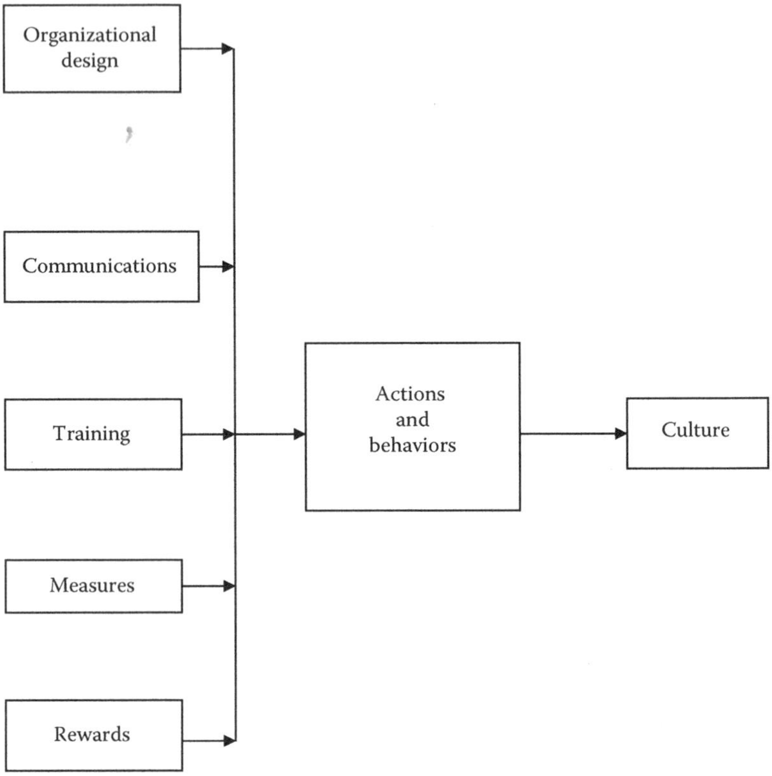

Control

Some of the previous forms are used here, such as 8, 12, 13, 15, 17, 21, and 23.

Miscellaneous Forms

25. Problem statement

According to...	Definition of the Problem	Unit of Measure	Source of Data / Data Type

26. Understanding the operational definition of the problem

Problem statement and project definition (original)	
Customer	CTQs
Nonconformance or variation (I will reduce…)	Data needed and units of measurement (relating to the nonconformance)
Current performance and reduction goal	
Potential benefits (COPQ)	
Project scope, limits, and boundaries	Potential team members
Problem statement and project definition (final)	

27. Gauge control plan

Gauge ID:	Department:	MSA reference:	Page __ of ___
Gauge name:	Gauge location:	Calibration procedure:	Document number:
Gauge type:	Storage:	Date:	Revision number:

					Calibration		GR&R		
Baseline	Date	Resolution	Bias	Linearity	Date	By	%R&R	P/T	Remarks

28. Work breakdown structures (WBS)

Task #	Task Name	Predecessor	Duration

29. Who does what in the project

		Relationship to the Project					
Stake holder	Wants, Needs, and Concerns	Impacted	Authority	Expertise	Influence	Resources	Other

30. How the work gets done

Who Needs Information	Message Forum	Message Frequency	Message Goal	Best Way to Present Message	Who Delivers the Message

31. Exploring the values of the organization

Priority	Description of What Is Valued

Typical Forms, in Addition to the Above, Used in Design for Six Sigma

32. Risk identification and mitigation

Risk Element	Risk Management	Risk Score	Mitigation Plan	

33. Competition matrix

Customer Needs	Priority	Competitor A	Competitor B	Best in Class	Best-in-Class Performance
Total Impact					

34. CTQ matrix and assessment of design risk

Customer Needs	Importance	Performance	Measurement	Target	USL	LSL	Date Type	Defect Rate	Map to Critical X	Impact on Customer	Impact on Company	Take Away

35. Design score card

CTQ	Measurement	Data Type	Target	USL	LSL	Sigma	DPMO

36. Pugh matrix

CTQs and Business Needs	Pugh Priority	Concept A	Concept B	Concept n
Sum of positives				
Sum of negatives				
Some of sames				
Weighted positives				
Weighted negatives				
Next score				

37. Scorecard: Critical to satisfaction (CTS) items

Project Description:									Date:				
CTSs	Units	Range		Contribution to Variability		Specs		Sample Statistics			z-score	
		Min	Max	Capability	%	LSL	USL	μ	σ	conf	σ shift	DPM

38. P-diagram

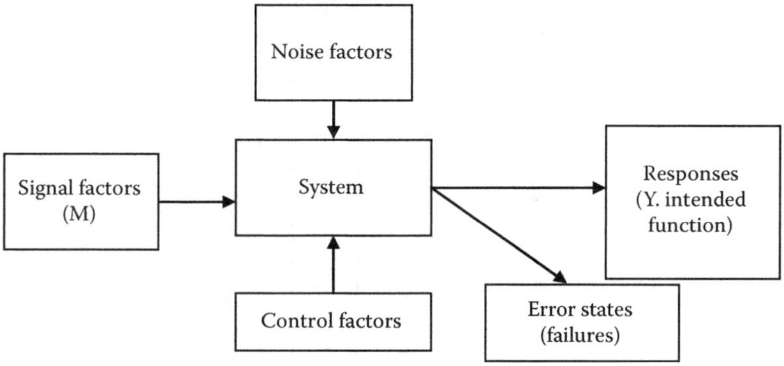

39. Translating language data to numeric data

Steps	Language Data	Numeric Data
Convert information to data Find the underlying message Identify the structure Evaluate importance plan appropriate actions		

40. Brand profiler

Attribute	Attribute Class	Priority	Primary Brand Positioning	Nameplate Brand Positioning	Program Specifics		Present Nameplate Entry
					Target Objectives	Status	
Usage experience							
Attribute differentiation:				Product attribute leadership strategy:			

41. Status of items critical to satisfaction and relationship to customer satisfaction (may be used at each stage of the DCOV model)

CTS or Surrogate	Units	T.F.? Y/N	Status		Competitor (BIC)		Target: Initial			Target: Aged			Customer Satisfaction Improvement
			μ	σ	μ	σ	μ	LSL	USL	μ	LSL	USL	

42. Customer dimensions

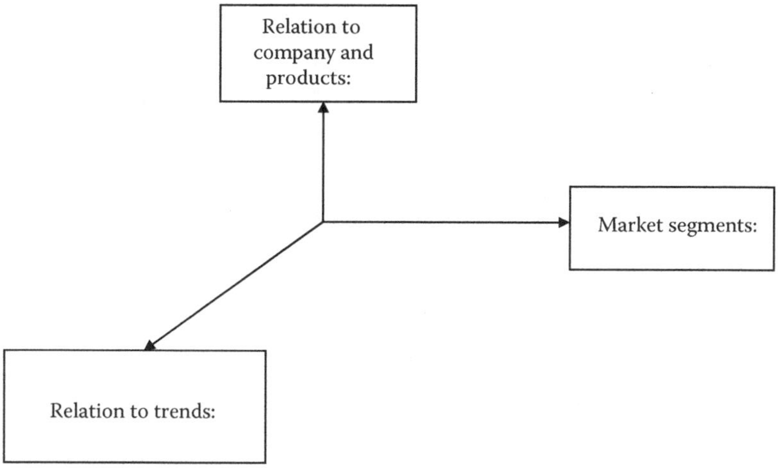

43. Reliability and robustness checklist

Functional Requirement Life Target (CTSs)	Design Parameters (Number and Name)	Tests (Number, Name, and Strategy)	Error States and Their Management

44. Design verification (DV) test plan

Test (Name, Type, and Frequency)	Objective	Expected Results	Real Outcome of Test

45. Triple 5-Why Analysis worksheet

For each root cause selected in Step 5, use this table to determine technical, detection, and systematic root causes.

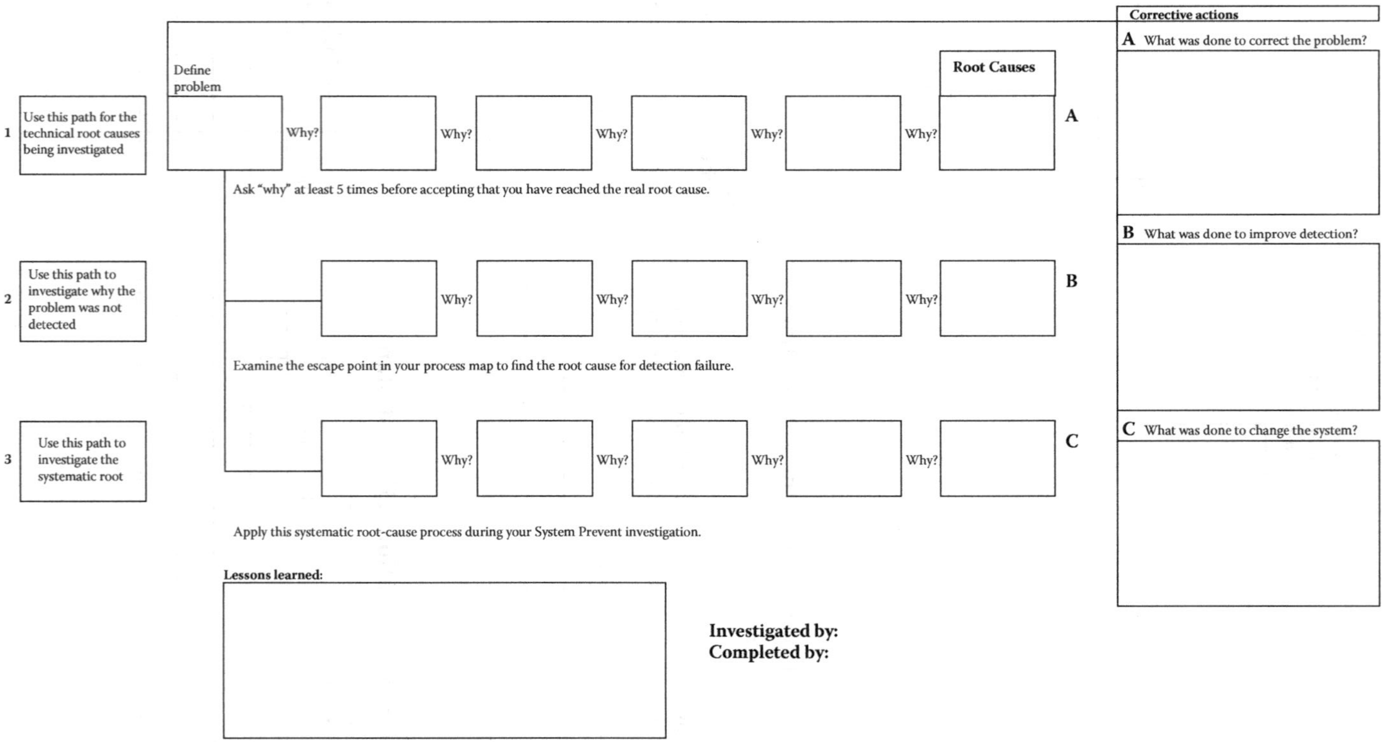

46. A selection of forms used in the process of identification and evaluation of a project

Customer-oriented goals

Unit: ____________________

Contact point	Customer reason for contact	Customer need, want or expectation	What is excellence?	Current status	Service or quality goal

Service / quality goals

Unit: ____________________

Non-price reason to buy	Customer need, want or expectation	What is excellence?	Current status	Service or quality goal

Team: ________

Date: ________
Project: ________

Idea Categorization Work Sheet
Function ______________________________

Category	Category	Category	Category
1	1	1	1
2	2	2	2
3	3	3	3
4	4	4	4
5	5	5	5
6	6	6	6
7	7	7	7
8	8	8	8
9	9	9	9
10	10	10	10
11	11	11	11

Blackbelts:

Workshop dates:
Project:

T-Chart Worksheet

Maximize good Advantages		Overcome bad Disadvantages		How can we make it work?
	Cost + −		Cost + −	
Net effect on cost		Net effect on cost		
Net cost impact:				

Team ____________________

____

____

Date: ____________

Project: ____________

____

Proposal selection worksheet

	Select appropriate criteria for evaluation and list below	Best alternatives from t-chart worksheets Rate from 0 to 10 (excellent)								
Relative importance	Criteria (Slight importance = 1, absolute must = 10)	Proposal 1	Proposal 2	Proposal 3	Proposal 4	Proposal 5	Proposal 6	Proposal 7	Proposal 8	Current Method
	Performance (basic function)									
	Reliability									
	Durability									
	Design aesthetics									
	Weight									
	Serviceability									
Total weighted rating - function:										

Piece cost $									
Value ratio = Function / Cost									

Investment $									
Engineering $									

Project Proposal Worksheet here

Proposal name:
Proposal #:

Black Belt:
Date:

Part name:	Carl lines affected:		Current function rating	Proposed function rating
Part number:	Volumes:		Current value ratio	Proposed value rating
Present condition (please describe).	Proposal description:		Advantages:	
			Disadvantages:	

(1) Cost savings per part-material	(2) Cost savings -labor	(3) Number of parts per vehicle	(4) Annual volume	(5) Lifetime volume	Weight reduction	Savings as % of piece cost		If all actions completed, implementation date:
(6) Tooling cost	(7) Engineering cost:	(8) Annual savings [(1+2)*3*4]	First year savings [8–(6+7)]	Lifetime savings [(1+2)*3*5–(6+7)]	Payback period: (mos.): [(6+7)/8]^12	Evaluation rank (Circle selection)	Immediately OK	Not accepted
							Needs examination/testing	Next iteration

47. Action plan worksheet

Workshop subject: ______________________ Date: __________

Proposal #: __________

	What needs to be done (Be specific)	Who does it (Name of person responsible)	Possible roadblocks (Names of people most likely to get in the way)	How to overcome	Completion date	Cost to complete
Design						
Samples						
Evaluation & engrg. judgement						
Tooling						
Other						

48. Cost / function worksheet

Black Belts: ____________________

Project: ____________________

Customer requirements: ____________________

Workshop dates: ____________________

#	Part/ operation	Cost	Function (active verb/measurable noun)									
			1	2	3	4	5	6	7	8	9	10

49. Process sequence flow chart

Black Belts: ______ ______ ______

Project: ______ ______ ______

Customer requirements: ______ ______ ______

Workshop dates: ______

#	What happens / Who does it									Cost per event	Time per event	Labor rate

50. Cost index worksheet

Team members: ____________________

Project: ____________________

Customer needs: ____________________

Workshop date: ____________________

Component / Model		Ours		Competitor 1		Competitor 2		Competitor 3		Competitor 4		Comments
Sketch →												
Part												
	Material Size Weight											

51. Manufacturer information gathering worksheet

Information required	Person responsible		Comment
	Manufacturer	O.E.M.	
Bill of materials (Complete parts list)			
Material cost			
Labor cost			
Freight & packaging costs			
Material flow/inventory costs			
Inspection costs			
Scrap & rework costs			
Any other cost components			
Original tooling costs			
Tooling capacity			
Standard Volume			
Process description with detailed costs			
Video of process (Mfg. & Ass'y.)			
Process flow charts			
Plant layout drawings			
Detail part drawing			
Material specifications			
Assembly drawing			
Photos of parts and/or processes			
Competitive parts			
Test / govt. requirements			
Warranty information			
List of suppliers' suppliers' phone #'s			
Invitation to employees & supervisors			
Conference room & equipment			

52. QFD block diagram

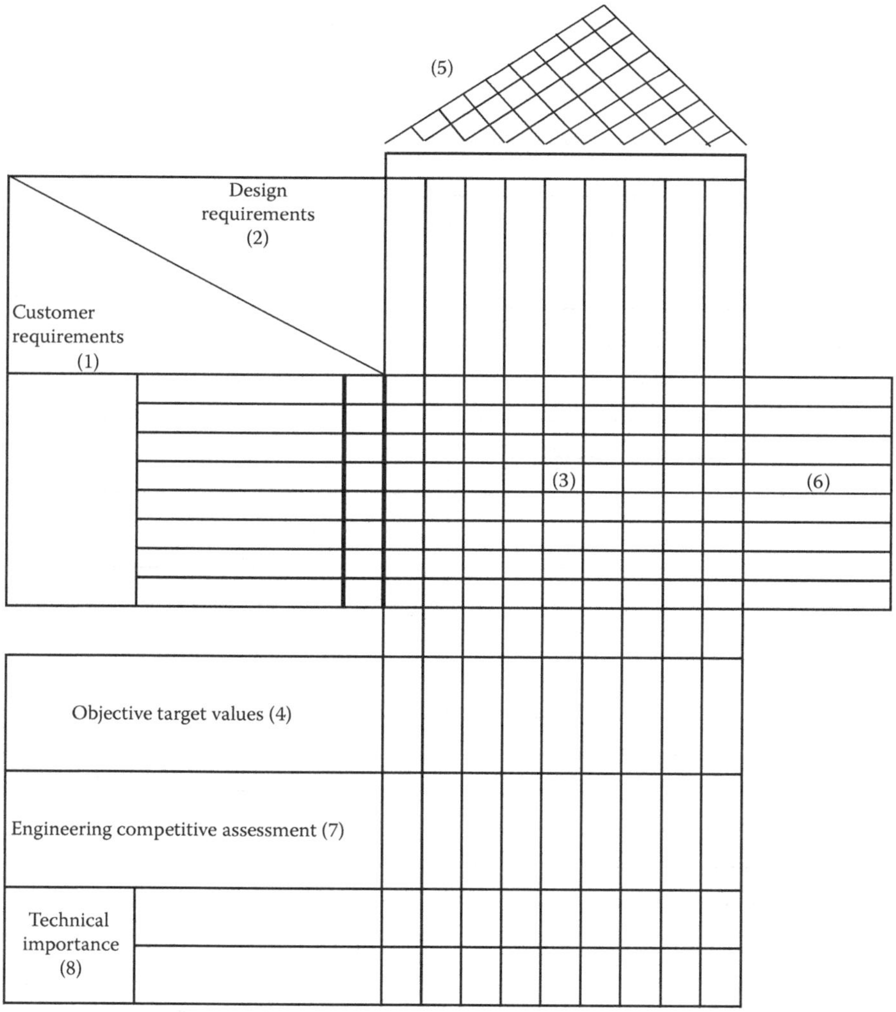

1. Customer requirements (focus on functionality). 2. Engineering requirements (based on customer's functionality). 3. Weighted relationship of customer functionalities. 4. Realistic target goals. 5. Relationship of engineering requirements. 6. Competitive cost relationship. 7. Engineering competitive evaluation. 8. Technical priority.

53. Function identification worksheet

Team members: ______________________________ Workshop dates: __________

______________________________ __________

Project: ______________________________

Listing of functions performed	
Active verb	**Measurable noun**

54. Function diagram

Black Belt(s): ____________________ Company: ____________________

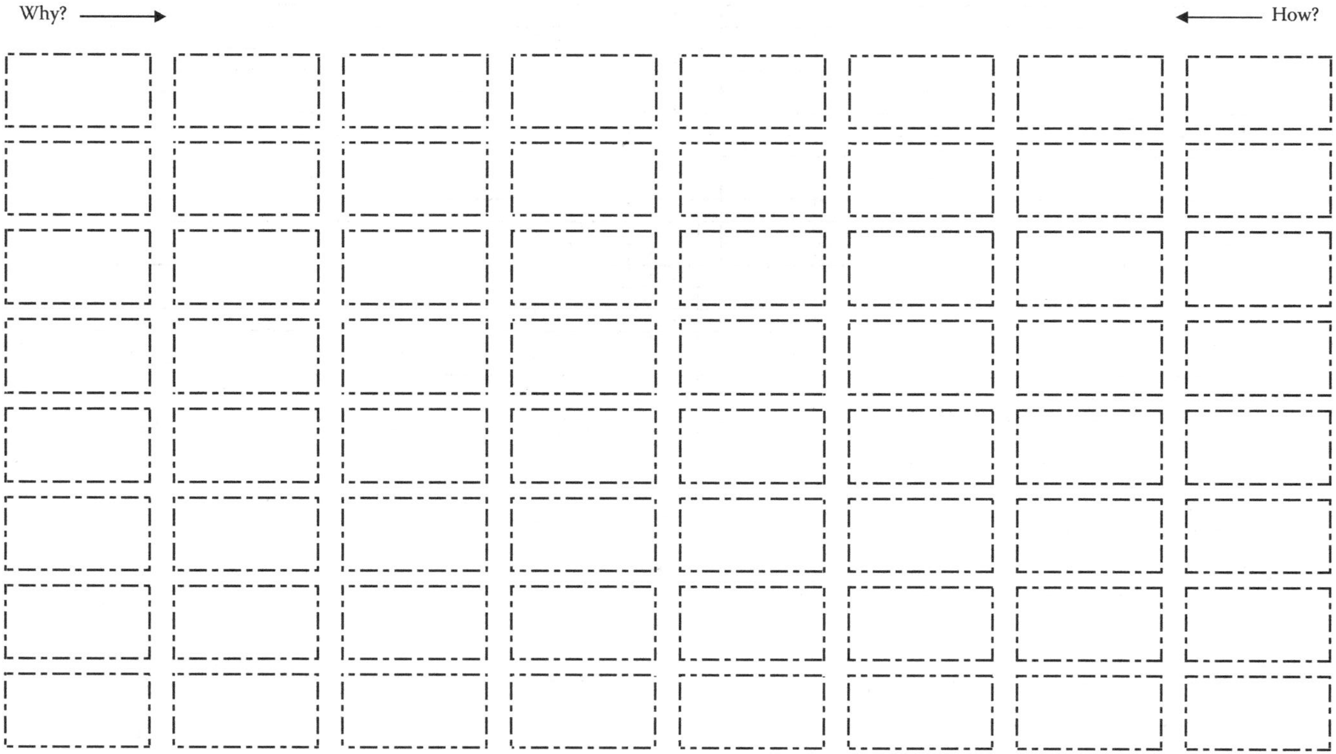

Project: ____________________ Date: ____________________

55. Alternative action plan

Legend

- **I** Initiate action and monitor performance
- **W** Perform work
- **C** Consultation required
- **R** Review/comment/summarize/recommend
- **A** Approve action or make decision
- **D** Make final decision if "A's" do not agree
- **N** Must be notified of decision

Title: ____________

Number: ____________

Objective: ____________

____

Prepared by: ____________

Date: ____________

		Characteristic of step taken												**Dates**		**Resources required or anticipated obstacles**	
Number	**Steps to be taken**	**A**	**B**	**C**	**D**	**E**	**F**	**G**	**H**	**I**	**J**	**K**	**L**	**Start**	**Stop**		

56. Competitive evaluation

Product/market: ______________________________

Please rank each company: 10 is best, 1 is worst

Key reasons to buy		**Weight**	**Company**	**Competitors**		
				1	**2**	**3**

57. Critical to satisfaction (cts) scorecard

Project description: ____________________

Status of critical to satisfaction (CTS) items and relationship to customer satisfaction															
CTS metric*	Units	T.F. ?	Status		Competitor/BIC		Target: Initial condition			Target: Aged			Associated cust sat improvement	Customer satisfaction graphical view
		Y/N	mean: μ	s.d.: σ	mean: μ	s.d.: σ	mean: μ	LSL	USL	mean: μ	LSL	USL		
														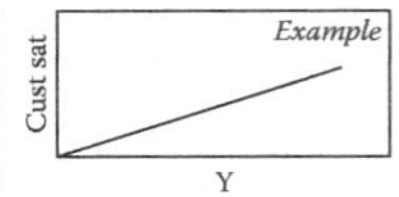
														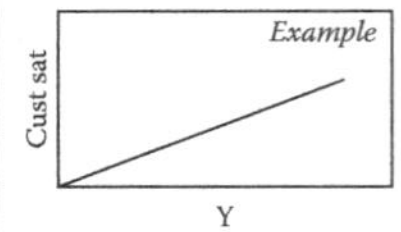
														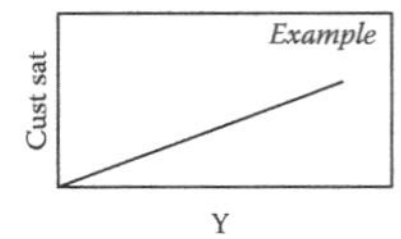

*Include reference to test or procedure for obtaining this measurement. Verification strategy will be captured in reliability & robustness checklist.

58. Design & Manufacturing Scorecard

Project Description: ____________________

Performance: Y or y		Transfer Function		Specification			Estimated Performance Capability			6σ Score			Actual Capability		
Characteristic	Units	Y/N	Formula (enter below)	Target	LSL	USL	mean: μ	s.d.: σ	Confidence/Comments	z_{st}	σ-shift	DPM_{LT}	mean: μ	s.d.: σ	Confidence/Comments
				0							1.50				

Variables: y's or x's			Range		Contribution to Variability		Specification		Sample/Database Statistics			6σ Score		
No.	Characteristic	Units	Min	Max	Sensitivity	%	LSL	USL	mean: μ	s.d.: σ	Confidence/Comments	z_{st}	σ-shift	DPM_{LT}
1													1.50	
2													1.50	
3													1.50	
4													1.50	
5													1.50	
6													1.50	
7													1.50	
8													1.50	
9													1.50	
10													1.50	
11													1.50	
12													1.50	
13													1.50	
14													1.50	
15													1.50	
Variables: n's, or signals, etc.														
17														
18														
19														
20														
21														

Capability: Graphical View

__Prediction? __Actual Result?

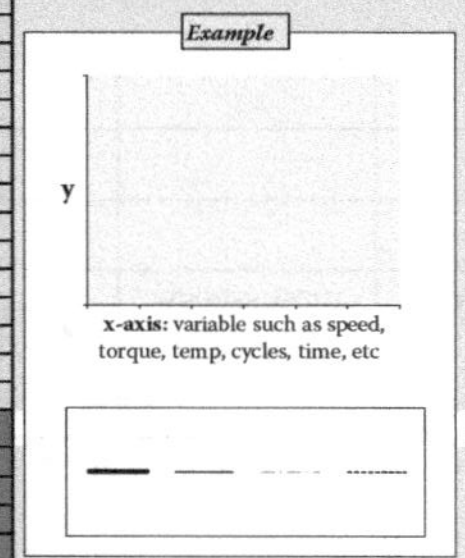

Cell Shading Key			
(white)	Enter Data	(light grey)	Enter formula (refer to cells J12, J13, ... representing x1, x2, ...)
(grey)	Do not enter data (Calculation)	(dark grey)	Do not enter data (Not applicable for Noise Factors)

Confidence Ratings	
High (H)	Mean & s.d. estimates based on customer-correlated model or same parts
Med (M)	Mean & s.d. estimates based on partial customer correlation or surrogate parts
Low (L)	Mean & s.d. estimates made without customer correlation or no process data available

Robustness Assessment						
Product		Range	mean: μ	s.d.: σ	S/N ratio	Beta
Present design						
Benchmark						
Process		Range	mean: μ	s.d.: σ	S/N ratio	Beta
Present design						
Benchmark						

Notes

Performance section

TransferFunction: Y/N	Has a transfer function been identified? Yes/No
Transfer Function: Enter Formula	Type in the Excel formula here, referring to the appropriate cells
Specification	Targets and spec limits based on customer needs
Estimated Performance Capability	Calculated using the transfer function and the listed means & s.d.'s of the variables

Variables section: y's: techni cal metrics; x's: control f actors; n's: noise factors

Characteristic	Name of variable being measured
Units	Units in which the characteristic is measured
Range: Min & Max	Design range (range in which nominal values of the characteristic can be set) or range of the noise space
Contribution to Variability:	
Sensitivity	Amount of change in performance with respect to change in variables, i.e., sensitivity of Δy to Δx
%	Contribution of the characteristic to variability in the performance (y)

Variables section, continued

USL	Upper Spec Limit
LSL	Lower Spec Limit
mean	Sample average or assumed population mean
s.d.	Standard deviation (may include mfg and/or usage variation)
Confidence/Comments	Qualitative confidence rating for data: see Conficence Ratings table. Short/Long-term indication.
z	z-score calculated from USL, LSL, mean and s.d.; z_{st} represents "unshifted," i.e. short term score
σ-shift	"Shift" of mean used in the DPM calculation; default is 1.5; (-1.5 for items with no USL)
DPM	Defects per Million, calculated using the σ-shift indicated to represent defects over the "long term"
Graphical View	If performance over the range of some variable is important, include a graph such as the one shown

Robustness Assessment

Results from steps D and E, indicate results using range, mean and s.d., or S/N and Beta

59. Reliability and robustness demonstration matrix

Program:				Project:						System:			
Date:				Team Lead:						Supplier Lead:			
Define Requirements						Design for Robustness				Verify Design			
Quality History Analysis				SDS and/or DVP		VMConditions(here we ID the high-impact noise factors)				Reliability Demonstration			
Criteria	Component	High-Impact Error State	VM Number	VM Description	VM Target	Number	Name	Critical Metric	Range	Demonstrated Result	Risk Assessment	Issues	Finish Date

60. Assessment of Six Sigma status

(The desired direction is always at the higher level. For your organization you may modify this to reflect your own needs)

6 Sigma category		Man. Commit.	Obsession with Excellence	Org. Is Customer	Customer Satisfaction	Training	Employee Involvement	Use of Incentives	Use of Tools	
DESIRED DIRECTION	5	Is continual improvement a natural behavior even for the routine tasks?	Is there a constant improvement in quality, cost, and productivity?	Is the primary goal to satisfy the customer?	Are customers maintaining a long-term relationship?	Is appropriate and applicable training available among employees?	Are employees proud to participate? Are employees self-directed? Are effective teams utilized in appraising and preventing problems	Are incentives appropriate and applicable for the entire team?	Is statistics used as a common language throughout the organization?	BENCHMARKING MATRIX
	4	Is focus on improving the system?	Are cross-functional teams used?	Is customer feedback used in decision-making?	Is improvement for value to the customer inherent in all routine behaviors of the process?	Is top management aware of the Six Sigma methodology and apply it?	Is manager the key decision with employees following his lead?	Are incentives appropriate and applicable for individuals in	Is at least some statistics and SPC used?	
	3	ARE appropriate resources applied to training?	Is Six Sigma supported by executives and managers?	Are appropriate and applicable tools and methodologies identified and used for identifying the wants, needs in the design?	Is there a verification of using the customer's feedback to improve processes and or complaints?	Are there ongoing training proposals?	Is the manager asking for input before a decision is made?	Is there a quality related selection and promotion criteria for employees?	Is at least some SPC used to reduce variation?	
	2	Is there a policy of balancing long-term goals with short-term objectives?	Is there an executive steering committee setup? Is there a champion for specific areas designated?	Are you sure that the customer's wants and needs are known?	Do you know how the customer is rating you?	Is there a big picture training plan developed?	Is the manager the key person for decision-making? Is the manager the person who decides and then asks his employees for ideas?	Is there an effective employee suggestion program in place?	Is at least some SPC used in key processes?	
	1	Traditional approach to quality control – very ineffective Emphasis is on inspection Quality is found only in manufacturing facilities								

Lean Functions for the Lean Enterprise

All activities fall into one of these three categories:

- Value-added
- Type I waste: adds no value but necessary, non-value-added work
- Type II waste:adds no value and avoidable

Categories of Action	Action	Examples
Value-added	Work that is worth something to the customer	Weld flange onto part Bolt muffler to vehicle
Type I waste	Non-value-added work	Pull down impact ranch Clamp and unclamp
Type II waste	Inspection Inventory	Sorting Waiting

Lean Deployment Guideline Sheet

Do you see this	Do you see this?	Do you see this?	Do you see this?
Visual Factory		Pull Systems	
Non-Lean	**Lean**	**Non-Lean**	**Lean**
No notification information (Andon) Boards No/little evidence of Lean metrics used Variations in work performance Standard procedures not being followed Dirty, cluttered, messy work areas	Used notification/ information (Andon) boards Demonstrate use of key Lean metrics Standardized work forms prominently displayed, workers knowledgeable Workers do the job the same way every time Sparkling clean equipment, tools, work areas, and bathrooms	High level of inventory on lines and in plant-receiving and shipping Materials waiting and stacked at lineside Delivery timing of materials unregulated Centralized schedules without Takt times Products pushed to the next area	Low level of inventory on lines and in plant Real Kanban system in use Material delivered frequently to lineside Use of small supermarket areas Localized/customer demand with Takt times

Do you see this	Do you see this?	Do you see this?	Do you see this?
Visual Factory		Pull Systems	
Non-Lean	**Lean**	**Non-Lean**	**Lean**
Messy bathrooms Materials piled everywhere Unmarked gauges and tools No min/max levels at lineside Empty shadow boards, tools missing, disorganized, dirty floors, overflowing drawers, cabinets and desks	Designated area for empty/full containers Marked gauges and key equipment Min/max levels clearly marked Shadow boards for tools that are used and maintained Shiny clean floor, painted		Use of load leveling
Evidence of Teams		*Returnable Container/Dunnage*	
No production data visible (or outdated) Lack of conference room for teams to use Suggestion campaign old, suggestion box not active, little implementation of suggestions accepted No problem-solving involvement at operator level 8D done in quality department Team not actively involved in standardized work activity or work flow	Available team meetings areas Team data prominently displayed • Quality data • Production real time data • Problem-solving evidence Training depth chart Teams implemented many of their own suggestions Proof of problem-solving activities many places in plant (fishbone diagrams filled out and displayed, PDCA cycle being followed, 5 Whys) Aggressive team standardized work input Six Sigma used to reduce variation	Cardboard containers Wooden pallets Large number of parts/containers Small number of parts/containers *Equipment Stability* Little visible evidence of preventive maintenance process Frequent breakdowns and no tracking No overall equipment effectiveness (OEE) data No study/analysis into major losses	Reusable containers (plastic, metal) Recycled between organization and suppliers Maintenance schedule posted and carried out; employees involved Breakdowns fixed quickly and prevented OEE at 85% or better (posted and tracked by teams)

(*Continued*)

Do you see this	Do you see this?	Do you see this?	Do you see this?
Visual Factory		Pull Systems	
Non-Lean	Lean	Non-Lean	Lean
Changeover		*Layout and Good Material Flow*	
Usually done by one person No tracking of changeover time Most work done when machine is down Lack of standardized methods; no specific work task/procedures identified	Changeover done by team Clear internal/external task identification Changeover chart tracks times, improvements, goals, ideas Use of changeover chart, visual organization used, standard methods	Insufficient, poorly spaced docks Cluttered staging areas Warehouse/storage poorly placed for continuous flow Designated scrap areas that are full Long conveyors, full of WIP	Point of use shipping and receiving docks Visually clear and simple staging areas Well-placed small market area Little scrap/use of mistake proofing Short conveyors, min/max levels marked
		Cell Design	
		Inefficient, assembly configurations Assembly is a long way from point of use Storage of completed materials (more than 1 day's supply)	U-shaped cells/no wasted motion Cells adjacent to point of use Subassembly and raw materials storage of 4 h to support production
		Lean Process Flow	
		Operation running much faster than customer requirements Large buffers/production push product Large offline repair bays	Operation tied to customer requirements Small buffers used sparingly to decouple major processes Production pulled from prior operation

First Time Through (FTT)

First time through (FTT) is the percentage of units that complete a process and meet quality guidelines the first time without being scraped, rerun, retested, diverted for offline repair, or returned. It is a measure of the quality of the manufacturing process.

$$\frac{\text{Units entering process} - (\text{scrap} + \text{reruns} + \text{retests} + \text{repaired offline} + \text{returns})}{\text{Units entering process}}$$

Stretch Objective: 100% FTT capability = zero defects made or passed on

Why

Improvements in FTT will result in

- Increased capacity
- Improved product quality to internal and external customers
- Reduced need for excess production inventory resulting in improved dock to dock time
- Improved ability to maintain sequence throughout the process resulting in improved build-to-schedule performance
- Increased quality input to constraint operations resulting in improved OEE
- Elimination of wastes due-to-scrap, repair, and excess inventory resulting in improved total cost

Where

FTT data should be collected at the end of processes.

When

FTT data should be collected and used, at minimum, by shift. FTT data must drive business practices behavior.

FFT Calculation

$$\frac{\text{Units entering process}\quad(\text{scrap} + \text{returns} + \text{repaired offline} + \text{retests})}{\text{Units entering process}}$$

Units entering	1000
Scrap	10
Returns	15
Retest	5
Repair offline	0
FTT%	1000 − (10 + 15 + 5 + 0)/1000
FTT%	970/1000
FTT%	97.0%

(Continued)

FTT multiple operations calculation examples	
FTT: Operation 1	92.87%
FTT: Operation 2	87.65%
FTT: Operation 3	82.34%
Final inspection: Operation 4	82.34%
Total FTT%	(0.9287 × 0.8765 × 0.6598 × 0.8234)
Total FTT%	44.22%

Dock to Dock (DTD)

Dock to dock is the elapsed time between the unloading of raw materials and the release of finished goods for shipment. DTD measures how fast raw materials are converted to finished goods and shipped. It is a measure (in hours) of the speed of material through the plant, not the speed of the processes.

Why

- Plants must eliminate waste and move toward customer pull systems.
- By building large buffers, you produce more parts (increase the speed of your process) but slow down the speed of materials.
- Decreasing inventories leads to less material handling and storage, which results in fewer opportunities to damage parts.

Where

- Raw material: Purchased part (control part). Begin counting when it is received in the plant, even on consignment, because it is already in its product form.
- Raw material that is changed to its product form.
- Work-in-process (WIP), including material in repair/rework areas, and all buffers.
- Finished goods end-item products: All finished goods containing the control part are included in the DTD calculation until they are physically shipped from the dock.

When

Data should be collected at the same time, which best represents the operating pattern of a product line.

Total DTD Calculation

$$\text{DTD(Single Operation)} = \frac{\text{Total Units of Control Part}}{\text{End of Line Rate}}$$

$$\text{DTD(Multiple Operation)} = \frac{\text{DTD (Operation 1)} + \text{DTD (Operation 2)} + \text{DTD (Operation 3)}\ldots}{\text{End of Line Rate}}$$

The end of line rate is calculated off the last operation in the service line for a product:

Total unit produced	730 (from last operation)	
Production hours	12 h (10 scheduled production hours = 2 unplanned overtime hours)	
End of line rate	730/12 = 60.83	
Units in area	Raw material + work-in-process (by operation) + finished goods	
Raw material	300	
Operation 1	181	
Operation 2	3	
Operation 3	3	
Finished goods	200	
DTD (raw material)	300/60.83	4.93 h
DTD (Operation 1)	181/60.83	2.98.h
DTD (Operation 2)	3/60.83	0.05 h
DTD (Operation 3)	3/60.83	0.05 h
DTD (finished goods)	200/60.83	3.29 h

Total DTD = 4.93 h + 2.98 h + 0.05 h + 0.05 h + 3.29 h

Total Dock to Dock time = 11.3 h

Overall Equipment Effectiveness (OEE)

OEE is a combination of three different calculations. It is a measure of the availability, performance efficiency, and quality rate of a given piece of equipment. OEE data will help identify several wastes caused by machine/process inefficiencies.

Stretch objective: Most common recommended standard, 85% OEE. The best OEE is 100%; however, it is not a realistic goal.

Why

- Higher throughput (increases capacity) decreasing total DTD times
- More stable processes improve production predictability, therefore improving build-to-schedule
- Higher throughput and lower rework/scrap leads to improved FTT and total cost

Where

OEE should be measured on machines in your constraint operations.

When

Data should be collected by shift and by product or job if applicable. OEE data should be reviewed daily, and trended over time to identify improvement opportunities and verify the effectiveness of process changes.

Special consideration: OEE data must not be used to compare plant-to-plant performance. Too many variables exist in physical environments, facility layout, and machine/process uniqueness, to compare on a valid basis.

OEE	Availability × Performance Efficiency × Quality
Availability	$\frac{\text{Operating Time}}{\text{Net Available Time}}$
Performance efficiency	$\frac{\text{Ideal Cycle Time} - \text{Total Parts Run}}{\text{Operating Time}}$
Quality	$\frac{\text{Total Parts Run} - \text{Total Defects}}{\text{Total Parts Run}}$

OEE Calculation Example

Availability	
Total schedule time (including overtime)	720min
Required downtime (e.g., 30 min lunch; 30 min breaks, etc.)	30 + 30 = 60 min
Net available time	720 – 60 = 660 min
All other downtime (e.g., machine downtime and die changes)	45 + 45 = 90 min

(Continued)

Operating time	$660 - 90 = 570$ min
Availability	$\frac{570}{660} = 86.3\%$
Performance Efficiency	
Ideal cycle time	0.33min/part
Total parts run	1440
Operating time	570 min
Performance	$(0.33 \times 1440) = 83.4\%$
Efficiency	570
Quality Rate	
Total parts run	1440
Rework	50
Scrap	40
Quality rate	$\frac{1440-(50+40)}{1440} = 93.8\%$
Overall Equipment Effectiveness	
Availability	86.3%
Performance efficiency	83.4%
Quality rate	93.8%
OEE	$0.863 \times 0.834 \times 0.938$
OEE	67.5%

Quick Changeover

Quick changeover is a method for people to analyze and significantly reduce setup and changeover time. Changeover time is the time between the last good piece off one production run and the first good piece off the next production run after the changeover.

Internal: Activities performed while the machine is shut down.

External: Activities performed while the machine is safely running.

<table>
<tr><td></td><td colspan="5">Changeover Time</td><td></td></tr>
<tr><td>1</td><td>Run A</td><td>External</td><td>Internal</td><td>External</td><td>Run B</td><td></td></tr>
<tr><td></td><td></td><td colspan="5">Start: Document current changeover elements</td></tr>
<tr><td>2</td><td>Run A</td><td colspan="2">Internal</td><td colspan="2">Run B</td><td>50% reduction</td></tr>
<tr><td></td><td></td><td colspan="5">Separate internal/external activities</td></tr>
<tr><td>3</td><td>Run A</td><td>Internal</td><td colspan="3">Run B</td><td>75% reduction</td></tr>
<tr><td></td><td></td><td colspan="5">Shift internal activity to external</td></tr>
<tr><td>4</td><td>Run A</td><td>Internal</td><td colspan="3">Run B</td><td>90% reduction</td></tr>
<tr><td></td><td></td><td colspan="5">Streamline internal and or external activities</td></tr>
</table>

Quick Changeover Process

1. Document current changeover process
   a. Document each element (consider photos/video)
   b. Time the process and the elements
   c. Establish baseline
2. Separate internal and external activities
   a. Internal activities must be performed while the machine is down
   b. External activities can be performed while the machine is running
3. Shift internal activities to external
   a. Analyze each element
   b. Ask 5 Whys
   c. Resolve if the element can be made external
4. Locate parallel activities
   a. Tasks that are independent of each other can be performed simultaneously
   b. Changeover time can be reduced significantly by doing jobs in parallel (simultaneously)
5. Streamline internal/external activities
   a. Adjustment elimination
   b. Fixtures and jigs
   c. Functional standards
   d. Quick die change hardware
6. Implement the revised plan
   a. Examine feasibility of all ideas

7. Validate the method/verify the results
   a. Develop draft procedures
   b. Test procedures
   c. Verify results with data
8. Document the new method on a standard worksheet
   a. One standard worksheet for external
   b. One for each set of parallel internal activities
9. Notify other areas

Visual Factory

A visual factory is a facility in which anyone can know in 5 min or less the who, what, where, when, how, and why of any work area, without talking to anyone, opening a book, or turning on a computer.

Visual Factory Levels

Visual Display

Level 1: Share information. Sharing all information about activities lets everyone see how closely performance conforms to expectations.

Level 2: Share standards at the site. Sharing information about standard specifications and methods lets everyone identify nonconformance as it occurs and helps to correct it.

Visual Controls

Level 3: Build standards into the workplace. Make the work environment (space limits, standard containers, equipment barriers) communicate established standards.

Level 4: Warn about abnormalities. Incorporate lights, alarms, bells, buzzers, and other devices that alert workers when an abnormality is detected, such as a defect generated or a shortage of parts.

Level 5: Stop abnormalities. Prevent defects from moving on after detection by stopping the process or rejecting the part. Devices are installed at the source of abnormalities.

Level 6: Prevent abnormalities. Level 6 means 100% prevention of abnormalities. Once the exact cause of a defect is known, an error-proofing device is installed to prevent the defect from occurring at all.

Five Steps to Workplace Organization

- Step 1: Sort (organization)
    - Distinguish between what is needed and what is not needed.
- Step 2: Stabilize (orderliness)
    - A place for everything and everything in its place.
    - Determine the best location for all necessary items.
- Step 3: Shine (cleanliness)
    - Eliminating dirt, dust, fluids, and other debris to make the work area clean.
    - Adopting cleaning as a form of inspection.
- Step 4: Standardize (adherence)
    - Maintain and monitor the first three S's.
    - Check/standardize/maintain/monitor/improve
- Step 5: Sustain (self-discipline)
    - Correct procedures have become habit.
    - The workplace is well ordered according to agreed-on procedures.
    - Manager, operator, and engineers are deeply committed to 5S.

Error Proofing

Error proofing is a process improvement to prevent a specific defect from occurring.

Three Sides of Error Proofing

Physical: Install hardware

Operational: Enforce procedure, sequence, and/or execution

Philosophical: Empowerment of workforce

Approaches to Error Proofing

Prevention: Prevents errors from creating defects

Detection: Detects defects and immediately initiates corrective action to prevent multiple defects from forming

Definition of a Defect

A defect is the result of any deviation from product specifications that may lead to customer dissatisfaction. To classify as a defect:

- The product has deviated from manufacturing or design specifications
- The product does not meet internal and/or external customer expectations

Definition of an Error

An error is any deviation from a specified manufacturing process. There can be an error without a defect. There cannot be a defect without an error first. The result of an error is demonstrated in a defect.

Error-Proofing Approach

Error-Proofing Teams

Error-proofing teams are composed of operators, supervisors, engineers, skilled trades, and support personnel. The observations, abilities, and creativity of each individual are combined to:

- Make changes to the work environment
- Increase production efficiency
- Take ownership in the manufacturing process
- Support continuous improvement in the workplace

Error-Proofing Method

1. Identify the problem
2. List possible errors
3. Determine the most likely error
4. Propose multiple solutions
5. Evaluate effectiveness, cost, complexity of solutions
6. Determine the best solution (include data analysis)

7. Develop implementation plan
8. Analyze preliminary benefits
9. Develop plan for long-term measure of benefits
10. Congratulate the team

Benefits of Error Proofing

- Business costs: Cost of additional energy and equipment required to manage a poor-quality production environment
- Liability costs: Cost of losing business due to poor product quality
- Cultural costs: Reduction of worker moral

Just-in-Time (JIT)

Just-in-Time (JIT) establishes a system to supply work (data, information, etc.) or services (patient care) at precisely the right time, in the correct amount, and without error. JIT is the heart of a Lean system and is an overriding theme for Lean healthcare. JIT is attained through the understanding and application of continuous flow, the pull system, and Kanbans.

Continuous Flow

Continuous flow is characterized by the ability of a process to replenish a single unit of work (or service capacity) when the customer has pulled it. The concept of continuous flow is used to move work, patients, or provide a service between processes with minimal or no wait (queue) time. It is further used to ensure that the process is performing the work required, no sooner or no later than requested, as well as in the correct quantity, with no defects (nonconformances). The goal is to not do any work or service that is not requested by the downstream process (or customer). It is synonymous with JIT. By focusing on continuous flow, the team will be able to

- Reduce or eliminate transport, delay, and motion waste
- Decrease lead times
- Reduce queue times
- Allow staff to identify and fix problems earlier

- Provide the needed flexibility in meeting demand changes
- Improve patient and staff satisfaction levels

True (100%) continuous flow in a manufacturing and service organization may not be achieved because of the staff and departmental interdependencies required in the various pathways of different processes. Therefore, the tools of in-process supermarkets, FIFO lanes, and Kanbans can be used at certain times when cycle time differences between processes exist and flow needs to be improved.

In-Process Supermarket

The in-process supermarket is a physical device between two processes that stores a certain quantity of work or service capacity that, when needed, is pulled by the downstream process. It originated in grocery stores with how clerks replenished food items. Supermarkets do not typically have large storage areas for many of their goods. Every item is received and transferred straight to the shelves (especially the perishable items), where it is made available to the customer. No costly storage. No large storage areas. No waiting. Minimum inventory. A good example of application in manufacturing is the tooling department. Tools need to be made on demand or have a very small inventory. Another example of in-process supermarket is the hospital pharmacy for making IVs. Due to specific requirements and time life considerations, the IVs have to be made practically on demand.

The in-process supermarket (i.e., the grocery store shelf between the supplier of the goods and the consumer) exists due to the variations in customer demand. The customer's lead time is minimal (the time it takes to remove something off the shelf); therefore, it would not be feasible to have just one item there. In order to establish the in-process supermarket, an ordering pattern, establishing minimum and maximum levels, needs to be determined to create a balance between the customer demand and the frequency of delivery to the store (and shelf). In-process supermarkets can also be explained in terms of cycle time. A supermarket exists between two processes to accommodate the differences in the cycle times of those processes (i.e., the time for the removal of a tool from the inventory to the time it takes to replenish that item). This is called the pull system. Only the amount that has been used by the customer for that day, week, or month (depending on the ordering pattern), or *pulled* from the shelf, is reordered. The maximum level is never exceeded. For in-process supermarkets to be successful, the following will need to be in place:

- A quantitative understanding of the downstream requirements
- Known cycle times for the downstream and upstream processes

- Communication system between the upstream and downstream processes
- Minimum and maximum number of work units or service capacity assigned to the supermarket
- Standard work created explaining the operation of the supermarket
- Training of all process workers on the supermarket operation
- Adequate documentation and visual controls communicating to the downstream process when there is a disruption in the flow
- A signal or Kanban triggers a pull from the downstream process to the upstream process to replenish with what was removed

For example, supplies in many departments can run out on an hourly, daily, or weekly basis. To minimize this disruption and maintain optimal functions, in-process supermarkets for unit/floor supply areas may provide value. The supermarkets would limit extra supplies throughout the area and prevent the searching for needed supplies at critical times. The supplies at these locations would be located in an in-process supermarket, meeting the hourly and daily needs of the staff. Replenishment of the supplies would be through Kanban cards when supplies reached a minimum ordering level.

It should be noted that information (i.e., charts, labs, consults, etc.) can also be considered a *supply* or *supermarket item* and the supermarket concept can be applied as such. Many organizations (manufacturing, service, healthcare, financial, defense, etc.) have found the supermarket to be the best alternative to scheduling upstream processes that cannot flow continuously. As you improve flow, the need for supermarkets should decrease.

Kanban

Kanban is a card or visual indicator that serves as a means of communicating to an upstream process precisely what is required at the specified time. In Japanese, Kanban means card, billboard, or sign. Kanban refers to the inventory control card used in a pull system. It is used to regulate the flow or work in and out of supermarkets as a visual control to trigger action.

Kanban is a form of visual control (information that allows a process to be controlled). This information states when, who, what, and how many work units are needed for movement. A Kanban can be anything from an actual index card, a file folder, or some type of electronic signal. There needs to be a mailbox or some repository for the Kanban to be deposited in, as well as the signal system identifying it is there.

The most common use for the in-process supermarket is the ordering of supplies. Kanbans for supplies will ensure that the dollars allocated for supplies will be at the minimum required. The Kanban system is used to create a pull of material, in this case a supply item, from the downstream process to the upstream process. The formal application to Kanban is an eight-step approach:

1. Conduct the supply survey: The team will need to create a standard list of supplies from which to draw. Distribute the list to all staff in the area requesting them to determine their usage level for a period of time (day/shift, week, or month). Examples are exam gloves, blue pads, general use syringes, and so on. A typical form will look like Table V.1.
2. Establish minimum/maximum levels: Once the list and special requests have been collected, gain a consensus on usage and establish minimum/maximum levels. Include the following:
   a. Type of standard supply (there are numerous glove sizes, so agreement must be made on a certain quantity for each size: small, medium, large, or one size fits all)
   b. Weekly or monthly usage for the items
   c. Establish a minimum quantity to have on hand
   d. Establish a maximum quantity to have on hand
3. Create the supply order form: The supply order form will list all the supplies represented by information obtained from the supply survey form (see Table V.2). The minimum quantity to have on hand is the quantity of items expected to be used during the time it takes to resupply that item (plus some X factor or buffer). The X factor or buffer is to ensure no stockout occurs at any time. The X factor is typically determined by the team and their experience (historical data) with the supplies. It may be one day's worth of a particular supply; it may be a week's worth, again depending on the experience of the

TABLE V.1

Typical Supply Survey Form

Supply Survey Form

Select from the list below the frequency of the items you will most likely use within a specific time (shift, week, month, and so on). If not on the list, please add the item at the bottom of the list. This information will be used in the Lean initiative to reduce or eliminate excess waste.

Item Description	*Quantity Used/Time Period*	*Comments*

TABLE V.2

Typical Order Form

Supply Order Form				
Fax this form to the central supply at extension XXXX by 16:00. Your supplies will be delivered by 06:30 the next day.				
Number	*Item Description*	*Item #*	*Quantity Ordered (Maximum – Minimum)*	*Unit Price*
1				
2				
3				

team. The maximum quantity to have on hand will be the minimum quantity plus the number of items that would be used during the supply or replenishment time.

4. Create the Kanban cards: There should be one Kanban card identified as the supply reorder Kanban for each supply item. It should be laminated and color-coded, differentiating more than one supply ordering location (i.e., production, tool room, central supply, housekeeping, etc.). Kanban cards should be affixed to the minimum reorder quantity item. For example, attach the Kanban card to the last box of exam gloves size (X) on the shelf. The card should be an appropriate size to visually convey the need. It is important here to note that in order to ensure that special items are identified, a special order Kanban card is created and is filled in by the staff for each of the special order items (e.g., welding gloves, latex-free band-aids). Monitor the special order Kanbans to determine if they should be included on the standard supply order form. This will be the case if the special items are used quite frequently and are part of the normal process. A typical card is shown in Table V.3.
5. Create standard work: Once the system has been designed and the reordering process has been determined, now the team is ready to begin creating the standard work chart. This should be posted in a

TABLE V.3

Typical Reorder Kanban Card

Supply Reorder Kanban Card
Item name:
Maximum quantity:
Minimum quantity:
Reorder quantity:
Supplier name:
Catalog page number:
Return this card to the Kanban envelope

place where it can be seen and used by the appropriate staff. It can also be used as a training vehicle for the new staff. A typical standard work may be a procedure and/or an instruction. It can be in a prose form or a process flow diagram.

6. Conduct the training: The training for the department should be done prior to implementation to ensure integrity of the system. The training should include the following:
   a. A brief explanation on the purpose of Kanbans
   b. Explanation on how the minimum/maximum levels were established and convey appreciation for everyone's input when the supply survey was conducted
   c. Explanation of how the system will work (distribute process flowcharts)
   d. Explanation of the two types of Kanbans: Supply Reorder and Special Order Kanban; a demonstration at the supply cabinet on how the Kanban system will work
   e. Acknowledgment of the key individuals within the team who contributed to this system
   f. Communication that this is a work-in-progress trial, and improvement ideas from all staff members will be welcomed
7. Implement the Kanban system: Training and implementation should occur simultaneously. Once the training has been completed, the Kanban for supplies will be ready to use.
8. Maintain the standards: After a month or two of usage, review the appropriate budget supply line item, determine the cost savings, congratulate the team, and convey the success of the initiative to management. At all times during the Kanban implementation, take suggestions on how the process can be improved. The responsibility to maintain the system should be rotated among the staff on a regular basis. This will help to ensure, as they take care of the system, that they will have an appreciation for how it works and, therefore, be more supportive of using the system as it was intended to be used. The benefits of implementing a Kanban system for supplies are
   a. Ensures minimum inventory
   b. Creates staff awareness of the cost of supplies
   c. Easy tool to implement and train
   d. Encourages teamwork
   e. Minimizes transactions on ordering supplies
   f. Reduces stress
   g. Reduces excess inventory waste

h. Allows staff to understand the concepts of flow and supermarkets
i. Reduces costs
j. Eliminates the waste of motion of searching through excess supplies

First-In First-Out (FIFO)

FIFO is a work-controlled method to ensure the oldest work upstream (first-in) is the first to be processed downstream (first-out). For example, in central supply areas, some materials are continuously rotated, replacing them with the up-and-coming expiration dates to the front of the shelf. This can also apply to work requests in production, laboratory, tool crib, employee-owned tools, and so on. The FIFO lane has the following attributes:

- Located between two processes (supply and demand).
- A maximum number of work units (written orders, lab requests, etc.) are placed in the FIFO lane and are visible.
- Is sequentially loaded and labeled.
- Has a signal system to notify the upstream process when the lane is full.
- Has visual rules and standards posted to ensure FIFO lane integrity.
- Has a process in place for assisting the downstream process when the lane is full and assistance is required.

The team can be creative in establishing the signal method, within the FIFO system, to indicate when the system is full. This could be a raised flag, a light, a pager code, an alert email, or text message to the upstream process. The important point is to ensure the signal established will work effectively. When the signal is released, the upstream worker lends support to the downstream worker until the work is caught up. There is no point in continuing to produce upstream when the downstream process is overloaded. When this happens, it becomes an overproduction waste, which is considered the greatest waste of all. A typical example of this is missing stock between processes and departments.

FIFO lanes help to ensure smooth workflow between processes with little or no interruption. Keep in mind that in-process supermarkets and FIFO lanes are compromises to true continuous flow. They should be viewed as part of a continuous improvement system and efforts should be made to find ways to reduce or eliminate them in the quest for continuous flow.

Pitch

Pitch is the time frame that represents the most efficient and practical work (or patient) flow throughout the value stream. It can be a multiple of Takt time. Since Takt time, for many organizations, typically is too small of a unit of time to move the work or information to the next process immediately, pitch is a solution that can be used. Pitch is the *optimal flow* of work at specific times through the value stream. Pitch is the adjusted Takt time (or multiple of) when Takt time is too short to realistically move something. Typically, each value stream (or process-to-process timed movement) will have its own pitch. Pitch will

- Assist in determining the optimal part or work flow
- Set the frequency for movement of the part or work to the next process
- Assist in reducing transport and motion waste
- Allow for immediate attention when interruptions to work flow arise
- Reduce wait (or queue) times

Very important note: Each value stream may require a separate pitch.

Pitch is used to reduce wait time and other wastes that exist within and between processes. For example, a production line was constantly calling to refill the stock bench every time a request from the supervisor had been called in that day. Pitch can be used as a tool in this case to reduce some of the wastes that existed in that process. The steps for calculating Pitch are

1. Calculate Takt time.
2. Determine the optimal number of parts or work units to move through the value stream (i.e., number of lab tests to be completed, number of parts to be processed within a specified time period, number of charts to be processed, etc.).
3. Multiply Takt time by the optimal number of work units. Pitch = Takt time × optimal number of work units.

For example, Lab tests requests: 20 tests per day. Time is an 8-h day (or 480 min).

Takt time: Time/Volume = 480 min/20 tests = 24 min. This means that approximately every half-hour someone is requesting someone from the production to call in a test for him or her. Optimal number of tests to be moved = 10 (10 tests can easily be called in at one time as agreed to by the staff). This means

Pitch = 24 min (Takt time) × 10 tests (optimal number of work units) = 240 min pitch or 4 h. This means every 4 h, 10 tests will be moved to the next process within the value stream.

Do not confuse *pitch* with the *cycle time*. In this example, it only takes approximately 1 min of cycle time (relaying the test to the lab technician). The total cycle time it takes to complete the 10 tests would be 10 min, plus 2 min to go to a phone and make the call, for a total cycle time of 12 min. If each test were called in separately, then there would be 20 min of going to the phone and making the call (each trip 2 min × 10 trips). From this example, the waste motion and transport of those other 9 times (or 18 min of staff time) would be eliminated. The cycle time would remain the same for each test. (The operator would have been notified of the time that the tests were to be called in.)

Pitch increments must be monitored to ensure they are being met. If an interruption of work arises, a system should be in place to address why the work is behind schedule. There are different ways this can occur:

- The group leader communicates the need for assistance when pitch cannot be met.
- The employee who cannot meet the next pitch work increment communicates the need for assistance.
- The runner (see next section) communicates the need for assistance.

The communication signal required for assistance can include:

- A pager code or text message to the supervisor or departmental manager
- An alert email, text message, and so on can be sent to all employees for a predefined round-robin type of support
- A phone call to the supervisor
- A physical meeting with the department leader or supervisor

Creating a visible pitch board will allow employees to think and work differently about the process that is being improved. A visible pitch board is a bulletin board, whiteboard, and so on displaying the pitch increments and/or associated work that is required to be done for that day or week. The simple fact of having a visual board at a location where the work is being done will be a motivating factor for employees to meet the goal (or pitch increment) that has been determined. This will create a sense of satisfaction each time the work (or pitch increment) is completed. The benefits of a visible pitch board are it

- Begins to remove the work from desks and to make it known
- Allows managers/supervisors to monitor hourly/daily progress and be aware of situations when work is getting behind

- Provides a foundation to advance to a more sophisticated scheduling system (i.e., leveling system) at a later date

If an employee gets behind (say more than 1 pitch increment) of what is required, there should be a standard procedure to follow (i.e., notify supervisor, request assistance from another staff member, etc.). The important point regarding pitch is to group work that can be aligned around a specified value stream, thereby reducing the queue time and other wastes that exist between two processes. Making the pitch increments as a visual cue is a great way to get everyone involved in a work rhythm (i.e., Takt time). It places emphasis on completing a certain quantity of work in an allotted time to meet customer demand and further provides the team a common purpose of maintaining this demand. After a specific period of time, usually a month or two of using visible pitch for a value stream, and as other value streams use the visible pitch board tool, then they all can be incorporated into a Heijunka or leveling system.

Workload Balancing

Workload balancing is the optimal distribution of work units throughout the value stream to maintain Takt time or pitch. Also known as employee/staff balancing or line balancing, workload balancing assures that no one worker is doing too much or too little work. Workload balancing begins with analyzing the current state of how work relative to the value stream is allocated and ends with an even and fair distribution of work, ensuring that customer demand is met with a continuous flow mentality. Workload balancing will accomplish the following:

- Determine the number of staff needed for a given demand (or value stream)
- Evenly distribute work units
- Ensure cycle times for each process are accurate
- Assist to standardize the process
- Assist in creating the future state value stream (or process) map
- Improve productivity
- Encourage teamwork through cross-training

The best tool to perform the workload balancing is an employee balance chart. The employee balance chart is a visual display, in the form of a bar chart, that represents the work elements, times, and workers for each process relative to the total value stream cycle time and Takt time (or pitch). The seven steps to work load balancing are

1. Visually display the list of processes from the current state value stream map: Be very clear about identifying each process, its beginning and end as well as the process's parameters. The team should have a good understanding of the processes from previously creating the current state value stream map.
2. Obtain individual cycle times for the various process activities: Cycle times should be derived from the current state value stream map and the document tagging worksheet and/or observation data. Revisit these times to ensure accuracy. Members of the team may want to use a stopwatch to time any process activities that may be questionable, evaluating the processes as they check the times. Take time to analyze the processes.
3. Add the individual cycle times to obtain the total cycle time for each process: Each process's total cycle time should match the process times located on the step graph at the bottom of the current state value stream map.
4. Create the employee balance chart of the current state: The employee balance chart is a bar chart identifying each process and staff, along with the various individual cycle times as derived from the current state value stream map. This may be the entire current state value stream map or sections of it where it is known that duties can be shared or consolidated to improve the overall flow. It is recommended that you visually display the chart on an easel with a flip chart so the team can review it as a group and comment. Use Post-it® Notes to represent the tasks associated with the processes. Make the Post-it® Notes proportional to the time element for each individual task. Draw a horizontal line to represent Takt time.
5. Determine the ideal number of staff for a value stream: When determining the ideal number of workers needed to operate the process or processes for the value stream, keep in mind that staff will most likely have multiple value streams for which they are responsible throughout the day. The ideal number of staff is determined by dividing the total process cycle time by the Takt time (or pitch). For example, in the lab, if an average test took 5 min (total cycle time) and the demand (Takt time) was determined to be 5 min (12 tests per hour), then the ideal number of staff needed would be 1 person (5 min total cycle time/5 min Takt time). However, it rarely comes out that evenly. Use this to assist in determining staffing levels. However, if the decimal number from the calculation is less than

X.5 workers required, balance the value stream to the lesser whole number. Ensure each worker is balanced to Takt time and allocate any excess time to one worker. Utilize the excess time to improve standard work procedures and conduct Kaizen activities, attempting to reduce additional wastes within the processes. Once the employee's efforts are no longer needed on the original project, that person can be placed in another continuous improvement capacity or position in the organization. *Lean is not about reducing the number of people,* it is about eliminating waste. Without this understanding, Lean will never be accepted. If the decimal number from the calculation is equal to or greater than X.5, then balance to the larger whole number.

6. Create the employee balance chart of the future state: Work with team members to move the Post-it® Notes around to balance the various work elements with each employee balanced to Takt (or pitch) time, while maintaining the flow of work.
7. Create standard work procedures and train staff: Once a consensus has been obtained on balancing, redistributing, and/or modifying the process activities, update all necessary standards and then train staff to those standards.

Standard Work

Standard work (standard operating procedures) establishes and controls the best way to complete a task without variation from the original intent. These tasks are then executed consistently, without variation from the original intent. Standard work offers a basis for providing consistent levels of healthcare productivity, quality, and safety, while promoting a positive work attitude based on well-documented work standards. Standard work, done properly, reduces all process variation. It is the basis for all continual improvement activities. Creating the standard work procedures is comprised of using two main tools: the standard work combination table and the standard work chart.

The standard work combination table is the visual representation displaying the sequential flow of all the activities related to a specific process. The standard work combination table will

- Document the exact time requirement (i.e., cycle times) for each work element or task in the process
- Indicate the flow (or sequence) of all the work in the process
- Display the work design sequence based on Takt time (ideally)

- Demonstrate the time relationship between physical work (patient care, dispersing meds, charting, etc.) to the movement of the patient or work (transporting patients, retrieving equipment, checking doctors' orders, etc.), queue times, and computer access time

The standard work combination table is an important tool for allocating work within the value stream when total cycle times are greater than Takt time. Capturing the motion of the process that is being reviewed is a good method to accurately document the times of each work element. The seven steps for creating the standard work combination table are:

1. Break the process into separate work elements (record motion to obtain accuracy).
2. Time each work element from Step 1 by observation or EMR/EMS database retrieval.
3. Complete the standard work combination table.
4. Review each work element (task). Question whether it should be eliminated or if the time can be improved.
5. Gain a consensus on any changes.
6. Create a new standard work combination table. Update the standard work procedures.
7. Train everyone on new standard work procedures and audit.

The standard work combination table is a powerful tool and should be used as the basis for all improvement activities. It does require time to thoroughly complete, but will be well worth the effort in the long run. It illustrates the sequence of the work performed and it provides a visual training aid for employees. Specifically, it (a) displays the work sequence, process layout, and work-in-process in relationship to each other,(b) displays the worker movement for each activity, task, or operation, and (c) identifies quality standards, safety concerns, and/or critical opportunities for errors. The process of creating the standard work chart is

- Draw the area layout or process flow on the chart. Label all items.
- Designate work element locations by number to correspond to items listed in (1).
- Use arrows to show movement of patients, information, staff, and any special instructions or safety concerns.
- Post the chart in the work area and use as basis for continual improvement.

It will do little good to improve a process if it is not from an existing standard. If you do not work from some type of standard, you will be attempting to improve to a moving target, one of process variation.

Column chart: Traditional approach to process balancing

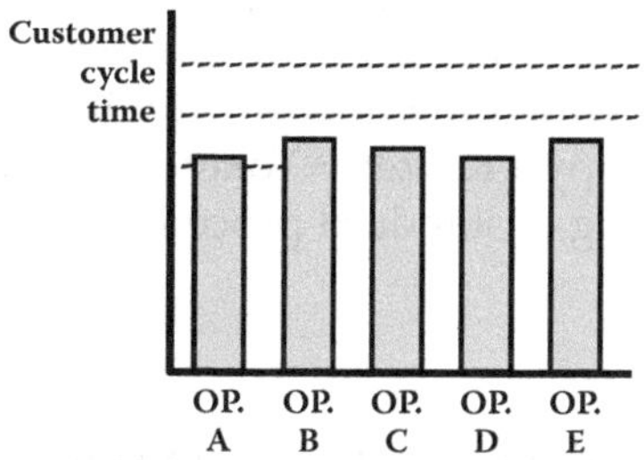

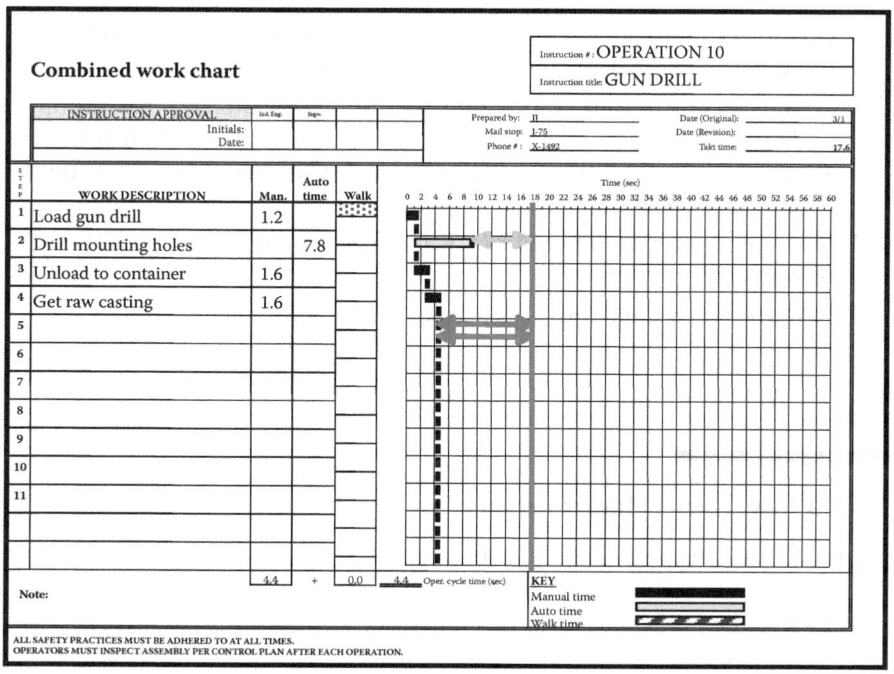

Yokoten

Yokoten means *best practice sharing* or *taking from one place to another.* It encompasses the methods of communicating, documenting, and distributing knowledge horizontally within an organization (peer-to-peer) about what works and what does not work from an improvement project (i.e., PDCA Kaizen event). Yokoten is a form of knowledge management. At its most basic level, Yokoten can be the notebook that a team keeps as a history of the group and problems/solutions encountered. Yokoten can be the library of A3 problem reports (storyboards) that a team or work group maintains for all to access. As a knowledge management device, the Yokoten process ensures information becomes part of the organizational knowledge base. For example, at Toyota there is an expectation that copying a good idea will be followed by some added Kaizen to that idea (copy + Kaizen = Yokoten). Yokoten standardizes a solution and shares it. Sharing of standard procedures across an organization is ideal.

Approach Yokotens as follows:

1. Create a standard improvement methodology that problem-solving, Lean, or Six Sigma teams will follow. Ensure adequate training is conducted.
2. Create standard forms and worksheets to be used. Inform employees where these forms are located on the local area network.
3. Assign a certain date and time each month (or quarter) for groups and/or departments to share their PDCA Kaizen event failures and successes. Allocate 10–15 min per group (or representatives) to share their improvement projects.
4. Document all completed PDCA Kaizen events on the network (LAN) or company intranet.
5. Ensure each team completes a Yokoten worksheet.

Basic Flowchart (Macro Level)

Approach creating this flowchart as follows:

1. Define the process boundaries with the beginning and ending points of the process.
2. Create no more than six boxes denoting what is occurring between the start and end points.

3. Review the flowchart with management to ensure everyone is on the same page with this macro view of the process or work flow.
4. Continue with more detailed flowcharting or value stream mapping (if appropriate) and follow the PDCA Kaizen event methodology.

Deployment Flowchart

Approach creating this flowchart as follows:

1. Identify the right people needed to develop the flowchart.
2. Define the process boundaries with the beginning and ending points of the process.
3. List the major steps (or time element) of the process vertically on the left sheet of paper.
4. List the responsible department (or process worker) across the top, each in a separate column.
5. List all the steps in their appropriate column and ensure they are connected by arrows.
6. Circulate the flowchart to other people within the process for input and/or clarification (if appropriate).
7. Identify areas for improvement.
8. Create the new flowchart and continue with the PDCA Kaizen event methodology.

Opportunity Flowchart

Approach creating this flowchart as follows:

1. Create a process flowchart (micro level).
2. Create separate columns on a flip chart or whiteboard. Label one value-added and the other one non-value-added (or cost-added only).
3. List each step from the process flowchart in either column and ensure they are connected by arrows. Expand the steps to show specific areas of concern, if needed.
4. Identify areas for improvement.
5. Create the new flowchart and continue with the PDCA Kaizen event methodology.

Spaghetti Diagram

Approach creating this flowchart as follows:

1. Define the beginning and ending of the process.
2. Obtain an engineering drawing or create a scale representation of the physical layout of the process.
3. List each step of the process on the flowchart (or on a separate piece of paper).
4. Label and draw each of the steps (3) sequentially as how the process flows. Connect each step with an arrow line denoting the flow.
5. Gain a consensus if the process is inefficient (i.e., too many touches, too many handoffs, too much travel, etc.).
6. Identify areas for improvement.

Create the new flowchart and continue with the PDCA Kaizen event methodology.

Process Flowchart (Micro Level)

Approach creating this flowchart as follows:

1. Identify the right people needed to develop the flowchart. This may require people from outside the PDCA Kaizen team for their expertise and knowledge.
2. Define the process boundaries with the beginning and ending points of the process.
3. Define the level of detail required.
4. Determine conditions and boundaries for the process flow.
5. List all the steps contained within the process flow. The Kaizen event team may need to walk the process.
6. Circulate the flowchart to other people within the process for input and/or clarification.
7. Identify areas for improvement.
8. Create the new flowchart and continue with the PDCA Kaizen event methodology.

Physical Layout

A Lean physical layout work area is a self-contained, well-ordered space that optimizes the flow of patients and work. Typically, many facilities have areas or departments separated by physical walls that may impede the efficiency of workflow. Eliminating walls may be difficult to achieve, but once people understand some of the efficiencies that can be gained, they become engaged in tackling the challenges. Many different methods can be used, such as the computer and visual controls simulating these walls coming down. It is important to realize that there are delays in many processes due to physical separations. Figure V.1 shows the comparison between the current physical arrangement with walls and a future Lean state designed without walls. Lean designs without walls remove all barriers that impact work flow, improving all types of communication and movement; no design should ever compromise patient safety or confidentiality.

In production, there are at least four possibilities in utilizing physical layout principles. They are

1. Product focused
   a. Equipment laid out in process order
   b. Straight stream flow

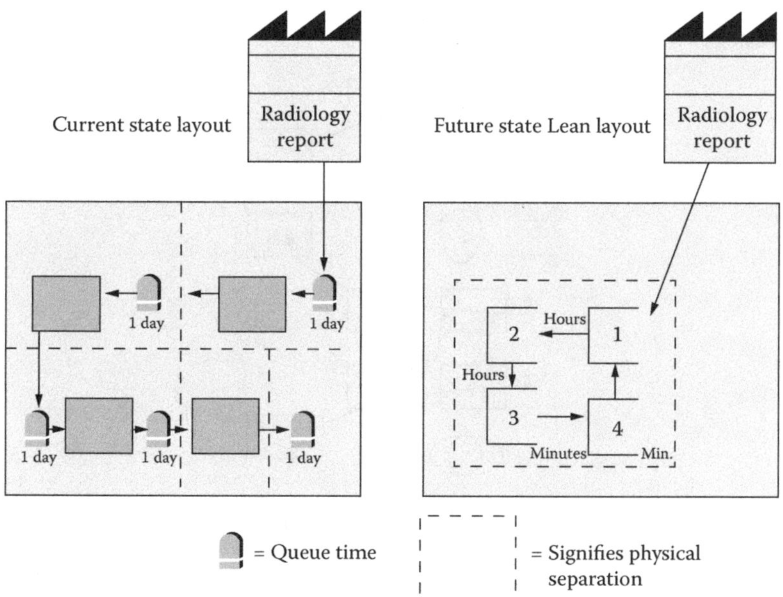

FIGURE V.1
Typical physical layout.

2. Machine closer (see Figure V.2)
   a. Facilitates one-piece flow
   b. Reduces non-value-added walking
3. U shape configuration (see Figure V.3)
   a. Allows volume flexibility with staffing
   b. Reduces non-value-added walking
   c. Improves communication and visual management
4. Customer focus
   a. Modules of capacity matched to major customer sites
   b. Allows operation to match customer needs

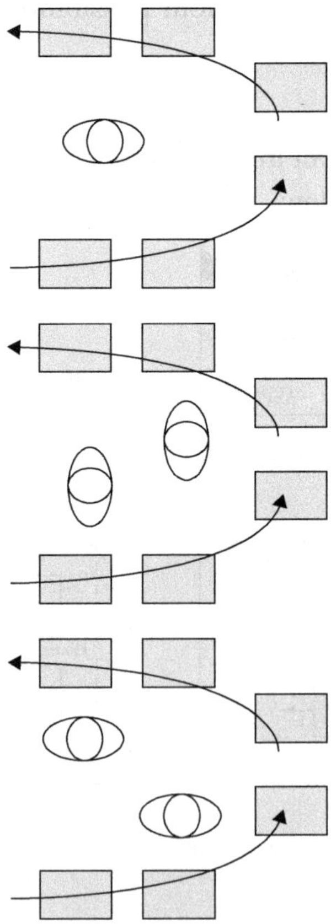

FIGURE V.2
Typical layout configuration.

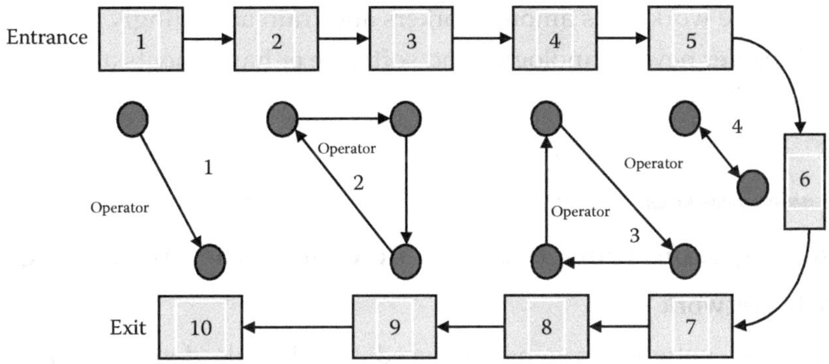

FIGURE V.3
Typical combination layout to take advantage of both equipment and operators.

The benefits for selecting a physical layout are

- Process knowledge can be better shared among the staff to allow everyone to be better trained
- Improved communications will occur between processes
- Wastes of delay, motion, and transport will be reduced or eliminated
- Most efficient use of equipment, people, and materials will be ensured
- Greater job flexibility will occur if someone is absent
- Less floor space will be required
- Work throughput will be increased as well as overall productivity

The following guidelines can be used when determining a new physical layout.

1. Review the current physical layout and associated process tasks to determine which wastes occur due to the current layout in terms of travel, motion, and delays.
2. Brainstorm to consolidate where to reduce or eliminate the wastes identified in (1). Processes may need to be modified or standardized and may require additional cross-training.
3. Determine if an in-process supermarket or FIFO lane is required.
4. Prepare a plan, including result expectations, to implement proposed changes.
5. Obtain management approval.
6. Implement new layout at a time when minimal disruption to the area would occur. Post any new standards when new layout has been completed.

7. Balance workloads among workers and train accordingly.
8. Consider new technologies and software enhancements as you continue to improve.

Various Specific Tools That May Be Used in the Process of CI

Activity Network

The activity network is generally used to schedule dependent activities within a plan. It can be used for describing and understanding the activities within a standard work process. The resulting diagram is useful for communicating the plan and risks to other people.

When to use it

- When planning any project or activity that is composed of a set of interdependent actions
- To calculate the earliest date the project can be completed, and to find ways of changing this
- To identify and address risk to completing a project on time

Affinity Diagram

The affinity diagram provides a visual method of structuring a large number of discrete pieces of information.

When to use it

- To bring order to fragmented and uncertain information and where there is no clear structure
- When information is subjective and emotive, to gain consensus while avoiding verbal argument
- When current opinions, typically about an existing system, obscure potential new solutions
- Rather than a relations diagram, when the situation calls more for creative organization than for logical organization

Bar chart

Bar charts are used to show the differences between related groups of measurements.

When to use it

- When a set of measurements can be split into discrete and comparable groups, to show the relative change between these groups

- When there are multiple sets of measurement groups, to show the relationship and change within and between groups
- Rather than a line graph, to display discrete quantities rather than continuing change
- Rather than a Pareto chart, when a consistent ordering of bars is wanted. This can ease recognition and comparison of current and previous charts

Brainstorming

Brainstorming is used to creatively generate new ideas. The creative synergy of a brainstorming session is also useful in helping a team bind together.

When to use it

- When new ideas are required, to generate a large list of possibilities
- When a solution to a problem cannot be logically deduced
- When information about a problem is confused and spread across several people, to gather the information in one place

Cause and Effect

The cause and effect diagram is used to identify and structure the causes of a given effect. Use it in preference to a relations diagram where there is one problem and causes are mostly hierarchical (this will be most cases).

When to use it

- When investigating a problem, to identify and select key problem causes to investigate or address
- When the primary symptom (or effect) of a problem is known, but possible causes are not all clear
- When working in a group, to gain a common understanding of problem causes and their relationship
- To find other causal relationships, such as potential risks or causes of desired effects

Check Sheet

The check sheet is used to manually collect data in a reliable, organized way.

When to use it

- When data is to be recorded manually, to ensure the data is accurately recorded and is easy to use later, either for direct interpretation or for transcription, for example, into a computer

- When the recording involves counting, classifying, checking, or locating
- When it is useful to check each measurement as it is recorded, for example, that it is within normal bounds
- When it is useful to see the distribution of measures as they are built up

Control Chart

The control chart is used to identify dynamic and special causes of variation in a repeating process. It is only practical to use it when regular measurements of a process can be made.

When to use it

- When investigating a process, to determine whether it is in a state of statistical control and thus whether actions are required to bring the process under control
- To differentiate between special and common causes of variation, identifying the special causes that need to be addressed first
- To detect statistically significant trends in measurements; for example, to identify when and by how much a change to the process has improved it
- As an ongoing health measure of a process, to help spot problems before they become significant

Decision Tree

The decision tree is used to select from a number of possible courses of action.

When to use it

- When making important or complex decisions, to identify the course of action that will give the best value according to a selected set of rules
- When decision making, to identify the effects of risks
- When making plans, to identify the effects of actions and possible alternative courses of action
- When there are chains of decisions, to determine the best decision at each point
- When data is available, or can reasonably be determined, on costs and probabilities of different outcomes

Design of Experiments (DOE)

Design of experiments is used to understand the effects of different factors in a given situation.

When to use it

- When investigating a situation where there are many variable factors, one or more of which may be causing problems
- When variable factors may be interacting to cause problems
- When testing a solution, to ensure that there are no unexpected side effects
- When there is not the time or money to try every combination of variables

Failure Mode and Effects Analysis (FMEA)

FMEA is used to identify and prioritize how items fail, and the effects of failure.

When to use it

- When designing products or processes, to identify and avoid failure-prone designs
- When investigating why existing systems have failed, to help identify possible causes and remedies
- When investigating possible solutions, to help select one with an acceptable risk for the known benefit of implementing it
- When planning actions, in order to identify risks in the plan and hence identify countermeasures

Fault Tree

Fault tree analysis is used to show combinations of failures that can cause overall system failure.

When to use it

- When the effect of a failure is known, to find how combinations of other failures might cause this
- When designing a solution, to identify ways it may fail and consequently find ways of making the solution more robust
- To identify risks in a system, and consequently identify risk reduction measures

- To find failures that can cause the failure of all parts of a fault-tolerant system

Flowchart

The flowchart is used to show the sequential steps within a process.

When to use it

- When analyzing or defining a process, to detail the actions and decisions within it
- When looking for potential problem points in a process
- When investigating the performance of a process, to help identify where and how it is best measured
- As a communication or training aid, to explain or agree on the detail of the process

Flow Process Chart

The flow process chart is used to record and illustrate the sequence of actions within a process.

When to use it

- When observing a physical process, to record actions as they happen, and thus get an accurate description of the process
- When analyzing the steps in a process, to help identify and eliminate waste
- Rather than a flowchart, when the process is mostly sequential, containing few decisions

Force Field Analysis

The force field diagram is used to weigh up the points for and against a potential action.

When to use it

- When decision making is hindered by a number of significant points for and against a decision
- When there is a lot of argument and indecision over a point, to clarify and agree on the balance of disagreement
- To help identify risks to a planned action and to develop a strategy for counteracting them
- To help identify the key causes of successful or unsuccessful actions

Gant Chart

Gantt charts are used to show the actual time to spend in tasks.

When to use it

- When doing any form of planning, to show the actual calendar time spent in each task
- When scheduling work for individuals, to control the balance between time spent during normal work hours and overtime work
- When planning work for several people, to ensure that those people who must work together are available at the same time
- For tracking progress of work against the scheduled activities
- When describing a regular process, to show who does what, and when
- To communicate the plan to other people

Histogram

The histogram is used to show the frequency distribution of a set of measurements.

When to use it

- To investigate the distribution of a set of measurements
- When it is suspected that there are multiple factors affecting a process, to see if this shows up in the distribution
- To help define reasonable specification limits for a process by investigating the actual distribution
- When you want to see the actual shape of the distribution, as opposed to calculating single figures like the mean or standard deviation

IDEFO (ICOM DEFinitions)

IDEFO is used to make a detailed and clear description of a process or system.

When to use it

- When formally describing a process, to ensure a detailed, clear, and accurate result
- When the process is complex, and other methods would result in diagrams that are more complex
- When mapping a wide variety of processes, as a consistent and scalable process description language (PDL)

- When there is time available to work on understanding and producing a complete and correct description of the process

Kano Model

Kano analysis is a quality measurement tool used to prioritize customer requirements based on their impact to customer satisfaction.

When to use

- Kano analysis is a tool that can be used to classify and prioritize customer needs. This is useful because customer needs are not all of the same kind, do not all have the same importance, and are different for different populations. The results can be used to prioritize your effort in satisfying different customers

Kappa Statistic

The Kappa is the ratio of the proportion of times the appraisers (see gauge R&R) did agree to the proportion of times the appraisers could agree. If you have a known standard for each rating, you can assess the correctness of all appraisers' ratings compared to the known standard. If Kappa = 1, then there is perfect agreement. If Kappa = 0, then there is no agreement. The higher the value of Kappa, the stronger the agreement. Negative values occur when agreement is weaker than expected by chance, but this rarely happens. Depending on the application, Kappa less than 0.7 indicates that your measurement system needs improvement. Kappa values greater than 0.9 are considered excellent.

When to use it

- When there is a need for comparison

Line Graph

The line graph is used to show patterns of change in a sequence of measurements.

When to use it

- When an item is repeatedly measured, to show changes across time
- When measuring several different items, which can be shown on the same scale, to show how they change relative to one another
- When measuring progress toward a goal, to show the relative improvement

- Rather than a bar chart, to show continuous change, rather than discrete measurements. It is also better when there are many measurements
- Rather than a control chart, when not measuring the degree of control of a process

Matrix Data Analysis Chart

The matrix data analysis chart (MDAC) is used to identify clusters of related items within a larger group.

When to use it

- When investigating factors that affect a number of different items, to determine common relationships
- To determine whether logically similar items also have similar factor effects
- To find groups of logically different items, which have similar factor effects

Matrix Diagram

The matrix diagram is used to identify the relationship between pairs of lists.

When to use it

- When comparing two lists to understand the many-to-many relationship between them (it is not useful if there is a simple one-to-one relationship)
- To determine the strength of the relationship between either single pairs of items or a single item and another complete list
- When the second list is generated because of the first list, to determine the success of that generation process. For example, customer requirements versus design specifications

Nominal Group Technique

The nominal group technique (NGT) is used to collect and prioritize the thoughts of a group on a given topic.

When to use it

- When a problem is well understood, but knowledge about it is dispersed among several people

- When a rapid consensus is required from a team, rather than a more detailed consideration
- When the team is stuck on an issue, for example, when they disagree about something
- When the group prefers a structured style of working together
- Rather than brainstorming, when a limited list of considered opinions is preferred to a long list of wild ideas, or when the group is not sufficiently comfortable together to be open and creative

Pareto Chart

The Pareto chart is used to show the relative importance of a set of measurements.

When to use it

- When selecting the most important things on which to focus, thus differentiating between the vital few and the trivial many
- After improving a process, to show the relative change in a measured item
- When sorting a set of measurements, to visually emphasize their relative sizes
- Rather than a bar chart or pie chart to show the relative priority of a set of numeric measurements

Prioritization Matrix

The prioritization matrix is used to sort a list of items into an order of importance.

When to use it

- To prioritize complex or unclear issues, where there are multiple criteria for deciding importance
- When there is data available to help score criteria and issues
- To help select items to be actioned from a larger list of possible items
- When used with a group, it will help to gain agreement on priorities and key issues
- Rather than simple voting, when the extra effort that is required to find a more confident selection is considered to be worthwhile

Process Capability

Process capability calculations indicate the ability of a process to meet specification limits.

When to use it

- When setting up a process, to ensure it can meet its specification limits
- When setting specification limits, to ensure they are neither too wide nor too narrow
- When investigating a process that is not meeting its specification limits
- When the process is stable and has a Normal distribution

Process Decision Program Chart (PDPC)

The process decision program chart (commonly just referred to as PDPC) is used to identify potential problems and countermeasures in a plan.

When to use it

- When making plans, to help identify potential risks to their successful completion
- When risks are identified, use it to help identify and select from a set of possible countermeasures
- Also to help plan for ways of avoiding and eliminating identified risks
- It is of best value when risks are nonobvious, such as in unfamiliar situations or in complex plans, and when the consequences of failure are serious

Relationship Diagram

The relationship diagram is used to clarify and understand complex relationships.

When to use it

- When analyzing complex situations where there are multiple interrelated issues
- Where the current problem is perceived as being a symptom of a more important underlying problem
- It is also useful in building consensus within groups

- It is commonly used to map cause–effect relationships, but also can be used to map any other type of relationship
- Rather than an affinity diagram, when there are logical, rather than subjective, relationships
- Rather than a cause and effect diagram, when causes are nonhierarchic or when there are complex issues

Scatter Diagram

The scatter diagram is used to show the type and degree of any causal relationship between two factors.

When to use it

- When it is suspected that the variation of two items is connected in some way, to show any actual correlation between the two
- When it is suspected that one item may be causing another, to build evidence for the connection between the two
- When both items being measured can be measured together, in pairs

Simulation in Lean

Simulation is a practical tool that changes complex parameters into mathematical, logical models, which are possible to analyze under different scenarios. Models can be classified from various aspects. They can be categorized as being mathematical or physical, static (Monte Carlo) or dynamic (shows a system that always changes), deterministic (includes no random variables) or stochastic (contains at least one random variable), discrete or continuous models. However, beside its benefits, the simulation also has some disadvantages.

When to use it

- Assessing hospital operations in different wards
- Estimating the number of needed doctors and nurses in the emergency ward
- Reducing the length of time that a patient has to stay in the emergency ward
- Predicating the required number of beds according to the patient interval

Advantages

1. The most important benefit of simulation is to get results at less cost
2. Assumptions about specific phenomena can be checked for feasibility

3. The simulation answers "what if' questions
4. Analysis of bottlenecks and finding the roots is easier
5. The system can be tested without allocating any resources

Disadvantages

1. Working with simulation modeling requires experience and specific training
2. Understanding the simulation results is difficult because they are based on random inputs

String Diagram

The string diagram is used to investigate the physical movement in a process.

When to use it

- When analyzing a manual or physical process that involves significant physical movement, in order to make movements easier and quicker. Movements may be of people, materials, or machines
- When designing the layout of a work area, to identify the optimum positioning of machines and furniture

Surveys

Surveys are used to gather information from people.

When to use it

- When information that is required is held by an identifiable and dispersed group of people
- To help decision making, by turning disparate qualitative data into useful quantitative information
- When the time and effort are available to complete the survey
- Rather than brainstorming or a nominal group technique, to gather real data about what a diverse group of people think (rather than an opinion of what they think)

Tables

Tables are used to organize and relate multiple pieces of information.

When to use it

- When gathering information, to help prompt for a complete set of data

- When information is disorganized, to help collate and understand it
- For summarizing information to make it easier to present and understand
- Specific table tools as frameworks for particular tasks, either to organize existing information or to prompt for specific categories of information

Tree Diagrams

The tree diagram is used to break down a topic into successive levels of detail.

When to use it

- When planning, to break down a task into manageable and assignable units
- When investigating a problem, to discover the detailed component parts of any complex topic
- When the problem can be broken down in a hierarchical manner
- Rather than a relations diagram, to break down a problem when the problem is hierarchical in nature

Value Analysis

Value analysis is used to determine and improve the value of a product or process.

When to use it

- When analyzing a product or process, to determine the real value of each component
- When looking for cost savings, to determine components that may be optimized
- When the item to be analyzed can be broken down into subcomponents and realistic costs and values allocated to these

Voting

Voting is used to prioritize and select from a list of items.

When to use it

- As a quick tool when a group must select one or more items from a list, for example, as generated by a brainstorming session
- When it is important that the group accept the result as fair

- When the knowledge to enable selection is spread within the group
- When the opinion of all group members is equally valued
- It can also be used to test the water, to determine opinions without committing to a final selection
- Rather than a prioritization matrix, when the added accuracy of the prioritization matrix is not worth the extra effort

Frequent Symbols for VSP

C/T	Cycle time
C/O	Changeover time
	Inventory
	Truck shipment
	External sources (suppliers, customers, etc.)
	Electronic information flow
	Movement of production material
	Supermarket (a controlled inventory of parts)
	Withdrawal (pull of materials, usually from a supermarket)
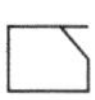	Production kanban (card or device that signals to a process how many of what to produce)
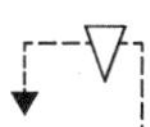	Signal kanban (shows when a batch of parts is needed)
	Kaizen starburst (identifies improvement needs)

Selected Bibliography

http://www.tec-ease.com/gdt-terms.php#sthash.kIZbaeuu.JTQQtteQ.dpuf. Retrieved on July 17, 2014.

http://www.toolingu.com/beta/definition-350200-5104-maximum-material-condition.html

____. (December 30, 2013). World watch. *Crain's Detroit Business*. 11–13.

Albert, M. (2014). Data driven manufacturing. *Modern Machine Shop*. 16.

Bangert, M. (2014). How to get the most out of lean. *Quality*. 40–42.

Bramble, K. (2009). *Geometric Boundaries II, Practical Guide to Interpretation and Application ASME Y14.5-2009*. Monroe, GA: Engineers Edge.

Briggs, S. (2012). (Re)visionary thinking. *Quality Progress*. 23–29.

Chaneski, W. (2013). The pulse of a company. *Modern Machine Shop*. 38–40.

Chaneski, W. (2014). Continuous improvement is vital for remaining competitive. Committing to it across the organization is equally important. *Modern Machine Shop*. 41–42.

Cox, H. (2014). Auditor's guide to calibration reports: A document without data is not much of a report. *Quality*. 13.

Creel, K. (2014). Working on the customer's behalf. *Quirk's Marketing Research Review*. 78–82.

Cressionnie, L. (2014). Revision runway: SIPOC implemented in process for updating AS9100 standards. *Quality Progress*. 50–52.

Dailey, K. (2003). *The Lean Manufacturing Pocket Handbook*. Tampa, FL: DW Publishing.

Daimler Chrysler Corporation. (2004). *Process Sign-Off (PSO)*, 5th ed. Rochester Hills, MI: Daimler-Chrysler Corporation.

De Mast, J., W. Schippers, J. Does, and E. Van Den Heuvel. (2000). Steps and strategies in process improvement. *Quality and Reliability Engineering International*. 16(4):301–311.

Drake Jr., P.J. (1999). *Dimensioning and Tolerancing Handbook*. New York: McGraw-Hill.

Dunmire, T. and G. Johnson. (2012). Pointed in the right direction. *Quality Progress*. 15–22.

Evans, J., Foster, S., and Linderman, K. (2014). A content analysis of research in quality management and a proposed agenda for future research. *Quality Management Journal*. 21(2):17–44.

Evers, A. (2014). The quest for consistent and comparable data. *World Trade 100*. 28–29.

Galsworth, G. (1997). *Visual Systems: Harnessing the Power of a Visual Workplace*. New York: American Management Association.

Gould, L. (2014). FEA: Vehicular development. *Automotive Design and Production*. 34–40.

Hayes, R. and S. Wheelwright. (1984). *Restoring Our Competitive Edge*. New York: John Wiley & Sons.

Henzold, G. (2006). *Geometrical Dimensioning and Tolerancing for Design, Manufacturing and Inspection*, 2nd ed. Oxford, UK: Elsevier.

Hinckley, C. (2001). *Make No Mistake*. Portland, OR: Productivity Press.

Hirano, H. (1995). *5 Pillars of the Visual Workplace: The Sourcebook for 5S Implementation*. Translated by Bruce Talbot. Portland, OR: Productivity Press.

Hirano, H. (1996). *5S for Operators: 5 Pillars of the Visual Workplace*. Portland, OR: Productivity Press.

Hollnagel, E. (2004). *Barriers and Accident Prevention*. Aldershot, UK: Ashgate Publishing.

Hollows, G. (2014). 5 tips to maximize imaging system performance. *Quality*. 1VS.

Howells, R. (2014). Companies need real-time analytics to cope with these new markets. *Worldtrade*. 31.

Hutchins, G. (2008). Risk management—The future of quality. *Quality Digest*. 49–51.

Juran, J. and F. Gryna. (1993). *Quality Planning and Analysis*, 3rd ed. New York: McGraw-Hill.

Littlefield, M. (2014). How closed-loop quality can drive: Near-perfect compliance in A & D. *Quality*. 19–20.

Liu, S. (2014). Catching fire: 7 strategies to ignite your team's creativity. *Quality Progress*. 18–24.

Lorden, A., Y. Zhang, S. Lin, and M. Cote. (2014). Measures of success: The role of human factors in lean implementation in healthcare. *Quality Management Journal*. 26–37.

Marion, R. (2000). *The Edge of Organization: Chaos and Complexity Theories of Formal Social Systems*. Thousand Oaks, CA: Sage Publications.

McCale, M. R. (1999). A conceptual data model of datum systems. *Journal of Research of the National Institute of Standards and Technology*. 104(4):349–400.

Merrill, P. (2014). Embracing change: Embedding innovation into your QMS via ISO 9001:2015. *Quality Progress*. 44–45.

Neumann, S. and A. Neumann. 2009. *GeoTol Pro: A Practical Guide to Geometric Tolerancing per ASME Y14.5-2009*. Dearborn, MI: Society of Manufacturing Engineers.

Nikkan, K. (1988). *Shimbun/Factory Magazine Poka-Yoke: Improving Product Quality by Preventing Defects*. Portland, OR: Productivity Press.

Norman, D. (1989). *The Design of Everyday Things*. New York: Doubleday.

Paterson, M. (1993). *Accelerating Innovation*. New York: Van Nostrand Reinhold.

Productivity Press Development Team. (2002). *Kanban for the Shopfloor*. New York: Productivity Press.

Rhinesmith, S. (1993). *A Manager's Guide to Globalization*. Alexandria, VA: The American Society for Training and Development; New York: Irwin Professional Publishing.

Rossett, A. and J. Gautier-Downes. (1991). *A Handbook of Job Aids*. San Francisco, CA: Jossey-Bass/Pfeiffer.

Schonberger, R. (2014). Quality management and lean: A symbiotic relationship. *Quality Management Journal*. 21(3):6–10.

Shacklett, M. (2014). Preparing the supply chain for the unexpected. *World Trade*. August:28–31.

Shacklett, M. (2014). SCM spells success in emerging markets. *Worldtrade*. June:28–31.

Shah, D. (2014). Measuring proficiency: Evaluating laboratory measurement performance. *Quality Progress*. 46–47.

Spath, P. (2000). Reducing errors through work system improvements. In: P.L. Spath, ed. *Error Reduction in Health Care*. Chicago, IL: AHA Press.

Srinivasan, V. (2008). Standardizing the specification, verification, and exchange of product geometry: Research, status and trends. *Computer-Aided Design*. 40(7):738–749.

Stewart, D. and S. Melnyk. (2000). Effective process improvement developing poka-yoke processes. *Production Inventory Management Journal*. 4148–4155.

Subbiah, S. and D. Bosik. (2014). Instead of big data, try value data. *Quirk's Marketing Research Review*. 22– 25.

Trunick, P. (2014). Compliance gets complex. *World Trade 100*. 20–23.

Tsuda, Y. (1993). Implications of fool proofing in the manufacturing process. In: W. Kuo, ed. *Quality through Engineering Design*. New York: Elsevier.

Tufte, E. 2001. *The Visual Display of Quantitative Information*. Cheshire, CT: Graphics Press.

Ulrich, K. and S. Eppinger. (1995). *Product Design and Development*, 2nd ed. Boston, MA: Irwin McGraw-Hill.

Wilson, B.A. (2005). *Design Dimensioning and Tolerancing*. Tinley Park, IL: Goodheart-Wilcox.

Worthington, J. (2014). How to manage change. *Quality Digest*. Electronic version. 1–2. http://www.qualitydigest.com/read/content_by_author/38243/6/17/14.

Yun, C., Y. Yong, and L. Loh. (1996). *The Quest for Global Quality*. Singapore: Addison-Wesley.

Zchuetz, G. (2014). The rule of thumb, part 1: Guidelines used to standardize the measuring process can provide a good basis for making gage. *Modern Machine Shop*. 56–58.

Zelinski, P. (2014). Improvement is not an option. *Modern Machine Shop*. 89–99.

Index

D

T